Natural Disaster at the Closing of the Dutch Golden Age

By the early eighteenth century, the economic primacy, cultural efflorescence, and geopolitical power of the Dutch Republic appeared to be waning. The end of this Golden Age was also an era of natural disasters. Between the late seventeenth and the mid-eighteenth century, Dutch communities weathered numerous calamities, including river and coastal floods, cattle plagues, and an outbreak of strange mollusks that threatened the literal foundations of the Republic. Adam Sundberg demonstrates that these disasters emerged out of longstanding changes in environment and society. They were also fundamental to the Dutch experience and understanding of eighteenth-century decline. Disasters provoked widespread suffering, but they also opened opportunities to retool management strategies, expand the scale of response, and to reconsider the ultimate meaning of catastrophe. This book reveals a dynamic and often resilient picture of a society coping with calamity at odds with historical assessments of eighteenth-century stagnation.

Adam Sundberg is an associate professor of History at Creighton University. His work has appeared in *Environmental History*, *Dutch Crossing*, and *The Low Countries Journal of Social and Economic History*.

Studies in Environment and History

Editors

J. R. McNeill, *Georgetown University*
Ling Zhang, *Boston College*

Editors Emeriti

Alfred W. Crosby, *University of Texas at Austin*
Edmund P. Russell, *Carnegie Mellon University*
Donald Worster, *University of Kansas*

Other Books in the Series

Germán Vergara *Fueling Mexico: Energy and Environment, 1850-1950*
David Moon *The American Steppes: The Unexpected Russian Roots of Great Plains Agriculture, 1870s-1930s*
James L. A. Webb, Jr. *The Guts of the Matter: A Global Environmental History of Human Waste and Infectious Intestinal Disease*
Maya K. Peterson *Pipe Dreams: Water and Empire in Central Asia's Aral Sea Basin*
Thomas M. Wickman, *Snowshoe Country: An Environmental and Cultural History of Winter in the Early American Northeast*
Debjani Bhattacharyya *Empire and Ecology in the Bengal Delta: The Making of Calcutta*
Chris Courtney *The Nature of Disaster in China: The 1931 Yangzi River Flood*
Dagomar Degroot *The Frigid Golden Age: Climate Change, the Little Ice Age, and the Dutch Republic, 1560–1720*
Edmund Russell *Greyhound Nation: A Coevolutionary History of England, 1200–1900*
Timothy J. LeCain *The Matter of History: How Things Create the Past*
Ling Zhang *The River, the Plain, and the State: An Environmental Drama in Northern Song China, 1048–1128*
Abraham H. Gibson *Feral Animals in the American South: An Evolutionary History*
Andy Bruno *The Nature of Soviet Power: An Arctic Environmental History*
David A. Bello *Across Forest, Steppe, and Mountain: Environment, Identity, and Empire in Qing China's Borderlands*
Erik Loomis *Empire of Timber: Labor Unions and the Pacific Northwest Forests*
Peter Thorsheim *Waste into Weapons: Recycling in Britain during the Second World War*
Kieko Matteson *Forests in Revolutionary France: Conservation, Community, and Conflict, 1669–1848*
Micah S. Muscolino *The Ecology of War in China: Henan Province, the Yellow River, and Beyond, 1938–1950*

George Colpitts *Pemmican Empire: Food, Trade, and the Last Bison Hunts in the North American Plains, 1780–1882*
John L. Brooke *Climate Change and the Course of Global History: A Rough Journey*
Paul Josephson et al. *An Environmental History of Russia*
Emmanuel Kreike *Environmental Infrastructure in African History: Examining the Myth of Natural Resource Management*
Gregory T. Cushman *Guano and the Opening of the Pacific World: A Global Ecological History*
Sam White *The Climate of Rebellion in the Early Modern Ottoman Empire*
Edmund Russell *Evolutionary History: Uniting History and Biology to Understand Life on Earth*
Alan Mikhail *Nature and Empire in Ottoman Egypt: An Environmental History*
Richard W. Judd *The Untilled Garden: Natural History and the Spirit of Conservation in America, 1740–1840*
James L. A. Webb, Jr. *Humanity's Burden: A Global History of Malaria*
Myrna I. Santiago *The Ecology of Oil: Environment, Labor, and the Mexican Revolution, 1900–1938*
Frank Uekoetter *The Green and the Brown: A History of Conservation in Nazi Germany*
Matthew D. Evenden *Fish versus Power: An Environmental History of the Fraser River*
Alfred W. Crosby *Ecological Imperialism: The Biological Expansion of Europe, 900–1900, second edition*
Nancy J. Jacobs *Environment, Power, and Injustice: A South African History*
Edmund Russell *War and Nature: Fighting Humans and Insects with Chemicals from World War I to Silent Spring*
Adam Rome *The Bulldozer in the Countryside: Suburban Sprawl and the Rise of American Environmentalism*
Judith Shapiro *Mao's War against Nature: Politics and the Environment in Revolutionary China*
Andrew Isenberg *The Destruction of the Bison: An Environmental History*
Thomas Dunlap *Nature and the English Diaspora*
Robert B. Marks *Tigers, Rice, Silk, and Silt: Environment and Economy in Late Imperial South China*
Mark Elvin and Tsui'jung Liu *Sediments of Time: Environment and Society in Chinese History*
Richard H. Grove *Green Imperialism: Colonial Expansion, Tropical Island Edens and the Origins of Environmentalism, 1600–1860*
Thorkild Kjærgaard *The Danish Revolution, 1500-1800: An Ecohistorical Interpretation*
Donald Worster *Nature's Economy: A History of Ecological Ideas, second edition*
Elinor G. K. Melville *A Plague of Sheep: Environmental Consequences of the Conquest of Mexico*
J. R. McNeill *The Mountains of the Mediterranean World: An Environmental History*

Theodore Steinberg *Nature Incorporated: Industrialization and the Waters of New England*
Timothy Silver *A New Face on the Countryside: Indians, Colonists, and Slaves in the South Atlantic Forests, 1500–1800*
Michael Williams *Americans and Their Forests: A Historical Geography*
Donald Worster *The Ends of the Earth: Perspectives on Modern Environmental History*
Robert Harms *Games against Nature: An Eco-Cultural History of the Nunu of Equatorial Africa*
Warren Dean *Brazil and the Struggle for Rubber: A Study in Environmental History*
Samuel P. Hays *Beauty, Health, and Permanence: Environmental Politics in the United States, 1955–1985*
Arthur F. McEvoy *The Fisherman's Problem: Ecology and Law in the California Fisheries, 1850–1980*
Kenneth F. Kiple *The Caribbean Slave: A Biological History*

Natural Disaster at the Closing of the Dutch Golden Age

Floods, Worms, and Cattle Plague

ADAM SUNDBERG
Creighton University

CAMBRIDGE
UNIVERSITY PRESS

University Printing House, Cambridge CB2 8BS, United Kingdom

One Liberty Plaza, 20th Floor, New York, NY 10006, USA

477 Williamstown Road, Port Melbourne, VIC 3207, Australia

314–321, 3rd Floor, Plot 3, Splendor Forum, Jasola District Centre, New Delhi – 110025, India

103 Penang Road, #05-06/07, Visioncrest Commercial, Singapore 238467

Cambridge University Press is part of the University of Cambridge.

It furthers the University's mission by disseminating knowledge in the pursuit of education, learning, and research at the highest international levels of excellence.

www.cambridge.org
Information on this title: www.cambridge.org/9781108831246
DOI: 10.1017/9781108923750

© Adam Sundberg 2022

This publication is in copyright. Subject to statutory exception and to the provisions of relevant collective licensing agreements, no reproduction of any part may take place without the written permission of Cambridge University Press.

First published 2022

A catalogue record for this publication is available from the British Library.

Library of Congress Cataloging-in-Publication Data
NAMES: Sundberg, Adam, 1984– author.
TITLE: Natural disaster at the closing of the Dutch golden age: floods, worms, and cattle plague / Adam Sundberg, Creighton University, Omaha.
DESCRIPTION: Cambridge, United Kingdom; New York: Cambridge University Press, [2022] | Series: Studies in environment and history | Includes bibliographical references and index.
IDENTIFIERS: LCCN 2021029006 (print) | LCCN 2021029007 (ebook) | ISBN 9781108831246 (hardback) | ISBN 9781108926591 (paperback) | ISBN 9781108923750 (epub)
SUBJECTS: LCSH: Natural disasters–Netherlands–History. | Floods–Netherlands–History. | Environmental degradation–Netherlands. | Rinderpest–Netherlands–History. | COVID-19 Pandemic, 2020--Netherlands. | BISAC: NATURE / General
CLASSIFICATION: LCC GB5011.54 .S86 2022 (print) | LCC GB5011.54 (ebook) | DDC 904/.5–dc23/eng/20211004
LC record available at https://lccn.loc.gov/2021029006
LC ebook record available at https://lccn.loc.gov/2021029007

ISBN 978-1-108-83124-6 Hardback

Cambridge University Press has no responsibility for the persistence or accuracy of URLs for external or third-party internet websites referred to in this publication and does not guarantee that any content on such websites is, or will remain, accurate or appropriate.

For Sara and Marsh

Contents

List of Figures	*page*	xiii
Acknowledgment		xvii
	Introduction	1
1	*Rampjaar* Reconsidered	26
2	"Disasters in the Year of Peace": The First Cattle Plague, 1713–1720	51
3	"The Fattened Land Turned to Salted Ground": The Christmas Flood of 1717 in Groningen	89
4	A Plague from the Sea: The Shipworm Epidemic, 1730–1735	122
5	"Increasingly Numerous and Higher Floods": The River Floods of 1740–1741	165
6	"From a Love of Humanity and Comfort for the Fatherland": The Second Cattle Plague, 1744–1764	212
7	The Twin Faces of Calamity: Lessons of Decline and Disaster	251
Bibliography		277
Index		325

Figures

I.1	The United Provinces of the Netherlands in 1710	page 8
1.1	The *Rampjaar* of 1672	27
1.2	Romeyn de Hooghe, *Ellenden klacht Van het Bedroefde Nederlandt. Sedert het Jaer 1672 tot den Aller-heyligen Vloet van het Jaer 1675*, 1675	29
2.1	Daniël Stopendaal, *Vuurwerk bij de viering van de Vrede van Utrecht*, 1713	52
2.2	Hendrick Hondius, *Koeien in een landschap*, 1644	61
2.3	Jan van de Velde after Willem Buytewech, *Terra* (The Elements), 1622	65
3.1	Christmas Flood mortality across the North Sea	92
3.2	Christmas Flood mortality in Groningen	113
3.3	Post-1717 dike design improvements	119
3.4	Diagrams of *kisthoofden*	120
4.1	Pieter Straat, *Nieuwe kaarte van het dijkgraafschap van 't Ooster Baljuwschap van West-Vriesland, genaamt Medenblick en de Vier Noorder Coggen*, 1730	123
4.2	Pieter Straat, *Nieuwe kaarte van het dijkgraafschap van 't Ooster Baljuwschap van West-Vriesland, genaamt Medenblick en de Vier Noorder Coggen*, 1730 (detail)	125
4.3	Early evidence of shipworm infestations	128
4.4	Jan Ruyter, *Studie van wormen*, 1726–44	130
4.5	Total number of VOC ship arrivals in Dutch waters	135
4.6	Spring and summer temperatures in central England and the Netherlands between 1659 and 1799	137

4.7	Anon., *De slechte toestand van de zeedijk vanaf Diemen*, 1705	141
4.8	M. Walraven, *Profiel van de Westvrieslandtsche Zee Dyk voor de Noorder Coggen*, 1732	142
4.9	Elias Baeck, *Abildung deren höchst schädliche unbekandten See-Würmer*, 1732	146
4.10	Pieter Straat and Pieter van der Deure, *Ontwerp tot een minst kostbaare zeekerste en schielykste herstelling van de zorgeliyke toestand der Westfriesche zeedyken*, 1733	160
5.1	Cornelis Velsen, *Kaart van den loop der rivieren de Rhyn, de Maas, de Waal, de Merwe, en de Lek, door de Provincien van Gelderland, Holland en Utrecht*, 1749	167
5.2	Map of territories affected by inundations of 1740–1	168
5.3	The Rhine–Meuse River System	172
5.4	Isaac Vincentsz. van der Vinne, *Het doorbreeken van den band-dyk voor Elden Anno 1740*, 1741	178
5.5	Melchior Bolstra, *Figurative kaart vande situatie van Gelderland, Holland, Uytrecht en OverYzel, ten regarde van zee, en rivieren*, 1751	186
5.6	Jan l'Admiral, *Bedroefde Watervloed, Voornamentlyk van Het Land van Heusden, van Althona, de Alblasserwaard, Crimperwaard, en Tielerwaard, waar door meer dan 33500 Morgen Lands onder water staan, na 't leven afgetekend*, 1741	189
5.7	Jan Smit, *Ware Afbeelding na het Leven, van het overstroomen, der Revieren, en het Doorbreeken van den Dyck, by Elden ... tot aen de Stat Cuylenburg ... Anno 1740*, 1741	192
5.8	Jan Smit, *Tweede plaat der overstroomingen vande provincien Gelderland en Holland in den jaare 1740 en 1741*, 1741	194
5.9	Jan Smit, *Derde Plaat der overstroominge inhoudende het Ryk Nimweegen ... met alle desselfs doorbraken in 't Jaar 1740 en 1741*, 1741	195
5.10	Jan Smit, *De Vierde Plaat van der overstroomingen der Landen ... waar by gevoegd is een nette afbeelding van de Kettingmoolen die gebruikt is; om het water uit de Stad te maalen in den Jaaren 1740 en 1741*, 1741	196
5.11	Winter and "long winter" temperatures between 1706 and 1800	201

5.12	Jan Caspar Philips, Title page from *Nederlands Water-Nood van den Jaare MDCCXL en MDCCXLI*, 1741	209
5.13	Jan Caspar Philips, *Nederlands Water-Nood en verscheide bezoekingen*, 1751 (detail)	210
6.1	Jan de Groot, *Afbeelding van de Staartsterren, verscheenen in de jaaren 1742 en 1744*	213
6.2	Jan Smit, *Gods slaandehand over Nederland door de pest-siekte onder het rund vee*, 1744	219
6.3	Cattle mortality by *gemeente* (North Holland) and *grietenij* (Friesland) between 1744 and 1745	221
6.4	(a and b) Jan Smit, *Gods slaandehand over Nederland door de pest-siekte onder het rund vee*, 1744 (detail)	231
7.1	The "Stone Man" of Friesland	270
7.2	The Janus face of Terminus	271

Acknowledgments

Like so many foreign visitors to the Netherlands, my first view of the Dutch landscape was from above. I was immediately struck by its rigid order as my plane descended to Schiphol Airport near Amsterdam. My window seat vantage framed tidy lines of trees that flanked narrow fields, all impossibly squeezed between roads and clusters of densely packed buildings. It seemed vast and flat – not altogether discomfiting for a plains Midwesterner – except water seemed to be everywhere! The view was a dissonant mix of the familiar and seemingly alien. It seemed at once insistently modern yet quietly timeworn. The balance of land and water appeared comfortable and controlled, yet I could not escape the feeling that if the landscape could speak, it might tell a different story. I had not yet heard of environmental history, and I would not find the language to articulate my thoughts about what I was seeing for some time, yet I found myself immediately wondering what had fashioned this unique place. I would spend much of the next year (and many more since) exploring the country and learning as much about the landscape, the people, and their shared history as I could. I could not have expected at the time where these travels would take me. This book is one destination.

It has been more than ten years since this first encounter, and although the journey to this book was far from direct, it seems a fitting place to begin because it threw me in the path of so many of the people that would make this work possible. It was during this trip to the Netherlands that I met Petra van Dam. Petra introduced me to Dutch environmental history as an undergraduate exchange student, sponsored me during my Fulbright year at the Free University Amsterdam as a graduate student, and she has championed my work ever since. Her encouragement and insights have been a constant source of support. Petra is a model advisor, a generous and critical reviewer, and an inspiring scholar and colleague.

My travels would eventually take me back to the Midwest and to the University of Kansas, where I would study under Greg Cushman. Greg's infectious enthusiasm for environmental history's messier stories was exciting. He encouraged me to engage the social and natural sciences and stretch the limits of what I then thought history could be. Greg encouraged me to think ambitiously, to work with the tools at my disposal, and develop those I didn't. I see his imprint throughout this book. I am also grateful to Don Worster for his mentorship, which extended from my first semester to my dissertation defense. The University of Kansas has long been fertile ground for environmental historians, and I count among my blessings the opportunities to work with and learn from many inspiring scholars in the discipline (and outside it), including Sara Gregg, Steven Epstein, Ed Russell, Johan Feddema, Joane Nagel, and Leslie Tuttle. My time at the University of Kansas would not have been half as meaningful had it not been for the friendship and support of my fellow graduate students, including Vaughn Scribner, Winchell Delano, Amanda Schlumpberger, Neil Schomaker, Andrés Lira-Noriega, Rebecca Crosthwait, and Alex Boynton. My particular thanks to Joshua Nygren, Nicholas Cunigan, Brian Rumsey, Brandon Luedtke, and P. J. Klinger who have all at some point reviewed the work at the center of this book.

I am also grateful to many other colleagues I have met and learned from in the process of this project. They include Toon Bosch, one of the earliest and most enthusiastic readers of my work who constantly encouraged me to think of the rivers; Harm Pieters, for his collegiality, friendship, and talent for sharing the histories of the landscapes we biked across; Filip van Roosbroeck and Tim Soens for showing me the wider world of social-economic and rural history in the Low Countries; and Michiel Bartels, for introducing me to dike archaeology and West Frisian hospitality. I'd also like to thank Dagomar Degroot, Adrie de Kraker, Christian Rohr, Chantal Camenisch, John McNeill and Ling Zhang, and the anonymous reviewers from Cambridge University Press who donated their time to read and review some or all of this work. I'd also like to thank Sam White, Derek Lee Nelsen, and Karel Davids who opened opportunities and encouraged me to think more deeply about many aspects of this book. My thanks also to Debbie Gershenowitz, who saw the promise of the work at an early stage, and Rachel Blaifeder, Lucy Rhymer, Melissa Ward, and Divya Arjunan for carrying it through to completion. Finally, I'm grateful to the many archivists and librarians across the Netherlands who made this work possible. I'm particular thankful to the staff at the Koninklijke Bibliotheek, Nationaal Archief, Westfries Archief, Tresoar, Stadsarchief Amsterdam, Zeeuws Archief, and Groninger Archieven, but I'd also like to acknowledge the assistance of archivists and librarians at institutions whose documentation appears less frequently or not at all in this book. This research journey was winding and took me in directions I did not expect. The present work is equally indebted to these detours, dead ends, and the assistance that made them possible.

This book project depended on the generous support I received from several organizations and institutions over the years, including the Fulbright Scholar Program, the National Science Foundation, the University of Kansas, The American Society for Environmental History, the Free University Amsterdam, and Creighton University. My thanks also to Wijnie de Groot and Frans Blom, whose Dutch paleography workshops at Columbia University are a unique and invaluable asset for young scholars. I also received helpful feedback from colleagues and opportunities to share my findings through the Stichting voor de Middeleeuwse Archeologie, Rijksuniversiteit Groningen, and Lotte Jensen's disaster research group at Radboud University.

Finally, I'd like to thank my family for their support. First and foremost, my partner Jess who has been my constant companion on this journey, even during the months when we were separated by an ocean and more. She's been my most resolute champion and a source of constant encouragement and inspiration. I'd also like to thank my siblings, Emma and Ike, who have cheered me on longer than almost anyone. I count among my family Errol Ertugruloglu, Laurens Bistervels, Kevin Mulder, David Horvath, and Paul Troost whose constant friendship have made the Netherlands a second home. If that second home has an address, it is certainly on the Willemsstraat because of Annie van den Oever and Pieter Troost's remarkable generosity. Lastly, I'd like to thank my parents, Sara and Marsh. They taught me that history and natural history are twin stories intertwined. Their encouragement is unflagging, and it is to them that I dedicate this book.

Introduction

The world is growing more hazardous. Natural disasters are increasing in frequency and severity, spurred in part by changes associated with a warming planet. In their 2020 joint report, the United Nations Office for Disaster Risk Reduction and the Centre for Research on the Epidemiology of Disasters found that the number of natural disasters rose precipitously since the 1980s, with each year bringing new human and economic losses.[1] Disasters affected 94.9 million people in 2019 alone, and 2020 brought a steady stream of record-breaking calamities, including super typhoons in Southeast Asia, historic wildfires in Australia and across the American West, locust swarms in East Africa and the Middle East, and a record-breaking Atlantic hurricane season. COVID-19 emerged as a global public-health emergency, which compounded the impacts of these and many other disasters.[2] The deadly

---

[1] CRED and UNDRR, *Human Cost of Disasters: An Overview of the Last 20 Years, 2000–2019*. https://cred.be/sites/default/files/CRED-Disaster-Report-Human-Cost2000–2019.pdf. These numbers may reflect, in part, better recording and reporting, and also the influence of climate change.

[2] CRED, *Natural Disasters 2019* (Brussels: CRED, 2020). Regine Cabato and Jason Samenow, "Super Typhoon Goni, World's Most Powerful Storm in Four Years, Smashes into the Philippines," *Washington Post*, November 1, 2020. Alexander I. Filkov, Tuan Ngo, Stuart Matthews et al. "Impact of Australia's Catastrophic 2019/20 Bushfire Season on Communities and Environment. Retrospective Analysis and Current Trends," *Journal of Safety Science and Resilience* 1, no. 1 (2020): 44–56. P. E. Higuera and J. T. Abatzoglou, "Record-Setting Climate Enabled the Extraordinary 2020 Fire Season in the Western United States," *Global Change Biology* 27 (2021): 1–2. Food and Agriculture Administration of the United Nations, "FAO Makes Gains in the Fight against Desert Locusts in East Africa and Yemen but Threat of a Food Security Crisis Remains," *FAO News*, November 5, 2020, www.fao.org/emergencies/fao-in-action/stor

consequences of the pandemic continue as of this writing. The burdens of catastrophes were and are endured unevenly around the world, often mirroring its inequalities, yet no region completely escaped their impacts. In the United States, the risk of hurricanes, wildfires, river floods, and droughts have intensified in recent decades, and the most recent US Climate Assessment warns of greater hazards in the future.[3] A dawning sense of urgency in the face of dramatic and accelerating socioeconomic and environmental change has produced a global clarion call for improved understanding of the roots, consequences, and response to disasters.

History informs this ambition. The past contains an archive of innumerable disaster experiences. Since its emergence in the 1980s, historical disaster studies have developed a variety of innovative strategies to utilize this information.[4] Historical analyses reveal trends in the frequency and severity of disasters, which provide baselines to assess the pace and scale of environmental and social change. History also serves as a laboratory to enrich our understanding of core concepts in disaster theory, such as the environmental and social construction of vulnerability, resilience, and adaptation.[5] More recently, historians have emphasized the importance of disaster perception. Disasters were complex, value-laden moments of profound dislocation that nevertheless remained moored to previous experience, cultural memory, and history. Past societies filtered their interpretation and response to calamity through these lenses, just as they do today. As a result, historical disaster scholarship is now a truly interdisciplinary endeavor, drawing insights from the social and physical sciences, memory studies and folklore, religious studies, as well as

---

ies/stories-detail/en/c/1275091/. Jeff Masters, "A Look Back at the Horrific 2020 Atlantic Hurricane Season," *Yale Climate Connections*, December 1, 2020, https://yaleclimateconnections.org/2020/12/a-look-back-at-the-horrific-2020-atlantic-hurricane-center/. Carly A. Phillips, Astrid Caldas, Rachel Cleetus et al., "Compound Climate Risks in the COVID-19 Pandemic," *Nature Climate Change* 10 (2020): 586–8.

[3] D. R. Reidmiller, C. W. Avery, and D. R. Easterling, *Impacts, Risks, and Adaptation in the United States: Fourth National Climate Assessment, Volume II* (Washington, DC: U.S. Global Change Research Program, 2018).

[4] Gerrit Jasper Schenk, "Historical Disaster Research. State of Research, Concepts, Methods and Case Studies," *Historical Social Research/Historische Sozialforschung* 32, no. 3 (2007): 10.

[5] Bas van Bavel, Daniel Curtis, Jessica Dijkman et al., *Disasters and History: The Vulnerability and Resilience of Past Societies* (Cambridge: Cambridge University Press, 2020). Greg Bankoff, "Comparing Vulnerabilities: Toward Charting an Historical Trajectory of Disasters," *Historical Social Research* 3, no. 32 (2007): 103–14.

material and visual culture.[6] Collectively, this work captures a richly variegated image of calamity's geophysical and human character. These stories enrich our understanding of the past and offer significant insights for the present.

Natural disasters are unique and powerful tools to evaluate historical change. They often result in an outpouring of documentation, whether from state and private institutions, news media and artists, or the personal accounts of the victims themselves. Historians have at their disposal a diverse array of sources that afford a uniquely high-resolution image of environments, cultures, and societies not otherwise available. This information can be mustered to evaluate the transformative power of disasters in history. The contingent nature of calamity resists any uniform characterization, however. Coastal and river floods may unearth buried social conflicts, but they might also engender solidarity. Epidemics may catalyze new medical treatments or state management efforts, but they might also entrench established strategies. Species introductions may pass unremarked, or they might motivate dramatic, even violent social or cultural, reaction. Context is critical. By unraveling the tangled strands of evidence that explain these divergent responses, historians enrich our understanding of environments in flux and communities in crisis.

Natural disasters also shed light on broader historical changes often deemed unconnected to calamity. Much of this research has focused on the modern era, but scholars are increasingly setting their sights on earlier periods. In early-modern Europe alone, research on disasters has contributed new interpretations of the emergence of witchcraft trials, the rise of Enlightened absolutism, and evolving interpretations of humanity's relationship with nature.[7] Disasters can be "totalizing" experiences, affecting

---

[6] Fred Krüger, Greg Bankoff, Benedikt Orlowski, and Terry Cannon, eds., *Cultures and Disasters: Understanding Cultural Framings in Disaster Risk Reduction* (New York: Routledge, 2014). Rasmus Dahlberg, Olivier Rubin, and Morten Thanning Vendelø, eds., *Disaster Research: Multidisciplinary and International Perspectives* (London: Routledge, 2016). Bernd Rieken, *"Nordsee ist Mordsee": Sturmfluten und ihre Bedeutung für die Mentalitätsgeschichte der Friesen* (Münster: Waxmann Verlag, 2005). Marco Folin and Monica Preti, *Wounded Cities: The Representation of Urban Disasters in European Art (14th–20th Centuries)* (Leiden: Brill, 2015). Monica Juneja and Gerrit Jasper Schenk, eds., *Disaster as Image: Iconographies and Media Strategies Across Europe and Asia* (Regensburg: Verlag Schnell & Steiner, 2014).

[7] Wolfgang Behringer, "Climatic Change and Witch-Hunting: The Impact of the Little Ice Age on Mentalities," *Climatic Change* 43, no. 1 (1999): 335–51. Behringer's thesis remains contested. Fredrik Charpentier Ljungqvist, Andrea Seim, and Heli Huhtamaa, "Climate and Society in European History," *Wiley Interdisciplinary Reviews: Climate Change* 12, no. 2 (2021): e691. Timothy D. Walker, "Enlightened Absolutism and the

all aspects of social structure and its relations to the environment.[8] Disasters, thus, offer scholars remarkable opportunities to open new inroads into well-trodden historical themes. Using disasters as a lens to explore broader social, cultural, and environmental change has become a hallmark of historical disaster research.

This book explores the historical experience of calamities to better understand their origins and meaning, just as it uses disasters to interpret an important historical transition – the decline of the Dutch Republic.[9] Between the last decades of the seventeenth century and the first half of the next, the Dutch Republic experienced a series of dramatic, destructive, and interconnected disasters. At the same time, the Dutch reckoned with the growing realization that their Golden Age of prosperity, security, and virtue was waning. Disasters and decline were connected, and this book explores their relationship. The Dutch had spent centuries developing social, economic, and technological systems to manage and exploit their dynamic rivers, coastlines, and landscapes. By the eighteenth century, each were changing in important ways. The disasters of the early eighteenth century were products of these transformations. Dutch perception and response to natural disasters also changed as the limits of their control grew increasingly defined. These responses, I argue, reinforced, transformed, and challenged Dutch interpretations of decline.

These twin approaches in disaster history – the one focusing on disasters as subjects, the other using disasters as tools to evaluate perceptions of social and environmental change – rarely intersect. To bridge these approaches, this book turns to environmental history. Many early, now classic studies reshaped our understanding of disasters in history, from

---

Lisbon Earthquake: Asserting State Dominance over Religious Sites and the Church in Eighteenth-Century Portugal," *Eighteenth-Century Studies* 48, no. 3 (2015): 307–28. Cindy Ermus, "The Spanish Plague that Never Was: Crisis and Exploitation in Cádiz during the Peste of Provence," *Eighteenth-Century Studies* 49, no. 2 (2016): 167–93. Michael Kempe, "Noah's Flood: The Genesis Story and Natural Disasters in Early Modern Times," *Environment and History* 9, no. 2 (2003): 151–71. Alessa Johns, ed., *Dreadful Visitations: Confronting Natural Catastrophe in the Age of Enlightenment* (New York: Routledge, 1999).

[8] Anthony Oliver-Smith, "'What Is a Disaster?': Anthropological Perspectives on a Persistent Question," in *The Angry Earth: Disaster in Anthropological Perspective*, eds. Anthony Oliver-Smith and Susanna Hoffman (London: Routledge, 1999), 20.

[9] Contemporaries referred to the Dutch Republic by several names, including the "Republic," "the Netherlands," and "the United Provinces." This book retains those conventions.

the Columbian encounter to the Dust Bowl.[10] The contributions of the field to historical disaster scholarship have only deepened in recent decades. The depth and consistency of engagement result from their close correspondence. Core themes in historical disaster scholarship parallel the central preoccupations of environmental history, whether reconstructing past social and physical environments, interrogating the social and economic systems that reorganized those environments, or exploring the ways that perception, ideas, and values informed those relationships.[11] Their cross-pollination has proved mutually beneficial as both fields have grown and matured.

They are also different in key ways. "Natural disasters" are increasingly understood to be social phenomena that reflect the characteristics of societies more than physical environmental shocks.[12] This explains why natural hazards often produce wildly uneven consequences for affected communities. Foregrounding unequal conditions of risk and opportunity revolutionized our understanding of disasters by re-centering attention on the social conditions that produce, exacerbate, and mitigate calamity. Perhaps as a result, however, "nature" is often a less-than-dynamic feature of disaster scholarship. Environmental history offers a hybrid approach that counterbalances this tendency.[13] Environmental histories of flooding, for instance, have explored the relationships between environmental phenomena, such as recurrent extreme events and long-term patterns in climate or coastal erosion, but also social conditions, including land ownership and settlement, industrial transformation of coasts or riverscapes, and civic or religious interpretations of environmental

---

[10] Alfred W. Crosby, *The Columbian Exchange: Biological and Cultural Consequences of 1492* (Westport, CT: Greenwood Publishing Group, 1972). Donald Worster, *Dust Bowl: The Southern Plains in the 1930s* (Oxford: Oxford University Press, 1979). Christof Mauch, "Introduction," in *Natural Disasters, Cultural Responses: Case Studies toward a Global Environmental History*, eds. Christof Mauch and Christian Pfister (New York: Lexington Books, 2009), 5–6.

[11] Donald Worster, "Transformations of the Earth: Toward an Agroecological Perspective in History," *The Journal of American History* 76, no. 4 (1990): 1087–106.

[12] Kenneth Hewitt, "The Idea of Calamity in a Technocratic Age," in *Interpretations of Calamity: From the Viewpoint of Human Ecology*, ed. Kenneth Hewitt (Boston: Allen & Unwin, 1983), 25. Terry Cannon, "Vulnerability Analysis and the Explanation of 'Natural Disasters,'" in *Disasters, Development and Environment*, ed. Ann Varley (Chichester: John Wiley and Sons, 1994), 13–30.

[13] Paul S. Sutter, "The World with Us: The State of American Environmental History," *Journal of American History* 100, no. 1 (2013): 94–119.

change.[14] Natural disasters emerged from these mutually reinforcing social and environmental changes and often influenced them in turn. The implications of this approach beyond the history of disaster are equally wide reaching. Environmental history cut its teeth reinterpreting historical causation and rewriting familiar historical narratives. Environmental histories of disaster present the same opportunities.

1.1 DISASTER AND THE INTERPRETATION OF HISTORY

The decline of the Dutch Republic is an historical narrative ripe for reinterpretation. Dutch decline in the eighteenth century was measured against its Golden Age, a period that roughly spanned the previous century. The Netherlands during the seventeenth century had been a vibrant center of global trade, a hub of empire, and the entrepôt of Europe.[15] It was during the Golden Age that the Dutch birthed the West (*WIC*) and East India Trading Companies (*VOC*), and Amsterdam emerged as a nexus for the exchange of information, goods, and capital and nurtured what some scholars consider the first "modern economy."[16] Dutch society was oligarchic, pluralistic, and strongly

---

[14] Greg Bankoff, "Constructing Vulnerability: The Historical, Natural and Social Generation of Flooding in Metropolitan Manila," *Disasters* 27, no. 3 (2003): 224–38. Tim Soens, "Flood Security in the Medieval and Early Modern North Sea Area: A Question of Entitlement?," *Environment and History* 19, no. 2 (2013): 209–32. Stéphane Castonguay, "The Production of Flood as Natural Catastrophe: Extreme Events and the Construction of Vulnerability in the Drainage Basin of the St. Francis River (Quebec), Mid-Nineteenth to Mid-Twentieth Century," *Environmental History* 12, no. 4 (2007): 820–44. John Emrys Morgan, "Understanding Flooding in Early Modern England," *Journal of Historical Geography* 50 (2015): 37–50. Grace Karskens, "Floods and Flood-Mindedness in Early Colonial Australia," *Environmental History* 21, no. 2 (2016): 315–42. James A. Galloway, "Coastal Flooding and Socioeconomic Change in Eastern England in the Later Middle Ages," *Environment and History* 19, no. 2 (2013): 173–207. Piet van Cruyningen, "Sharing the Cost of Dike Maintenance in the South-Western Netherlands: Comparing 'Calamitous Polders' in Three 'States,' 1715–1795," *Environment and History* 23, no. 3 (2017): 363–83.

[15] David Onnekink and Gijs Rommelse, *The Dutch in the Early Modern World: A History of a Global Power* (Cambridge: Cambridge University Press, 2019).

[16] Oscar Gelderblom, Abe De Jong, and Joost Jonker, "The Formative Years of the Modern Corporation: The Dutch East India Company VOC, 1602–1623," *The Journal of Economic History* 73, no. 4 (2013): 1050–76. Clé Lesger, *The Rise of the Amsterdam Market and Information Exchange: Merchants, Commercial Expansion and Change in the Spatial Economy of the Low Countries, c. 1550–1630* (Surrey: Ashgate, 2006). Jan de Vries and Ad van der Woude, *The First Modern Economy: Success, Failure, and Perseverance of the Dutch Economy, 1500–1815* (Cambridge: Cambridge University Press, 1997).

divided in politics and religion, but it nevertheless nurtured luminaries in European art, science, and philosophy. This was the era of Rembrandt and Vermeer, Van Leeuwenhoek and Huygens, Spinoza and the renowned printing culture that fostered Descartes, Bayle, and Locke. The Republic, whose borders largely correspond to the modern Netherlands, was a unique and powerful geopolitical entity in the early-modern world. In an age of state centralization, the Republic was less a single country than a union of seven largely sovereign provinces. Each exhibited stark differences in urbanization, economic development, and political interest (with Holland by far the wealthiest, most populous, and powerful), yet the Dutch Republic proved remarkably stable and adept at exerting its influence in Europe and around the globe[17] (Figure I.1).

The very existence of the Republic was perhaps its greatest and most surprising accomplishment. After defeating the Spanish Empire in an eighty-year-long revolt, this loose confederacy of provinces, born of necessity, bordered by superior military rivals and seemingly at odds with its own amphibious geography, remained a dominant European power for over a century. In an era defined by the adverse climatic conditions of the Little Ice Age and a "seventeenth-century crisis" that featured widespread conflict, economic dislocation, demographic stagnation, and political instability across much of Europe and the globe, the Dutch appeared the great European exception.[18] Indeed, Dutch success was both perplexing and enviable to their neighbors. "There grows nothing in Holland," one anonymous observer declared, "yet there is the wealth of the

---

[17] Maarten Prak, *The Dutch Republic in the Seventeenth Century: The Golden Age* (New York: Cambridge University Press, 2005). Oscar Gelderblom, ed., *The Political Economy of the Dutch Republic* (Surrey: Ashgate, 2016).

[18] Jan de Vries, *The Economy of Europe in an Age of Crisis, 1600–1750* (Cambridge: Cambridge University Press, 1976), 1–29. Niels Steensgaard, "The Seventeenth-Century Crisis," in *The General Crisis of the Seventeenth Century*, eds. Geoffrey and Lesley M. Smith Parker (London: Routledge, 1985), 26–56. Dagomar Degroot, *The Frigid Golden Age: Climate Change, The Little Ice Age, and the Dutch Republic, 1560–1720* (Cambridge: Cambridge University Press, 2018). Ivo Schöffer, "Did Holland's Golden Age Coincide with a Period of Crisis?," in *The General Crisis of the Seventeenth Century*, eds. Geoffrey and Lesley M. Smith Parker (London: Routledge, 1985), 83–109. The extra-European implications of this phenomenon have been explored in Geoffrey Parker, *Global Crisis: War, Climate Change and Catastrophe in the Seventeenth Century* (New Haven, CT: Yale University Press, 2013). Sam White, *A Cold Welcome: The Little Ice Age and Europe's Encounter with North America* (Cambridge: Harvard University Press, 2017). Sam White, *The Climate of Rebellion in the Early Modern Ottoman Empire* (Cambridge: Cambridge University Press, 2011).

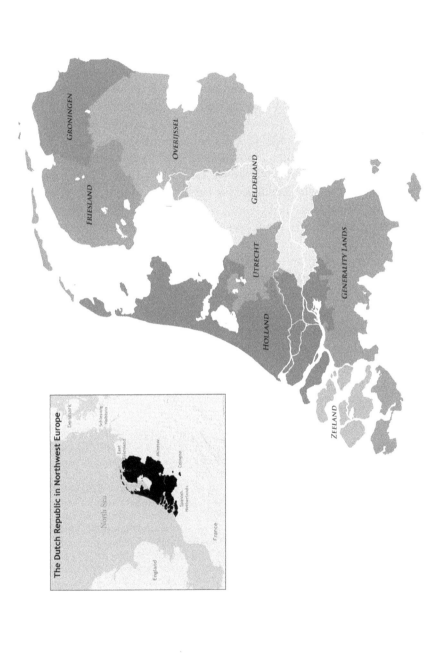

FIGURE 1.1 The United Provinces of the Netherlands in 1710. The Dutch Republic consisted of Holland, Utrecht, Friesland, Groningen, Overijssel, Gelderland, and Zeeland. The Estates General controlled several additional territories, called Generality lands. Provincial boundaries adapted from Dr. O. W. A. Boonstra (2007). NLGis shapefiles. DANS. http://dx.doi.org/10.17026/dans-xb9-t677.

world."[19] The country had flourished amidst social, religious, political, and environmental upheaval, which made its reversal of fortune all the more striking.

By the dawn of the eighteenth century, the Dutch Golden Age of wealth, power, and prestige appeared to be waning. The Republic remained among the wealthiest states in Europe, but core sectors of its economy experienced relative, if not absolute, decline. Between 1650 and 1750, Dutch agriculture endured a prolonged recession, the Dutch herring fishery collapsed, and industry and urbanization flatlined. Dutch primacy in overseas trade likewise diminished. After 1713, the Dutch no longer ranked in the top tier of geopolitical powers in Europe. By the last decades of the eighteenth century, contemporaries in the Republic and abroad widely described Dutch decline in universalizing language. In the words of eminent political historian E. H. Kossmann, decline "was thought to be total, on all levels, in all human endeavor – moral, economic, social, cultural, political."[20] This characterization remains influential today.

The origins and meaning of Dutch decline have provoked sustained debate since the eighteenth century.[21] Its enduring appeal as an historical question speaks to the complexity and significance of the subject. Some contemporaries pointed to the decay of Dutch morality and civic virtue. Others highlighted mercantilist trade policies, an inefficient tax system overwhelmed by foreign wars, or a general conservatism born of past prosperity. Nineteenth-century historians wrote disparagingly of the previous century's decadence and stagnation and termed the era the Periwig Period (*Pruikentijd*).[22] Named after the French fashion of wearing powdered wigs, it evoked an atmosphere of indulgent malaise. Historians employed the term well into the twentieth century. No less a figure than cultural historian Johan Huizinga still described the "collapse

---

[19] *The Dutch Drawn to Life* (London: Tho. Johnson and H. Marsh, 1664), introduction.
[20] E. H. Kossmann, "The Dutch Republic in the Eighteenth Century," in *The Dutch Republic in the Eighteenth Century: Decline, Enlightenment, and Revolution*, eds. Margaret C. Jacob and Wijnand W. Mijnhardt (Ithaca, NY: Cornell University Press, 1992), 28.
[21] Modern interest in decline was stimulated by Johannes de Vries, *De economische achteruitgang der Republiek in de achttiende eeuw* (Leiden: Stenfert Kroese, 1968). De Vries's arguments find some earlier grounding in J. G. Van Dillen, "Omstandigheden en psychische factoren in de economische geschiedenis van Nederland," in *Mensen en achtergronden*, ed. J. G. Van Dillen (Groningen: J. B. Wolters, 1964), 53–79.
[22] Wyger R. E. Velema, *Republicans: Essays on Eighteenth-Century Dutch Political Thought* (Leiden: Brill, 2007).

of Dutch culture in the eighteenth century" in 1941. With more than a touch of irony, Huizinga termed the eighteenth century the true "golden age" because "wood and steel, pitch and tar, colour and ink, pluck and piety, fire and imagination" more aptly described the dynamism of the seventeenth century.[23]

Recent interpretations of decline vary, and explanations range from geopolitical competition and protectionism to structural changes in the socioeconomic fabric of Dutch society. Most economic historians maintain nuanced positions that contrast with earlier historians' breathless proclamations of universal, national decay. The trajectory of decline varied over time and manifested differently by region, city, and sector of the economy. The deurbanization of densely populated parts of coastal Holland, for instance, was among the most dramatic consequences of this change. Leiden, formerly a center of textile production with 70,000 inhabitants in the last quarter of the seventeenth century, dropped to half that number by the third quarter of the eighteenth. Other areas of the Republic, especially its more thinly populated rural regions, experienced less of this economic and social tumult. Even in the aggregate, the Dutch Republic remained perhaps the wealthiest country in Europe with per-capita incomes well in excessive of its rivals. The Republic certainly experienced a loss in geopolitical stature and economic power, but that decline was most apparent when compared to the growing influence of competitors such as England and France.[24]

---

[23] Johan H. Huizinga, *Dutch Civilisation in the Seventeenth Century: And Other Essays* (New York: Harper & Row, 1969), 100, 104.

[24] Herman Diedriks, "Economic Decline and the Urban Elite in Eighteenth-Century Dutch Towns: A Review Essay," *Urban History Yearbook* 16 (1989): 78–81. De Vries and Van der Woude, *First Modern*, 52–5. Jonathan Israel, *The Dutch Republic: Its Rise, Greatness, and Fall, 1477–1806* (New York: Oxford University Press, 1995), 1007. J. L. van Zanden, "De economie van Holland in de periode 1650–1805: Groei of achteruitgang? Een overzicht van bronnen, problemen en resultaten," *BMGN: Low Countries Historical Review* 102, no.4 (1987): 562–609. J. C. Riley, "The Dutch Economy after 1650: Decline or Growth?," *Journal of European Economic History* 13 (1984): 521–69. Jan de Vries, "The Decline and Rise of the Dutch Economy, 1675–1900," in *Technique, Spirit, and Form in the Making of the Modern Economies: Essays in Honor of William N. Parker*, eds. Gary R. Saxonhouse and Gavin Wright (London: JAI Press, 1984), 149–89. Jonathan Israel, *Dutch Primacy in World Trade, 1585–1740* (New York: Oxford University Press, 1989), 392. Maarten Prak and J. L. van Zanden, *Nederland en het poldermodel: Sociaal-economische geschiedenis van Nederland, 1000–2000* (Amsterdam: Bert Bakker, 2013), 165. Patrick O'Brien, "Mercantilism and Imperialism in the Rise and Decline of the Dutch and British Economies, 1585–1815," *De Economist* 148, no. 4 (2000): 469–501. W. Frijhoff, J. Kloek, and M. Spies, *Dutch Culture in a European Perspective: 1800, Blueprints for*

Social and cultural historians have also revisited the relationship between the Golden Age and eighteenth-century decline. The characterization of the seventeenth century as an idyllic moment of (usually Holland's) prosperity and virtue was largely fashioned by eighteenth- and nineteenth-century moralists and historians who required an Edenic myth to explain their own perceived fall from grace. As we will see, these Golden Age myths supplied ample fuel for social and moral critique in the eighteenth century, but they also reified an enduring vision of the seventeenth century that glossed over the undeniably uneven and unequal experiences of the era.[25] By contrast, the dynamism of the eighteenth century is often obscured within its gilded shadow. The reputation of eighteenth-century culture, for instance, has long suffered by comparison to its forbearer. In the words of one historian, the era "cultivated no great spirits. Neither the arts nor the sciences produced any dizzyingly high achievements. Its people were mediocre and so were their accomplishments."[26] Yet just as with social and economic history, scholars have gradually begun to revisit even the most established evidence of cultural decay. Scholarship on eighteenth-century Dutch literature and the visual arts, for instance, have enjoyed a minor renaissance in recent decades.[27] Historians are no longer comfortable with universalizing characterizations of either the glittering seventeenth or stagnant eighteenth centuries, and, as a result, their work is yielding a richly figured image of the era.

Environmental changes have thus far made little impression in this evolving picture of the Dutch eighteenth century even though the period featured dramatic transformation of what many contemporaries considered the material foundations of the Republic's prosperity and cultural

---

a National Community (Assen: Royal Van Gorcum, 2004), 16–19, 31–4, 69. Onnekink and Rommelse, *The Dutch in the Early Modern World*, 203–10.

[25] Perhaps the most pernicious consequence is the tendency to ignore or downplay the roots of Dutch prosperity in colonialism and imperialism. Helmer J. Helmers and Geert H. Janssen, "Introduction: Understanding the Dutch Golden Age," in *The Cambridge Companion to the Dutch Golden Age*, eds. Helmer J. Helmers and Geert H. Janssen (Cambridge: Cambridge University Press, 2018), 1–12.

[26] A. T. van Deursen, "The Dutch Republic, 1588–1780," in *History of the Low Countries*, eds. J. C. H. Blom and E. Lamberts (New York: Berghahn Books, 2006), 215.

[27] Inger Leemans and Gert-Jan Johannes, *Worm en donder: Geschiedenis van de Nederlandse literatuur, 1700–1800: De Republiek* (Amsterdam: Bert Bakker, 2013). Frans Grijzenhout, "A Myth of Decline," in *The Dutch Republic in the Eighteenth Century: Decline, Enlightenment, and Revolution*, eds. Margaret C. Jacob and Wijnand W. Mijnhardt (Ithaca: Cornell University Press, 1992), 324–37. Piet Bakker, "Crisis? Welke crisis? Kanttekeningen bij het economisch verval van de schilderkunst in Leiden na 1660," *De Zeventiende Eeuw* 27, no. 2 (2012): 232.

identity. Despite its high degree of urbanization, the Dutch Republic remained an agrarian society in the eighteenth century. Its highly specialized agricultural sector worked symbiotically with commercial and industrial activity.[28] The changing environmental conditions of rural society thus affected Dutch prosperity as a whole. The seventeenth century certainly featured its own suite of challenges in the countryside, yet in the context of decline, natural disasters seemed unique, increasingly common, and far more consequential. Cattle plague outbreaks, for instance, repeatedly threatened dairy production and an oxen trade that linked rural environments to urban consumers. These assaults on rural productivity were also tests of identity and character. Cattle were among the most potent cultural and economic embodiments of Golden Age wealth and Arcadian fertility. Their loss, whether through disease, floods, or other natural disasters, registered in both financial and moral terms.

The Republic's prosperity was also undeniably tied to water. The North Sea served as a vital link to overseas markets and a seemingly boundless resource for exploitation. The sea also posed great risks and inspired primal fear. Rivers were instruments of industry, agriculture, and transportation, yet by the eighteenth century they appeared increasingly unstable. Much of the Republic's population, major cities, and most valuable agricultural land lay below sea level or exposed to riverine inundation. A vast water management infrastructure protected these resources, yet in the context of climatic change, the slow erosion and subsidence of the coast, as well as short-term changes such as storm surges, ice dams, and the arrival of shipworms, the Dutch encountered new and increasingly troubling threats. The natural disasters of the eighteenth century pushed these concerns into public view. For a society that so closely linked its prosperity, virtue, and security to the mastery of nature, disasters seemed evidence of decline.

Decline scholarship has largely ignored these environmental challenges for two reasons. First, they disproportionately affected the countryside. Most interpretations of decline focus on trade or geopolitical changes rather than rural concerns. The *Propositie* of 1751, one of the earliest state-sponsored efforts to assess the roots of economic decline, emphasized its commercial roots.[29] This document heavily influenced

---

[28] De Vries and Van der Woude, *First Modern*, 195–6.
[29] *Propositie van syne hoogheid ter vergaderingen van haar Hoog Mogende en haar Edele Groot Mog. gedaan, tot redres en verbeeteringe van den koophandel in de Republicq* ('s Gravenhage: Jacobus Scheltus, 1751).

eighteenth-century discussions of the subject and later scholarship continued to favor interpretations that privileged trade and the interests of urban elites. A perspective that emphasizes rural society and includes the views of ministers, poets, printmakers, farmers, and surveyors reveals a different picture. Their accounts of adversity were often first-person and deeply affecting. They revealed economic ambition, religious conviction, and anxieties about the future. Their voices did not represent every segment of rural society, yet they speak to important changes in rural environments that have thus far remained marginal in accounts of decline.

Second, environmental catastrophes make few appearances in the literature because direct, causal relationships between disaster and long-term decline are lacking. This is due in part to the different scales of the phenomena. Relative to decades-long processes of stagnation, historians tend to characterize natural disasters as moments of intense, albeit temporary setbacks.[30] Disasters may amplify crises for specific populations during specific moments in history, but they fade into the background as our scope widens. Historians have also demonstrated that the relationship between disasters and social crises are complex and contingent. Even when clustered, disasters did not necessarily result in long-term social or economic problems. During the seventeenth century, the Republic endured its fair share of calamity and flourished. Disaster outcomes depended upon far too many variables to establish simple connections. Even for the Dutch Republic, which was relatively resilient in the face of environmental changes, disasters produced both crisis and opportunity.[31]

---

[30] Jan de Vries, "Measuring the Impact of Climate on History: The Search for Appropriate Methodologies," *The Journal of Interdisciplinary History* 10, no. 4 (1980): 599–630. Johan de Vries, *Economische Achteruitgang*, 151. A. M. van der Woude, *Het Noorderkwartier: een regionaal historisch onderzoek in de demografische en economische geschiedenis van westelijk Nederland van de late middeleeuwen tot het begin van de negentiende eeuw* (Wageningen: H. Veenman en zonen, 1983), 594. De Vries, *Economy of Europe in an Age of Crisis*, 73. De Vries and Van der Woude, *First Modern*, 123, 212, 302. Marjolein 't Hart, "Town and Country in the Netherlands, 1550–1750," in *Town and Country in Europe, 1300–1800*, ed. S. R. Epstein (Cambridge: Cambridge University Press, 2001), 103. C. R. Boxer, "The Dutch Economic Decline," in *The Economic Decline of Empires*, ed. Carlo Cipolla (London: Taylor & Francis, 1970), 253. J. Aalbers, "Holland's Financial Problems (1713–1733) and the Wars against Louis XIV," in *Britain and the Netherlands* Vol. 6, eds. A. C. Duke and C. A. Tamse (The Hague: Martinus Nijhoff, 1978), 81. J. L. Price, "The Dutch Republic," in *A Companion to Eighteenth-Century Europe*, ed. Peter H. Wilson (Oxford: Wiley, 2009), 298.

[31] I use the term "resilience" broadly. In disaster studies, it remains a contested concept and has been interpreted in a variety of ways, from "buffering capacities" to the ability of social systems to "bounce back" from shocks, to the likelihood of adaptation. Van Bavel et al., *Disasters and History*, 35–7.

Emphasizing direct causal relationships deflects attention from the ways that natural disasters continued to shape Dutch society and culture long after the environmental hazards receded. Their long lives were codified in institutional reorganization; materialized in rebuilt infrastructure; memorialized in poems, maps, and monuments; or etched onto the landscape itself. Disasters left enduring impressions in social and physical environments, especially if they compounded other calamities. Disasters likewise influenced the perception of seemingly disconnected crises, including those associated with Dutch decline. Disasters forced the faithful to square their Golden Age past with their diminished present. The rift between the two widened over the course of the eighteenth century. Natural disasters did not cause decline. This does not mean that disasters were unimportant, nor does it mean they were unrelated. Anxieties about decline shaped the perception of disaster origins; it influenced disaster decision-making and encouraged people to connect calamities to broader cultural, environmental, and social concerns.

Finally, natural disasters highlight the inadequacy of decline as a unifying framework for understanding the collective experience of the Dutch eighteenth century. Social, economic, and environmental challenges varied by region, changed over time, and provoked mixed interpretations of their meaning and consequences. Disasters thus map the multiplicity of declensionist narratives. Disasters resulted in trauma and at times existential despair, but also self-reflection and calls for improvement. River systems could be modified, coastal infrastructure redesigned, and disease could be controlled with enhanced state management and novel medical treatments. By the mid-eighteenth century, early Enlightenment thinkers linked moral, social, and environmental adversity together in a progressive program of national improvement directed at decline, which they considered the primary problematic of the age. Work that examines this constellation of crisis and response tends to emphasize the latter half of the century when declensionist anxieties appeared fully formed, yet the natural disasters covered in this book showcase its formation in the decades before.[32] What emerge are often dynamic visions of

---

[32] Jan Willem Buisman, *Tussen Vroomheid en Verlichting: een cultuurhistorische en- sociologisch onderzoek naar enkele aspecten van de Verlichting in Nederland (1755–1810)* (Zwolle: Waanders, 1992). Disasters make few, if any, appearances in Dutch Early Enlightenment scholarship. Wiep van Bunge, ed., *The Early Enlightenment in the Dutch Republic, 1650–1750* (Leiden: Brill, 2003). Jonathan Israel, *Radical Enlightenment: Philosophy and the Making of Modernity, 1650–1750* (Oxford: Oxford University Press, 2001). W. Frijhoff and M. Spies, *Dutch Culture in a*

Dutch society and culture at odds with historical assessments of early eighteenth-century stagnation. Natural disasters were trials of faith and reason. Each calamitous event presented new opportunities for moral, economic, and social recovery. Disasters, thus, paint a different, more resilient portrait of the Dutch at the end of their Golden Age.

I.2 HISTORY AND THE INTERPRETATION OF DISASTER

The eighteenth-century Dutch Republic is an ideal time and place to explore the environmental history of disasters. Natural disasters not only manifested frequently but also coincided with profound changes in the Republic's social and physical environment. No western European territory had been so completely transformed by human influence – an influence that both limited and contributed to disasters throughout the earlymodern period. In the best of times, the centuries-long investment in coastal flood defenses and river management strategies kept the waters at bay and opened new landscapes to productive use. Technologies of environmental control likewise fashioned new vulnerabilities. Dike adaptations that promised greater protection against erosion and storm surges simultaneously invited outbreaks of wood-boring mollusks. Reclamation and commercialization of the countryside encouraged agricultural productivity, and networks of trade drew the agricultural wealth of northwest Europe into the Dutch Republic. In the process, these changes exposed Dutch landscapes to perennial flood risk and panzootic outbreak. Natural disasters today reflect the resource-driven exploitation of environments with even greater clarity. The manufacturing of risk at the interface of natural and technological systems is often associated with modernity itself.[33] The Dutch eighteenth century encourages us to explore its premodern genealogy.

European Perspective: 1650, Hard-Won Unity (Assen: Uitgeverij Van Gorcum, 2004). N. C. F. van Sas, *De metamorfose van Nederland* (Amsterdam: Amsterdam University Press, 2005). Margaret C. Jacob and Wijnand W. Mijnhardt, eds., *The Dutch Republic in the Eighteenth Century: Decline, Enlightenment, and Revolution* (Ithaca, NY: Cornell University Press, 1992).

[33] Ian Burton, Robert W. Kates, and Gilbert F. White, *The Environment as Hazard* (New York: Guilford Publications, 1993). Simon Dalby, "Anthropocene Formations: Environmental Security, Geopolitics and Disaster," *Theory, Culture & Society* 34, no. 2–3 (2017): 233–52. Ulrich Beck, *Risk Society: Towards a New Modernity*, trans. Mark Ritter (London: Sage, 1992). Franz Mauelshagen, "Defining Catastrophes," in *Catastrophe and Catharsis: Perspectives on Disaster and Redemption in German*

Natural disasters were also a formative part of Dutch identities during the early-modern period. The struggle with water in particular left deep impressions in culture, political institutions, and religious beliefs.[34] The amphibious physical and human geography of the Netherlands exposed most of the population to the risk of flooding, yet deep historical experience with disaster also produced unique cultures of coping that included impressive management strategies to mitigate those challenges. The legacies of these adaptations resonate today. Flood disasters in recent decades have invariably provoked calls in even the most developed states to "learn from the Dutch," whether their technocratic approaches to the control of water or more recent challenges to that paradigm.[35] Dutch disasters offer insight into the way history and memory molded this learning process.

The world of the eighteenth century, of course, presented far different challenges than those experienced in the modern Netherlands. Yet in those differences, other opportunities for comparison emerge. Unlike today, disasters were a "frequent life experience" in the eighteenth century. Between 1713 and 1745 alone, the Dutch Republic experienced numerous natural disasters. River and coastal floods repeatedly inundated towns and the countryside. Cattle plague swept over the rural landscape on two separate occasions. Bitterly cold winters provoked widespread dearth, and droughts led to the outbreak of mollusks in the wooden infrastructure that protected Dutch dikes. In recent years, social scientists and historians have begun emphasizing the cumulative outcome of reoccurring disasters in cultural coping strategies and the perception of

---

Culture and Beyond, eds. Katharina Gerstenberger and Tanja Nusser (Rochester, NY: Camden House, 2015), 179–84.

[34] Lotte Jensen, *Wij tegen het water: Een eeuwenoude strijd* (Nijmegen: Uitgeverij Vantilt, 2018).

[35] Martin Reuss, "Learning from the Dutch: Technology, Management, and Water Resources Development," *Technology and Culture* 43, no. 3 (2002): 465–72. John McQuaid, "What the Dutch Can Teach Us about Weathering the Next Katrina," *Mother Jones*, August 28, 2007, www.motherjones.com/environment/2007/08/what-dutch-can-teach-us-about-weathering-next-katrina/. Russell Shorto, "How to Think Like the Dutch in a Post-Sandy World," *The New York Times*, April 9, 2014. www.nytimes.com/2014/04/13/magazine/how-to-think-like-the-dutch-in-a-post-sandy-world.html. Kiah Collier, "Can the 'Masters of the Flood' Help Texas Protect Its Coast from Hurricanes?," *The Texas Tribune*, July 15, 2019, www.texastribune.org/2019/07/15/can-masters-flood-help-texas-protect-its-coast-hurricanes/.

risk. The repeated disasters that marked the Dutch eighteenth century invite comparisons to other "regions of risk" in the past and present.[36]

The disasters of the eighteenth century seem increasingly relevant in one other way. Despite their long experience with hazards such as storm surges, extreme river discharge, and outbreaks of disease in animal populations (epizootics), the catastrophes of the early eighteenth century seemed unnervingly exceptional to the Dutch. Cattle plague manifested with unmatched intensity, and river floods appeared more severe and frequent. The deep cultural memory of coastal flooding provided few analogues for the devastation wrought by the Christmas Flood of 1717. The shipworm epidemic of the 1730s appeared utterly novel. In an era already marked by hardship, unprecedented natural disasters appeared the new normal. In reality, each of these disasters had historical precedent, yet exceptionality served as a dominant frame for Dutch perception. Exceptionality incentivized unique technological responses, justified translating moral interpretations of decline across scale, and revealed the subtle ways that the memory and history of environmental change could be molded to suit diverse needs. Learning from disaster was a political, technological, and cultural act refracted through the lens of perception. Focusing on exceptionality contributes to a growing literature on the ways that disaster perception encourages and constrains response.[37] In a warming future that threatens increasingly frequent and severe disasters, these stories assume still greater relevance.

1.3 AN ENVIRONMENTAL HISTORY OF DISASTER AND DECLINE

This book makes the case that the environmental history of the Dutch Republic offers compelling lessons for our understanding of natural disaster, and those disasters inform our understanding of its decline. It grounds this argument in four interpretive frameworks, building upon themes at the nexus of environmental history and disaster studies. These frameworks help explain individual disaster origins and outcomes, as well

---

[36] Greg Bankoff, *Cultures of Disaster: Society and Natural Hazards in the Philippines* (London: Routledge, 2003). Kenneth Hewitt, *Regions of Risk: A Geographical Introduction to Disasters* (Essex: Longman, 1997).

[37] Christian Pfister, "Climatic Extremes, Recurrent Crises and Witch Hunts: Strategies of European Societies in Coping with Exogenous Shocks in the Late Sixteenth and Early Seventeenth Centuries," *The Medieval History Journal* 10, no. 1-2 (2006): 33-73. W. Neil Adger, Suraje Dessai, Marisa Goulden et al., "Are There Social Limits to Adaptation to Climate Change?," *Climatic Change* 93, no. 3-4 (2009): 335-54.

as the way perception molded experience into response. First, disasters were hybrid phenomena that blur distinction between social and environmental conditions of adversity. The "natural"-ness of natural disasters is no longer assumed in scholarship. Rather than describing disastrous events as singular and historically inexplicable acts of God, historians and social scientists now interpret them as socially and culturally constructed phenomena.[38] "De-naturalizing" disaster should not diminish the role of physical environmental influences, nor the role of people in shaping them. In the words of anthropologist Anthony Oliver-Smith, disasters "occur at the intersection of nature and culture and illustrate, often dramatically, the mutuality of each in the constitution of the other."[39] Dutch catastrophes bridged nature and culture. They produced cultural and environmental consequences, just as they emerged out of mutually reinforcing social vulnerabilities and physical hazards.

Dutch victims and observers understood the multidimensionality of disasters, although they never would have described them in this way. Disaster prints and maps, sermons and farmers' journals, state resolutions and medical treatises all depicted catastrophes as totalizing phenomena. Documents that described the "Disaster Year" of 1672, for instance, integrated the cumulative impacts of war and social unrest, and also coastal floods and windstorms. This hybrid interpretation of adversity connected disaster to decline, linking broken dikes, conquered cities, and toppled buildings to more general themes of impoverishment and loss of geopolitical stature. Many historians would later argue the Disaster Year inaugurated Dutch decline, but its social-environmental character has been largely ignored. Emphasizing disaster hybridity performs the valuable task of deconstructing categories historians have tended to impose upon calamity and brings new insight into the relationship between disaster and decline.

Second, disasters were both events and cumulative processes. They were moments of intense violence and social-environmental developments

---

[38] Anthony Oliver-Smith and Susanna Hoffman, eds., *The Angry Earth: Disaster in Anthropological Perspective* (London: Routledge, 1999). Ted Steinberg, *Acts of God: The Unnatural History of Natural Disaster in America* (Oxford: Oxford University Press, 2000).

[39] Anthony Oliver-Smith, "Theorizing Disasters: Nature, Power, and Culture," in *Catastrophe & Culture: The Anthropology of Disaster*, eds. Susanna M. Hoffman and Anthony Oliver-Smith (Oxford: James Currey, 2002), 24. Piers Blaikie, Terry Cannon, Ian Davis, and Ben Wisner, *At Risk: Natural Hazards, People's Vulnerability and Disasters* (New York: Routledge, 1994).

that slowly played out over centuries. It can be tempting to limit the study of disasters to moments of high drama, limited in space and time. Much of the early historiography of disaster identified catastrophic events as pivots in history.[40] The early-modern period seems particularly amenable to this tendency. A rich tradition of event-based research of disasters continues today, of which work on the 1703 London storm, the 1720 Marseilles plague outbreak, and the Lisbon earthquake of 1755 are oft-cited examples.[41] Historian Greg Bankoff critiques this preoccupation with disasters "as a 'trigger' or 'key moment' in determining or changing the course of human history."[42] Disasters were also cumulative processes. When disasters struck simultaneously or in quick succession, their impacts compounded and interacted in complex ways that disaster scholars term "cascades." The legacy of cascading disasters often continued long after the acute phase of disorder abated.[43] The interpretation of disasters as process pushes historical scholarship to the center of disaster studies. Catastrophes, past and present, resolve as "the outcome of processes that change over time and whose genesis lies in the past."[44]

Both process and event-based approaches have value. Concepts like cultural memory, historical trends like the growth of hydraulic expertise, and environmental changes like sedimentation, subsidence, and shifts in climate only acquire definition when considered in the long term.[45] Event-based approaches on the other hand foreground the environmental and

---

[40] Kenneth Hewitt identified this trend as early as 1983, arguing that this event-centered, near-determinist model of disaster analysis overemphasizes the uniqueness of disaster events and hides their roots in "normal everyday life." Hewitt, *Interpretations of Calamity*, 1983. Research into "focusing events" or episodes of dramatic, catastrophic change that catalyze policy action, media attention, and community mobilization support this trend. Thomas A. Birkland, *After Disaster: Agenda Setting, Public Policy, and Focusing Events* (Washington, DC: Georgetown University Press, 1997).
[41] Johns, *Dreadful Visitations*.
[42] Greg Bankoff, "Time Is of the Essence: Disaster, Vulnerability, and History," *International Journal of Mass Emergencies and Disasters* 22, no. 3 (2004): 23–42.
[43] Disaster cascades are processes triggered by natural hazards, amplified by preexisting vulnerabilities, that work through interrelated environmental, social, and technological systems. Susan L. Cutter, "Compound, Cascading, or Complex Disasters: What's in a Name?," *Environment: Science and Policy for Sustainable Development* 60, no. 6 (2018): 21.
[44] Bankoff, "Comparing Vulnerabilities," 110.
[45] I use "cultural memory" in the tradition of Jan Assman. Cultural memory is a form of collective identity production, as well as a process of preserving and/or forgetting events. Jan Assmann, "Communicative and Cultural Memory," in *Cultural Memory Studies: An International and Interdisciplinary Handbook*, eds. Astrid Erll and Ansgar Nünning (Berlin: Walter de Gruyter, 2008), 109–18.

social contingencies in their development.[46] Catastrophic events are a necessary point of departure for disaster history, but environmental history stands to gain by incorporating these events into coevolutionary stories of human response and natural change. This book treats disasters as both events and processes to link catastrophe with decline. Declensionist anxieties, like disasters, emerged and evolved in the long term, yet they remained fundamentally tied to these moments of profound unsettling.

Third, disaster response varied across scale. In disaster scholarship, the uneven spatial extent of disaster impacts and their varied interpretation across scale have emerged as important determinants of response.[47] The asymmetry of exposure to hazards and the differential vulnerability of communities produced widely variable outcomes. Cattle plague affected every province, for instance, but some far more than others. This presented diverse options for response. The decentralized character of Dutch governance, where most intervention occurred on the local and provincial rather than the "national" scale, amplified the significance of these differences. Local responses to cattle plague included efforts to "cleanse" polluted pastures and stables, whereas provinces enacted import and export restrictions. On the scale of the Republic at large, meanwhile, attention focused on moral improvement. Divergent responses reflected different and sometimes competing interests as well as unequal exposure to hazards.

The perception of those distinctions evolved over time, informed by shifting environmental conditions. By the 1740s, river floods increasingly overwhelmed local relief strategies. Victims thus sought outside assistance by appealing to Dutch solidarity and civic virtue while framing the disasters as regional events or as calamities that affected the entire state. Similarly, hydraulic experts began applying an interprovincial perspective

---

[46] Wolfgang Behringer argues event-centered perspectives should be the "first commandment" of disaster research. Wolfgang Behringer, "Der Krise von 1570. Ein Beitrag zur Krisengeschichte der Neuzeit," in *Um Himmels Willen: Religion in Katastrophenzeiten*, eds. Manfred Jakubowski-Tiessen and Hartmut Lehmann (Göttingen: Vandenhoeck & Ruprecht, 2003), 153.

[47] Birgine Refslund Sørensen and Kristoffer Albris, "The Social Life of Disasters: An Anthropological Approach," in *Disaster Research: Multidisciplinary and International Perspectives*, eds. Rasmus Dahlberg, Olivier Rubin, and Morten Thanning Vendelø (London: Routledge, 2016), 73–4. Greg Bankoff, "Bordering on Danger: An Introduction," in *Natural Hazards and Peoples in the Indian Ocean World: Bordering on Danger*, eds. Greg Bankoff and Joseph Christensen (New York: Palgrave, 2016), 1–30.

to the problem of flooding because they perceived that disasters were growing more intense and more frequent. Interpretations of scale were dynamic and drew upon multiple (contested) arguments about the ultimate meaning of disaster. These arguments, or causal stories, provided essential rhetorical evidence that supported adaptive strategies and linked disaster to decline.[48] As we shall see, the early eighteenth century presented no shortage of dramatic disasters, each variable in scale and severity.

Finally, the past was a critical tool for disaster interpretation and response. It supplied valuable templates for action, baselines to measure regression or improvement, and causal stories that defined responsibility. Historians and disaster scholars frequently argue that disaster events catalyzed new social, scientific, and technological adaptations or retrenched populations in longstanding socio-environmental relationships.[49] Dutch disasters demonstrate that these narratives defied clear distinction. Past experience and the promise of the future proved equally powerful motivation. Disasters are thus ideal theaters to explore the interplay between the seemingly contradictory narratives of innovation and reliance upon tradition. Dike authorities, for instance, tapped both reservoirs of legitimacy to promote rebuilding strategies in the wake of coastal floods. Following the Christmas Flood of 1717, they couched reconstruction in the language of improvement. These proposals only gained traction, however, because they accommodated the longstanding customs and traditions of dike governance. Diverse interpretations of disasters past, especially their frequency and exceptionality, encouraged victims and observers to mold history, memory, and tradition to suit their purposes.

Dutch disasters also revealed the limits of history and memory. Shipworm novelty encouraged investigation into their natural origins, just as it reinforced the moral conviction they signaled decline. The river floods of the 1740s and 1750s produced a new, public awareness of a changing fluvial environment depicted in maps and described in state

---

[48] The "causal story" comes from the political science of agenda setting and is a useful model to define and explain methods of translating disaster events into action. Deborah A. Stone, "Causal Stories and the Formation of Policy Agendas," *Political Science Quarterly* 104, no. 2 (1989): 281–300.

[49] Karel Davids, *The Rise and Decline of Dutch Technological Leadership: Technology, Economy and Culture in the Netherlands, 1350–1800* (Leiden: Brill, 2008), 69. Jason D. Rivera, "Resistance to Change: Understanding Why Disaster Response and Recovery Institutions Are Set in Their Ways," *Journal of Critical Incident Analysis* 4, no. 1 (2014): 44–65.

publications. The belief that river floods were increasing in frequency and severity reinforced anxieties about decline, just as they prompted new investigations into their cause. The picture that emerges from these intersections reveals a population tentatively testing the boundaries of remembered experience and their mastery of nature, while at the same time gradually coming to grips with social and environmental challenges many could only characterize as decline.

The following chapters unpack these arguments by focusing on five natural disasters that occurred amidst the social, economic, and cultural transformation of the Dutch Republic between the late-seventeenth and mid-eighteenth centuries. This period encompassed the years historians broadly recognize as the beginning of Dutch decline. Along the way, I will introduce interpretive themes to aid in disaster interpretation: cultural memory, novelty, causal stories, and disaster cascades. Each offer valuable insights into the nature of disaster and shifting perceptions of decline. This approach encourages new readings of published and unpublished source material as well as historical and scientific reconstructions of past environments. It also encourages novel interpretations of familiar and historically obscure disasters. These were by no means the only natural disasters (or even the only types of disasters) to take place during this era, yet they count among the most catastrophic and consequential. Financial accounting of the loss of cattle, farmers' journals revealing the trauma of coastal inundation, and maps of river flooding rarely feature in histories of Dutch decline, yet they offer important insights into how the Dutch understood and coped with their changing reality. Dutch contemporaries frequently acknowledged the centrality of environmental forces in their experience of eighteenth-century adversity. These sources reveal new connections between disasters long considered separate or peripheral to strictly socioeconomic or cultural investigations of decline.

1.4 CHAPTER SUMMARY

Chapter 1 begins during the "Disaster Year" (*Rampjaar*) of 1672 when the French, the English, and their allies invaded and nearly toppled the Dutch Republic. To many observers and later historians, the *Rampjaar* signaled the end of the Golden Age. It is thus a useful moment to begin our investigation of the relationship between disaster and decline. This chapter introduces the Dutch Republic and proposes several ways that an environmental history of disasters enriches our understanding of the development and meaning of decline. It explores these interventions through a deep reading of the print "Miserable Cries of the Sorrowful Netherlands" (*Ellenden klacht van het bedroefde Nederlandt*). This image visually merges the political and military disasters of 1672 with the floods and windstorms that followed. It condenses time, works across scale, and frames the collective environmental, cultural, social, and economic consequences of the *Rampjaar* as a breach with the past. *Ellenden klacht* reads like a founding document of the Dutch decline narrative, but it also contained visual clues that pointed to alternative interpretations. This chapter argues that disasters, especially natural disasters, were traumatic and they challenged the moral, economic, and political standing of the Dutch Republic. At the same time, disasters could yield opportunities for adaptation, recovery, and growth.

Chapter 2 opens in 1713 at the conclusion of the War of the Spanish Succession. The Dutch Republic emerged from this conflict a weakened state and declensionist anxieties expanded as a result. That same year, an outbreak of cattle plague emerged in the Republic. Originating in the eastern European steppes, this panzootic spread slowly across Europe following networks of war and trade. Centuries of landscape transformation in the Netherlands set the stage for this disaster, and weather associated with a changing climate conditioned its severity. The disease killed hundreds of thousands of cattle in the Republic, impacting Dutch urban and rural livelihoods. Between 1713 and 1720, state authorities and the public struggled to understand and manage the disaster. This chapter investigates the social and environmental origins of cattle plague, as well as cultural and state response. State authorities based their strategies in environmentalist and contagionist theories of disease transmission that varied across scale. Its impacts were far from uniform, but moralists framed cattle plague as a problem that affected the entire country, which reinforced narratives of Dutch decline. This chapter argues that *causal stories* explaining the origins and meaning of the epizootic both reinforced pessimistic decline narratives and prompted a universalist approach to medical responses.

Chapter 3 explores the Christmas Flood of 1717 – likely the deadliest coastal flood in North Sea history. The storm impacted the entire southern coast of the North Sea basin, but the majority of its more than 13,000 victims lived in marginalized communities in the northern Netherlands and coastal Germany. This chapter investigates the origins, impact, and response to the Christmas Flood on the province of Groningen. The Netherlands had a long history of coping with coastal flooding and moralists, state officials, and dike authorities exploited the *cultural memory* of previous floods to advocate solutions. The city of Groningen and its rural hinterlands wielded the past to divergent ends in their efforts to reframe financial responsibility for reconstruction. Provincial technocrats balanced tradition with the rhetoric of improvement to build support for new and improved seawalls. Moralists emphasized the unprecedented severity of the flood to scale up its significance and embed it in broader decline narratives. The Christmas Flood revealed the diverse ways that the past could be wielded to promote and resist change.

A fourth disaster began in the late fall of 1730 after a coastal flood hit the Dutch island Walcheren. Inside the broken wooden revetments strewn across its beaches, dike authorities noticed peculiar, tiny holes. These holes contained a species of shipworms (*Teredo navalis*), a marine mollusk that bored into the wooden infrastructure that protected coastal dikes. This discovery prompted the most significant redevelopment and rebuilding of coastal flood defenses in the early-modern period. Chapter 4 investigates the origins, interpretation, and response to the "shipworm epidemic" of the 1730s. It argues that the perception of shipworm *novelty* influenced this dramatic change. In contrast to animal plagues or coastal floods, the cultural memory of disaster presented no ready solutions for shipworms. Shipworms' perceived novelty catalyzed new natural historical investigations of the animals, as well as innovative dike designs. Shipworms also produced new connections to decline. Pietist ministers and enlightened spectatorial journalists united in their condemnation of the moral decay of the Dutch Republic by linking shipworms to a wave of sodomy trials. Their novelty connected shipworms to an unprecedented period of persecution.

By the mid-eighteenth century, river flooding superseded coastal floods in severity and frequency. Observers worried that floods seemed to be increasingly numerous and severe. Chapter 5 evaluates the origins, interpretations, and consequences of the 1740–1 river floods. Victims could not help but interpret these floods in the context of recent years of dearth and disaster. The historically bitter winter of 1739–40 had catalyzed a *disaster cascade* in the hardest hit areas of the riverlands that amplified the impacts of inundation and expanded its consequences. At the same time, Dutch surveyors and hydraulic engineers, ministers, and state authorities promoted a discourse of increasing moral and geographic risk of inundation. In contrast to the Christmas Flood, where technocrats grounded dike innovations in the cultural memory of prior inundations, river floods forced observers to consider problematic futures. Surveyors and cartographers mapped flood risk in the Dutch riverlands and warned of potential consequences should the state ignore their new river management strategies. The floods of 1740–1 and narratives of increasing risk added to distress and anxiety about decline, but they also prompted the first proto-national flood-relief efforts and increased emphasis on the systemic, interprovincial nature of Dutch river challenges.

The return of cattle plague to the Netherlands in 1744 was the nadir of the eighteenth-century era of disaster. Hardly a generation removed from the first outbreak, it returned to the Republic with far greater intensity. It lasted over twice as long and resulted in over a million cattle deaths. Chapter 6 compares the second outbreak of cattle plague to the first, assessing changing response. Like the first outbreak, the disease emerged and spread in the context of conflict and extreme weather. Unlike the previous episode, it interacted with an ongoing disaster cascade that amplified and prolonged its consequences. Popular and state response showed remarkable continuity. Cattle plague was not novel and prior experience proved beneficial as provinces tapped the cultural and institutional memory of the previous outbreak. Provincial decrees quickly reinstituted bans on cattle importation, enacted quarantines, and issued certificates of health. Pamphlet literature once again highlighted the human tragedy of the animal disease and bemoaned its moral implications. The extensive geographic scope and duration of this outbreak attracted new attention from an international network of medical scholars and practitioners. Its increased severity prompted novel medical responses, including the first inoculation trials for cattle plague in the 1750s. These trials reveal the diffusion of declensionist fears into the economic and social program of the Dutch Enlightenment.

The concluding chapter, Chapter 7, steps back to assess the changing perception of Dutch decline across the first half of the eighteenth century. Declensionism did not emerge fully formed in 1672, nor any other date. Rather, it developed over time. What do natural disasters reveal about this process? The chapter also considers what this era of decline can teach us about disasters more broadly. Disasters were events and processes that manifested at the intersection of natural and cultural change. They produced differential consequences for Dutch society across scale, just as they do today. These conditions influenced Dutch perception of disaster and affected their response. The Golden Age past was key to learning from these disasters – whether as a model to emulate or a baseline to measure progress. Dutch "decline" and the natural disasters that punctuated it served as social and cultural tools that resolved in the long term.

I

Rampjaar Reconsidered

In 1672, the Dutch Republic seemed on the brink of collapse. For more than a century since its successful revolt against Habsburg Spain, the Dutch Republic had successfully defended its borders against foreign invasion and built a Golden Age of security and prosperity. Disaster struck when the combined forces of Louis XIV of France, King Charles II of England, and the prince-bishops of Münster and Cologne declared war on the Republic in spring 1672. The bishop of Cologne attacked from the east, while the bishop of Münster laid siege to the northern Dutch city of Groningen. A combined Anglo-French fleet attacked from the sea, choking off ports and trade. The French king invaded from the south with a force of over 130,000 men, four times the number of Dutch defenders. His army quickly overwhelmed Dutch defenses in the borderlands, conquering the less-populated eastern provinces and much of Limburg. By June, French forces swept into the heart of the Republic intent on capturing the large, prosperous city of Utrecht. "So desperate and in such confusion" were its city leaders that they capitulated without a fight[1] (Figure 1.1).

In response to the shocking speed of the invasion, the provincial assembly of Holland (called the Estates of Holland) ordered the countryside flooded to halt the enemy advance, essentially conceding the eastern half of the country.[2] These strategic inundations would stretch from the

---

[1] Unless otherwise indicated, all Dutch has been translated to English by the author. Judith Brouwer, *Levenstekens: Gekaapte brieven uit het Rampjaar 1672* (Hilversum: Verloren, 2014), 19. Note 19.

[2] Petra Dreiskämper, *Redeloos, radeloos, reddeloos: de geschiedenis van het Rampjaar* (Hilversum: Verloren, 1998), 45–6. Israel, *Dutch Republic*, 797–8.

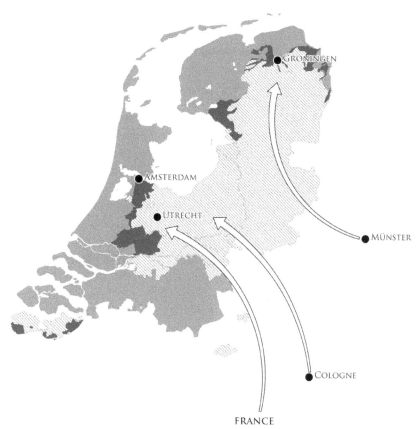

FIGURE 1.1. The *Rampjaar* of 1672. The water lines (dark grey) were critical defenses against the invading forces of Louis XIV of France and the bishops of Cologne and Münster. Hashed grey areas indicate occupied territories. Water line boundaries adapted from Gottschalk, *Stormvloeden*, 236, 240, 247.

Southern Sea (*Zuiderzee*) northeast of Amsterdam to the Merwede River in southern Holland. The Dutch had institutionalized this tactic in the sixteenth century during the defense of the Northern Netherlands from Spanish forces. River and polder dikes in several provinces could be breached, providing multiple layers of protection.[3] The "Holland water line" (*Hollandse waterlinie*) protected the wealthiest and most populous

---

[3] In 1672, Inundations took place in Holland, Zeeland, Friesland, Groningen, and Zeelandic Flanders, and offensive inundations occurred in Overijssel and Gelderland. M. K. E. Gottschalk, *Stormvloeden en rivieroverstromingen in Nederland*, Vol. 3 (Amsterdam: Van Gorcum, 1977), 235–43.

province of the Republic. Inundation, ordinarily an enemy, seemed in 1672 a potential ally. Unfortunately, the preceding months of drought reduced available river water, limiting its effectiveness.[4]

By mid-summer, the stunning success of the enemy advance prompted widespread social and political unrest in Dutch cities. An angry mob violently unseated Holland's chief statesman, the Grand Pensionary (*raadpensionaris*) Johan de Witt, later killing him and his brother. Burghers rioted in cities across the Netherlands, looting houses and attacking the regents they deemed responsible for the defeats. By August, the Dutch Republic had lost its leadership and two-thirds of its land area. The Dutch refer to 1672 as the "Disaster Year" (*Rampjaar*) – an appropriate encapsulation of this calamitous moment of social and political upheaval. Historians often characterize the Dutch Republic's geopolitical situation during the *Rampjaar* as *radeloos* (desperate), their government as *redeloos* (irrational), and the country itself *reddeloos* (irretrievable).[5] The *Rampjaar* represented a crisis of dramatic and revolutionary consequence.

Few artists captured the climate of anxiety and helplessness during the *Rampjaar* as powerfully as Romeyn de Hooghe. The remarkably active artist produced over two dozen etchings of the invasion and its aftermath.[6] His print "Miserable Cries of the Sorrowful Netherlands" (*Ellenden klacht van het bedroefde Nederlandt*) is among the most striking examples. It presents a frenzied vision of calamity (Figure 1.2). The central female figure in panel nine symbolizes the Republic. She clasps her hands in desperation, surrounded by images of wartime atrocities and social unrest. Panel two depicts the invasion of the French army and the subsequent "misery" inflicted on the Dutch populace. Below in panel three, De Hooghe illustrates the "civilian and farmer struggle" (*Borger- en boeren-krijgh*) that followed the invasion, which pitted "the rich against the rich, the holy against the holy, everyone against each other." These scenes move the viewer though the opening months of the invasion into its darkest moments of social turmoil.[7]

---

[4] Ibid., 240–1. [5] Dreiskämper, *Redeloos*, 69–73. Israel, *Dutch Republic*, 799–804.
[6] Henk Van Nierop, "Romeyn de Hooghe and the Imagination of Dutch Foreign Policy," in *Ideology and Foreign Policy in Early Modern Europe (1650–1750)*, eds. David Onnekink and Gijs Rommelse (Surrey: Ashgate, 2011), 197–214.
[7] Romeyn de Hooghe, *Ellenden klacht van het bedroefde Nederland te sedert het jaer 1672 tot den Aller-heyligen Vloet van het jaer 1675* (Amsterdam: Romeyn de Hooge, 1675).

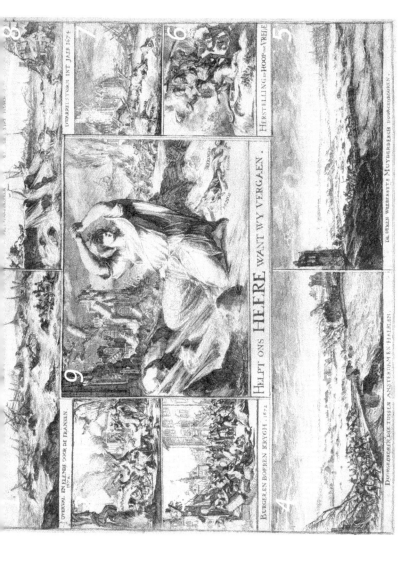

FIGURE 1.2. This image depicts the disastrous events of the period between 1672 and 1675, here unified into a period of disaster. 1. Dike breach at Den Helder (1675); 2. Pillaging of the invading French Army (1672); 3. Citizen uprising (1672); 4. Dike breach between Amsterdam and Haarlem (1675); 5. Dike breach at Muiderberg (1675); 6. Restoration and hope of peace with William III (1672); 7. Windstorm in Utrecht (1674); 8. Dike breach at Hoorn (1675); 9. Floods, warfare, and windstorms surround the maid of Holland. Source: Romeyn de Hooghe, *Ellenden klacht Van het Bedroefde Nederlandt. Sedert het Jaer 1672 tot den Aller-heyligen Vloet van het Jaer 1675*, Print, 1675, Rijksmuseum, Amsterdam, http://hdl.handle.net/10934/RM0001.COLLECT.466803.

Ellenden klacht's purpose extended beyond documenting social and political disorder, however. De Hooghe dedicated most of the space in this print to natural disasters. Floods and storms swirl around the central figure, and framing images depict broken dikes and buildings toppled by storm winds. Panel seven on the right depicts the windstorm (*dwarrelstorm*) that destroyed part of the Cathedral of Utrecht (*Domkerk*) in 1674.[8] Panels one, four, five, and eight depict inundations during the Second All Saints' Day Flood (*Tweede Allerheiligenvloed*) that occurred near Amsterdam and in the north of Holland, a region called West Friesland and the Northern Quarter.[9] De Hooghe downplays any distinction between these localized "natural" disasters and catastrophes resulting from social or military disorder. The *Rampjaar* was a hybrid phenomenon of national significance. He also condenses time, visually merging calamities that occurred across multiple years. De Hooghe's *Rampjaar* stretched from the invasion of 1672 to the floods of 1675. His portrayal of nature's violence mirrored social and military devastation, underlining the interconnectedness both he and his contemporaries found between natural and social disasters. Once again, the central panel underscores this point. The broken staff of the god Mercury represents wealth and lies at the feet of the central figure, the *Domkerk* crumbles behind her, the inscriptions "poverty" (*armoe*) and "decay of commerce" (*neeringloosheid*) swirl in the floodwaters around her feet. The large central inscription underscores the figure's distress, giving voice to the gravity of this collective misfortune. "Help us Lord," she pleads, "because we perish."

De Hooghe's *Ellenden klacht* so effectively captures the relationship between natural disaster and Dutch decline in the late seventeenth and early eighteenth century because it conveys two seemingly contradictory narratives. The first centers on the drama, chaos, and terror of 1672. It connects this moment of trauma with social disorder and natural disasters that continued years after the military outcomes concluded. From this

---

[8] Katrin Hauer, "Wahrnehmung, Deutung und Bewältigung von Starkwinden. Der Ostalpenraum und Holland im Vergleich (1600–1750)" (PhD diss., Universität Salzburg, 2008). The storm of 1674 was a rare straight-line windstorm called a "derecho." G. van der Schrier and R. Groenland, "A Reconstruction of 1 August 1674 Thunderstorms over the Low Countries," *Natural Hazards Earth Systems Science*, 17, no. 2 (2017): 157–70.

[9] This "Second All-Saints Day Flood" was even more extensive than this print acknowledges. Flood damage extended from the Wadden Islands and the province of Friesland in the north of the Netherlands, south to Antwerp in present-day Belgium. Gottschalk, *Stormvloeden*, 260.

perspective, the *Rampjaar* did not end in 1673, nor were its hazards limited to foreign armies or unruly burghers. Water and wind proved equally dangerous foes. Contemporary disaster commentary had long found political, social, and moral meaning in natural disasters, but the *Rampjaar* seemed different because its threats appeared so transparently existential. The print clearly expresses this new condition. It contrasts the historic, blessed condition of the Netherlands with ongoing social and natural disasters of 1672–75. The accompanying text gives voice to the personified Netherlands. "I was an arbiter of crowns, the darling of prosperity, a mirror of the world," it proclaims, "now heaven, earth, and sea rise up against me."[10] De Hooghe's depiction of the *Rampjaar* resonated with viewers because the scenes of destruction were such a reversal of fortune. The Golden Age of wealth and security seemed past. In this light, it is easy to see why observers might read this image as commentary on Dutch decline.

At the same time, *Ellenden klacht* subverts this declensionist message. Despite its many visual cues to the contrary, this document remains optimistic. It embraces disaster as a trial of faith (*beproeving*) and projects confidence about the future. Indeed, cultural historian Simon Schama argues that this type of optimism became a "formative part" of Dutch Golden Age culture.[11] "Take heart in her disasters," the text resolutely declares, "[s]he [the Dutch Republic] moves her hands and make right the hardship, saves her commerce and best polders, and she will have power enough left to force her enemies to a fair peace."[12] She reassures her audience that the disasters of the 1670s are merely temporary hardships, not indicative of God's lasting disfavor. The final line of *Ellenden klacht* quotes Virgil and underscores this optimistic Golden Age mentality. "*O Passi graviora dabit Deus his quoque finem.*" (O friends and fellow sufferers, who have sustained severer ills than these, to these, too, God will grant a happy period.)[13] According to this alternate reading, the central figure is not wrenching her hands in fear so much as she is extending them heavenward in a prayer of submission. *Ellenden klacht* vividly displays the shock and trauma of the *Rampjaar*, but its underlying message is one of piety and resilience. Calamity and perseverance, both

---

[10] De Hooghe, *Ellenden klacht*.
[11] Simon Schama, *The Embarrassment of Riches: An Interpretation of Dutch Culture in the Golden Age* (New York: Knopf, 1987), 25.
[12] De Hooghe, *Ellenden klacht*.
[13] Translation from *The Works of Virgil Translated into English Prose* (London: Geo B. Whittaker et al., 1826), 197.

themes apparent in De Hooghe's print, defined the Dutch eighteenth century as the brilliance of the Golden Age faded.

1.1 PRELUDE TO DISASTER

Disasters reflect the environments and societies that produce them. On the eve of the crisis in 1672, the Dutch Republic was already a society in transition. The conclusion of its eighty-year-long revolt against Habsburg Spain in 1648 confirmed the Republic as a major political power and the commercial and financial center of Europe. For much of the previous century, the Republic had prospered due to its unique synergy of urban growth, maritime trade, and highly commercialized agriculture and fishing. The roots of these developments were regionally diverse and began long before the founding of the Republic, but they accelerated between the late sixteenth and early seventeenth centuries.[14] Waves of immigration, driven by warfare and religious persecution or enticed by opportunity, swelled Dutch towns and cities and provided capital, labor, and connections instrumental to the expansion of Dutch industry and overseas trade.[15] The most dynamic changes occurred in low-lying maritime areas of the country, especially the many cities and towns of Holland. In the western countryside, urban capital, high agricultural prices, new drainage technologies, and a complex system of local and regional water management institutions transformed the bogs and peat lakes of low-lying coastal regions into productive commercial farms.[16] Foreign visitors marveled at the Dutch capacity to transform these landscapes and also the "miracle" of economic growth driven by Dutch overseas commerce. "By Meanes of their shipping," one English author noted in 1640, "they are

---

[14] Jan Bieleman, *Geschiedenis van de landbouw in Nederland 1500–1950: Veranderingen en verscheidenheid* (Meppel: Boom, 1992), 34–9. De Vries and van der Woude, *First Modern*, 195–210. Bo Poulsen, *Dutch Herring: An Environmental History, c. 1600–1860* (Amsterdam: Aksant, 2008), 82–105. Bas van Bavel and Jan Luiten van Zanden, "The Jump-Start of the Holland Economy during the Late-Medieval Crisis, c. 1350–c. 1500," *The Economic History Review* 57, no. 3 (2004): 503–32.

[15] Jelle van Lottum, *Across the North Sea: The Impact of the Dutch Republic on International Labour Migration, c. 1550–1850* (Amsterdam: Aksant, 2007), 58–60.

[16] De Vries and van der Woude, *First Modern*, 27–31. Davids, *Rise and Decline*, 72–7. P. H. Nienhuis, *Environmental History of the Rhine-Meuse Delta: An Ecological Story on Evolving Human-environmental Relations Coping with Climate Change and Sea-Level Rise* (Dordrecht: Springer, 2008), 82–94. Milja van Tielhof and Petra J. E. M. van Dam, *Waterstaat in stedenland: Het hoogheemraadschap van Rijnland voor 1857* (Utrecht: Matrijs, 2006), 182–4.

plentifully suplied with whatt the earth affoards For the use of Man ... with which supplying other Countries, they More and More enritche their owne [sic]."[17] The shipping and re-shipping of "bulk" trade items such as Baltic grain, North Sea herring, and later "luxury" goods from Asia, Africa, and the Americas integrated Dutch ports into an emerging global economy with Amsterdam at its center. By the third quarter of the seventeenth century, the Republic had become a dominant force in world commerce.[18]

During the 1650s, important elements of this picture began to change. This was partly the result of changing geopolitics. Scarcely four years removed from the Peace of Münster that ended the Eighty Years' War, the Republic again found itself embroiled in conflict, this time the result of commercial competition and festering political and ideological tensions with England. The Republic fought two Anglo-Dutch Wars (1652–54, 1665–7) to protect their interests prior to 1672. Although the Dutch maintained their position as a major political and military power throughout the seventeenth century, the cost of war and rising tide of protectionist trade policies during this era was immense.[19]

Perhaps the most fundamental shift occurred in the countryside, which experienced a century-long slump in agricultural prices beginning in the 1660s. In the context of this secular trend, agricultural revenue declined even as water management and labor costs increased.[20] The close connection between overseas trade and agricultural exports ensured the broad impact of these changes, especially in Holland. Economic inequalities,

---

[17] Peter Mundy, *The Travels of Peter Mundy in Europe and Asia, 1608–1667*, Vol. IV, ed. Richard Carnac Temple (Cambridge: Hakluyt Society, 1907), 72.

[18] Milja van Tielhof, *The "Mother of All Trades": The Baltic Grain Trade in Amsterdam from the Late 16th to the Early 19th Century* (Boston: Brill, 2002), 1–8. Christiaan van Bochove, "The 'Golden Mountain': An Economic Analysis of Holland's Early Modern Herring Fisheries," in *Beyond the Catch: Fisheries of the North Atlantic, the North Sea and the Baltic, 900–1850*, eds. Louis and Darlene Abreu-Ferreira Sicking (Leiden: Brill, 2009), 209–44. Richard W. Unger, "Dutch Herring, Technology, and International Trade in the Seventeenth Century," *The Journal of Economic History* 40, no. 2 (1980): 253. De Vries and van der Woude, *First Modern*, 402–9.

[19] David Ormrod and Gijs Rommelse, "Introduction: Anglo-Dutch Conflict in the North Sea and Beyond," in *War, Trade and the State: Anglo-Dutch Conflict, 1652–89*, eds. David Ormrod and Gijs Rommelse (Rochester, NY: Boydell and Brewer, 2020), 21–9. Gijs Rommelse, "Prizes and Profits: Dutch Maritime Trade during the Second Anglo-Dutch War," *International Journal of Maritime History* 19, no. 2 (2007): 139–60.

[20] De Vries and van der Woude, *First Modern*, 39–40. Jan Bieleman, *Five Centuries of Farming: A Short History of Dutch Agriculture, 1500–2000* (Wageningen: Academic, 2010), 23.

present during periods of explosive growth, increased.[21] The economic picture was far from grim, however. The West India Company would soon lose its colonial grip on Brazil and New Netherland, but the Dutch Atlantic economy expanded. The luxury trades continued to flourish as the VOC expanded its influence in the East Indies and Southern Africa. These gains often exacted a brutal human cost, especially as Dutch participation in the Atlantic and Asian slave trades expanded after midcentury. Closer to home, export-oriented industry diversified, populations grew in most areas of the Republic, and urbanization in many of its largest cities continued. Amsterdam became a city of 200,000 people by 1672. It was a bustling, cosmopolitan city and symbol of the Republic's continued dynamism. Although some historians would retrospectively point to the mid-seventeenth century as its economic high watermark, in the estimation of many contemporaries at home and abroad on the eve of invasion, the Republic was experiencing a Golden Age.[22]

The period between 1650 and 1672 also featured dramatic cultural and political change. The Reformed (Calvinist) faith enjoyed a position of privilege as the public Church and victory over the Spanish seemed confirmation that the United Provinces enjoyed divine favor. Devout Calvinists fashioned these beliefs into a self-image of the Republic as a second "Israel."[23] After midcentury, the Further Reformation – a movement within Dutch Pietism that called for the purification of the Church and society and a deepening of personal morality – gained influence as

---

[21] Guido Alfani and Wouter Ryckbosch, "Growing Apart in Early Modern Europe? A Comparison of Inequality Trends in Italy and the Low Countries, 1500–1800," *Explorations in Economic History* 62 (2016): 145–7. Anne E. C. McCants, "Inequality among the Poor of Eighteenth Century Amsterdam," *Explorations in Economic History* 44, no. 1 (2007): 1–4.

[22] Gert Oostindie and Jessica Vance Roitman, "Repositioning the Dutch in the Atlantic, 1680–1800," *Itinerario* 36, no. 2 (2012): 136–9. Israel, *Dutch Republic*, 611, 620. Wim Klooster, *The Dutch Moment: War, Trade, and Settlement in the Seventeenth-Century Atlantic World* (Ithaca, NY: Cornell University Press. 2016), 164. Matthias van Rossum, "'Vervloekte goudzugt': De VOC, slavenhandel en slavernij in Azië," *TSEG/Low Countries Journal of Social and Economic History* 12, no. 4 (2015): 42. Frijhoff and Spies, *Hard-Won Unity*, 125–31. Koenraad Walter Swart, *The Miracle of the Dutch Republic as Seen in the Seventeenth Century: An Inaugural Lecture Delivered at Univ. Coll. London, 6 Nov. 1967* (London: H. K. Lewis, 1969).

[23] Frijhoff and Spies, *Hard-Won Unity*, 128–30, 357–67. G. Groenhuis, "Calvinism and National Consciousness: The Dutch Republic as the New Israel," in *Britain and the Netherlands: Vol VII, Church and State Since the Reformation*, eds. A. C. Duke and C. A. Tamse (The Hague: Martinus Nijhoff, 1981), 118–33. Cornelis Huisman, *Neerlands Israël: Het natiebesef der traditioneel-gereformeerden in de achttiende eeuw* (Dordrecht: J. P. van den Tol, 1983).

well. This notion of the Dutch as a chosen people in an imperfect relationship with God would remain an important cultural framework to interpret disasters in the eighteenth century.

Despite the power of the Reformed Church, the decentralized character of Dutch governance fostered religious pluriformity and a high degree of tolerance by European standards. Confessional conflicts nevertheless remained subjects of intense theological and political debate.[24] Although the peace celebrations following the Treaty of Münster encouraged conciliation between these groups and promised a new age of concord, many of the internal social and religious divisions within the Republic hardened soon thereafter.[25] These conflicts served up consistent challenges for the state, which likewise navigated tensions between its cities and provinces.[26] Particularism was embedded in every level of governance, from foreign affairs to water management. During disasters, the decentralized, pluralistic character of Dutch society and culture meant that the highest level of state involvement often resided with the provinces, but regional and city interests weighed heavily on disaster response as well.

On the scale of the Republic at large, political conflict also arose out of the contradictory ambitions of its two most powerful players: Holland and the House of Orange. Relative to other provinces, Holland was first among equals by virtue of its wealth, sizeable urban population, and disproportionate share of tax responsibility. The Princes of Orange, meanwhile, exerted considerable influence in Dutch politics based on their claims as governors (*stadhouder*) of provinces. Stadhouders traditionally commanded the Dutch Army during wartime. William ("the Silent") of Orange had led the Dutch Revolt and his descendants commanded Dutch forces throughout the eighty-year-long conflict with Spain. During this period, the balance of power between the Princes and Holland had waxed and waned but reached a new inflection point in 1650 when William II staged a coup that failed after he died of smallpox. This left the Republic without a unitary stadhouder. The province of Holland, under Johan de Witt, capitalized on this opportunity to assert Holland's dominance in political affairs and culture and institute what his republican allies labeled

---

[24] Israel, *Dutch Republic*, 676.
[25] Lotte Jensen, *Celebrating Peace: The Emergence of Dutch Identity, 1648–1815* (Nijmegen: Vantilt, 2017), 41–2.
[26] Frijhoff and Spies, *Hard-Won Unity*, 73, 124–5, 139.

an era of "True Freedom." This "first stadhouderless period" continued until the *Rampjaar* in 1672.[27]

The dynamic environmental conditions of the third quarter of the seventeenth century added yet another layer of complexity to this changing social and political landscape. The diverse physical geography of the Republic was heavily influenced by water, whether the North Sea along its northern and western borders or large European rivers such as the Rhine, Meuse, and Scheldt that bounded and bifurcated the country. These fluid elements modified highly erodible peat and sandy soils, producing a diverse and variegated landscape. Long-term human interaction with these physical features, whether via agriculture, peat extraction, drainage, or canal building, produced intensively managed cultural landscapes. These processes were already well under way in the Middle Ages, though the degree and character of influence varied sharply by region.[28] By the mid-seventeenth century, the Dutch had manufactured cityscapes and landscapes highly tailored to their needs through a combination of technological, social, and institutional ingenuity.

These efforts to harness nature yielded considerable rewards, but also presented unforeseen problems. Many of the most serious and persistent challenges resulted from the management of water. Centuries of embankment, sedimentation, and storm surges shifted the distribution of river water along the main branches of the Rhine, which created problems in the Dutch riverlands for flood defense, navigation, and settlement. Further intervention accelerated sedimentation and the silting up of river mouths and harbors. Drainage-induced subsidence and peat extraction for energy lowered the level of the land relative to rivers, lakes, and seas. Each of these conditions influenced the frequency and severity of flooding.[29] Dredging harbors, draining or pumping excess water, and

---

[27] Israel, *Dutch Republic*, 595–609. Frijhoff and Spies, *Hard-Won Unity*, 105.
[28] Guus J. Borger and Willem A. Ligtendag, "The Role of Water in the Development of the Netherlands – A Historical Perspective," *Journal of Coastal Conservation* 4, no. 2 (1998): 109-14. G. P. van de Ven, *Man-made Lowlands: History of Water Management and Land Reclamation in the Netherlands* (Utrecht: Matrijs, 2004), 15-20, 24-6, 30-3. Bas van Bavel, *Manors and Markets: Economy and Society in the Low Countries, 500–1600* (Oxford: Oxford University Press, 2010), 15-28. De Vries and van der Woude, *First Modern*, 33-40.
[29] Van de Ven, *Man-made Lowlands*, 165-70, 179-80. Toon Bosch, *Om de macht over het water: de nationale waterstaatsdienst tussen staat en samenleving, 1798–1849* (Zaltbommel: Europese Bibliotheek, 2000), 21-4. Bertus Wouda, *Een stijgende stand met een zinkend land: Waterbeheersingssystemen in polder Nieuw-Reijerwaard, 1441–1880* (Hilversum: Verloren, 2009), 65-9. William H. TeBrake, "Taming the

building and maintaining flood infrastructure mitigated some of these risks but exacerbated others. Once begun, labor and capital-intensive water management necessitated continuous, costly investment. These costs, combined with rising inequality and inflexible water management institutions, meant that flood vulnerability grew in many areas of the Republic throughout this era.[30]

Climate change presented another important environmental influence during the mid-seventeenth century. Beginning in the 1640s, the climate of northwest Europe entered the "Maunder Minimum" – a low point in solar output and one of four great cooling phases of the Little Ice Age. The term "Little Ice Age" refers to a period between approximately 1300 and 1850 when average annual temperatures around much of the world declined. Within this general trend, significant variation took place and even included periods of remarkable warmth.[31] Expressions of climate varied dramatically by region and shifted in the medium and short term, due in part to natural oscillations in atmospheric and oceanic circulation, volcanic eruptions, fluctuating solar radiation, and possibly land use change.[32] Beyond temperature, historical climatologists have linked the Little Ice Age conditions to regional changes in average precipitation,

---

Waterwolf: Hydraulic Engineering and Water Management in the Netherlands during the Middle Ages," *Technology and Culture* 43, no.3 (2002): 489–98.

[30] H. J. A. Berendsen, "Birds-Eye View of the Rhine-Meuse Delta (The Netherlands)," *Journal of Coastal Research* 14, no. 3 (1998): 740–52. De Vries and van der Woude, *First Modern*, 40–5. Bas van Bavel, Daniel R. Curtis, and Tim Soens, "Economic Inequality and Institutional Adaptation in Response to Flood Hazards: A Historical Analysis," *Ecology and Society* 23, no. 4 (2018): 30. P. D. J. van Iterson, "Havens," in *Maritieme geschiedenis der Nederlanden*, Vol. 3, eds. G. Asaert, J. van Beylen, and H. P. H. Jansen (Brussum: De Boer Maritiem, 1977), 59–91.

[31] Sam White, "The Real Little Ice Age," *Journal of Interdisciplinary History* 44, no. 3 (2014): 327–52. The "Global Little Ice Age," however, shows a remarkable synchronicity between northern and southern hemispheres during only one extended period, which lasted from 1594 to 1677. Raphael Neukom, Joëlle Gergis, David J. Karoly, et al., "Inter-Hemispheric Temperature Variability over the Past Millennium," *Nature Climate Change* 4, no. 5 (2014): 362–7.

[32] Heinz Wanner, Jürg Beer, Jonathan Bütikofer et al., "Mid-to Late Holocene Climate Change: An Overview," *Quaternary Science Reviews* 27, no. 19–20 (2008): 1791–828. Eduardo Moreno-Chamarro, Davide Zanchettin, Katja Lohmann et al., "Winter Amplification of the European Little Ice Age Cooling by the Subpolar Gyre," *Scientific Reports* 7, no. 1 (2017): 1–8. Michael Sigl, Mai Winstrup, Joseph R. McConnell et al., "Timing and Climate Forcing of Volcanic Eruptions for the Past 2,500 Years," *Nature* 523, no. 7562 (2015): 543–9. David J. Reynolds, J. D. Scourse, P. R. Halloran et al., "Annually Resolved North Atlantic Marine Climate over the Last Millennium," *Nature Communications* 7, no. 1 (2016): 1–11. White, *A Cold Welcome*, 21.

wind speed and direction, and storminess.[33] In the northern hemisphere, the most pronounced cooling occurred during the Grindelwald Fluctuation (1560–1628) and the Maunder Minimum (1645–1720). As the Dutch entered the second half of the seventeenth century, therefore, climate was again shifting. Despite intense variability, average winter temperatures in northwest Europe declined, summers grew wetter, and storminess likely increased.[34]

This evolving climate brought weather that impacted Dutch society in multiple and complex ways, which complicates efforts to connect them to discrete disaster events. Climate changes played out over decades or centuries. Within longer climatic trends, annual and seasonal weather varied intensely, especially during the coldest periods of the Little Ice Age. Scholars thus construct causal claims that relate shared human experiences to similar weather events, whether the prevalence of subsistence crises during years of shortened growing seasons, epizootics following wet summers and cold winters, or the likelihood of flooding during periods of elevated storminess.[35] The contingent nature of social

---

[33] Javier Mellado-Cano, David Barriopedro, Ricardo García-Herrera et al., "Euro-Atlantic Atmospheric Circulation during the Late Maunder Minimum," *Journal of Climate* 31, no. 10 (2018): 3849–63. Atle Nesje and Svein Olaf Dahl, "The 'Little Ice Age' – Only Temperature?," *The Holocene* 13, no. 1 (2003): 139–45. Ricardo García-Herrera, David Barriopedro, David Gallego et al., "Understanding Weather and Climate of the Last 300 years from Ships' Logbooks," *Wiley Interdisciplinary Reviews: Climate Change* 9, no. 6 (2018): e544. Dennis Wheeler, Ricardo Garcia-Herrera, Clive W. Wilkinson, and Catharine Ward, "Atmospheric Circulation and Storminess Derived from Royal Navy Logbooks: 1685 to 1750," *Climatic Change* 101, no. 1–2 (2010): 257–80. Dagomar Degroot, "'Never Such Weather Known in These Seas': Climatic Fluctuations and the Anglo-Dutch Wars of the Seventeenth Century, 1652–1674," *Environment and History* 20, no. 2 (2014): 247–73.

[34] The timing of the Little Ice Age and the dating of its coldest periods vary based on historical climatological sources and dating techniques, geographic emphases, and expression of climate (glaciation, summer temperature), and whether cultural impact factors into chronology. Chantal Camenisch and Christian Rohr, "When the Weather Turned Bad. The Research of Climate Impacts on Society and Economy during the Little Ice Age in Europe: An Overview," *Cuadernos de Investigación Geográfica* 44, no. 1 (2018): 99. This work follows Dagomar Degroot's chronology. Degroot, *Frigid Golden Age*, 2, 43.

[35] Bruce M. S. Campbell, "The European Mortality Crises of 1346–52 and Advent of the Little Ice Age," in *Famines During the 'Little Ice Age' (1300–1800): Socionatural Entanglements in Premodern Societies*, eds. Dominik Collet and Maximilian Schuh (Cham: Springer International Publishing, 2018), 19–41. Chantal Camenisch, "Two Decades of Crisis: Famine and Dearth during the 1480s and 1490s in Western and Central Europe," in *Famines During the 'Little Ice Age' (1300–1800): Socionatural Entanglements in Premodern Societies*, eds. Dominik Collet and Maximilian Schuh (Cham: Springer, 2018), 72. Christian Pfister, "Little Ice Age-Type Impacts and the

response to weather and extreme events means that many conclusions must be couched in the language of probability.[36] Human decisions, technologies, and institutions likewise conditioned disaster impacts. Climate was important but by no means the only variable in play.

Complicating matters further, the experience of weather was always filtered through the lens of perception, prior experience, and underlying social and economic vulnerabilities. These factors influenced the interpretation of events, subsequent outcomes, and the likelihood they would be recorded. Extreme events like storm surges, river flooding, and extreme winters produced far more documentation than relatively benign conditions.[37] Despite these challenges, scholars have developed robust and nuanced strategies to parse the relationship between climate, weather, and human history.[38] Historians and historical climatologists have demonstrated that across large regions of the world, climate change during the coldest and most erratic periods of the Little Ice Age increased the likelihood of harvest failures, epidemics, and other disasters. In combination with a variety of other factors, this contributed to widespread violence,

---

Mitigation of Social Vulnerability to Climate in the Swiss Canton of Bern prior to 1800," in *Sustainability of Collapse? An Integrated History and Future of People on Earth*, eds. Robert Costanza, Lisa J. Graumilch, and Will Steffen (Cambridge: MIT Press, 2007), 197–212. Sam White, "Animals, Climate Change, and History," *Environmental History* 19, no. 2 (2014): 319–28. Philip Slavin, "The Great Bovine Pestilence and Its Economic and Environmental Consequences in England and Wales, 1318–50," *The Economic History Review* 65 (2012): 1239–66. H. H. Lamb and Knud Frydendahl, *Historic Storms of the North Sea, British Isles and Northwest Europe* (New York: Cambridge University Press, 1991). Adriaan M. J. de Kraker. "Reconstruction of Storm Frequency in the North Sea Area of the Pre-industrial Period, 1400–1625 and the Connection with Reconstructed Time Series of Temperatures," *History of Meteorology* 2 (2005): 51–69.

[36] Dagomar Degroot, "Climate Change and Society in the 15th to 18th Centuries," *Wiley Interdisciplinary Reviews: Climate Change* 9, no. 3 (2018): 4.

[37] Pfister, "Climatic Extremes." Georgina H. Endfield, Sarah J. Davies, Isabel Fernández-Tejedo, Sarah E. Metcalfe, and Sarah L. O'Hara, "Documenting Disaster: Archival Investigations of Climate, Crisis, and Catastrophe in Colonial Mexico," in *Natural Disasters, Cultural Responses: Case Studies toward a Global Environmental History*, eds. Christof Mauch and Christian Pfister (New York: Lexington Books, 2009), 305–25. Mark Carey, "Climate and History: A Critical Review of Historical Climatology and Climate Change Historiography," *Wiley Interdisciplinary Reviews: Climate Change* 3, no. 3 (2012): 239–41. Soens, "Flood Security," 209–17.

[38] Degroot, *Frigid Golden Age*, 15. Chantal Camenisch, Kathrin M. Keller, Melanie Salvisberg et al., "The 1430s: A Cold Period of Extraordinary Internal Climate Variability during the Early Spörer Minimum with Social and Economic Impacts in North-western and Central Europe," *Climate of the Past* 12, no. 11 (2016): 2107–26. Bruce M. S. Campbell, *The Great Transition: Climate, Disease and Society in the Late-Medieval World* (Cambridge: Cambridge University Press, 2016).

social unrest, and political instability. Some of the worst impacts in Europe occurred during the Maunder Minimum, which encompassed the period leading up to 1672.[39]

The Dutch experience of climate change during the Maunder Minimum was no less complex than in other territories, yet compared to its many neighbors struggling amidst adversity, the Dutch enjoyed relative stability and strong economic growth.[40] Recent research by historian Dagomar Degroot has demonstrated that, relative to other European powers, Little Ice Age conditions presented more benefits than disadvantages to the Republic. The reasons were multifaceted – and included the Dutch capacity to capitalize on the circulation of people, goods, and information – to tap energy sources such as wind and, in some cases, avoid the worst consequences of disaster experienced elsewhere.[41] The Dutch Republic was certainly not immune to environmental setbacks during this era, including natural disasters. Communities continued to experience epidemics in cities and the countryside, devastating coastal and river floods, and harvest failures. The uneven consequences of these disasters demonstrated that environmental conditions, including climate change, rarely determined outcomes; rather, they worked through a suite of social, economic, and political structures that likewise remained in flux.[42] As temperatures descended into a new nadir in the 1650s that

---

[39] Parker, *Global Crisis*. White, *The Climate of Rebellion*. Georgina H. Endfield and Sarah L. O'Hara, "Conflicts over Water in 'The Little Drought Age' in Central Mexico," *Environment and History* 3, no. 3 (1997): 255–72. David D. Zhang, Harry F. Lee, Cong Wang et al., "The Causality Analysis of Climate Change and Large-Scale Human Crisis," *Proceedings of the National Academy of Sciences* 108, no. 42 (2011): 17296–301. The "Global Crisis" thesis has also been criticized in recent years. Paul Warde, "Global Crisis or Global Coincidence?," *Past & Present* 228, no. 1 (2015): 287–301.

[40] Jan de Vries, "The Crisis of the Seventeenth Century: The Little Ice Age and the Mystery of the 'Great Divergence,'" *Journal of Interdisciplinary History* 44, no. 3 (2013): 374–5.

[41] Degroot, *Frigid Golden Age*, 300–4.

[42] Daniel R. Curtis, "Was Plague an Exclusively Urban Phenomenon? Plague Mortality in the Seventeenth-Century Low Countries," *Journal of Interdisciplinary History* 47, no. 2 (2016): 139–70. Daniel R. Curtis and Jessica Dijkman, "The Escape from Famine in the Northern Netherlands: A Reconsideration using the 1690s Harvest Failures and a Broader Northwest European Perspective," *The Seventeenth Century* 34, no. 2 (2019): 229–58. Jessica Dijkman, "Feeding the Hungry: Poor Relief and Famine in Northwestern Europe, 1500–1700," in *An Economic History of Famine Resilience*, eds. Jessica Dijkman and Bas van Leeuwen (New York: Routledge, 2019): 93–111. Tim Soens, "Resilience in Historical Disaster Studies: Pitfalls and Opportunities," in *Strategies, Dispositions and Resources of Social Resilience: A Dialogue Between Medieval Studies*

would last through the invasion of 1672, changing environmental conditions would again yield benefits, but also challenges for the Republic.

To later critics and historians of the Republic searching for signs of waning influence and prosperity, the second half of the seventeenth century presented a clouded picture. The many economic, political, and environmental changes of the era produced little in the way of absolute rupture, but their influence would gradually reshape Dutch culture and society. On their own, none determined the course of later events, yet each exerted an important influence. The interconnection between these structural changes in society and environment, not to mention the influence of culture and individual decisions are difficult to parse for historians and would have been virtually impossible for contemporaries. During and following disasters such as the *Rampjaar*, however, greater degrees of this complexity gained definition. The richness of written and visual disaster documentation and the tendency of contemporaries to self-assess, critique, and reflect on the meaning of calamities present ideal opportunities to reconstruct these relationships and serve as a reminder that perception governed interpretation and response. The trauma of the *Rampjaar* and its attendant social and natural disasters presented one such moment of heightened awareness.

1.2 THE *RAMPJAAR* AND THE BEGINNING OF DUTCH DECLINE

The *Rampjaar* sparked a profound reaction in Dutch society and culture. The trials of that year fundamentally altered the status quo, and their consequences reverberated across subsequent decades. "Who ever lived in more remarkable times," one pamphleteer remarked in 1672, "when were there ever days with as much change as ours?"[43] As with disasters before and since, opportunists welcomed them as moments of creative destruction. De Hooghe's *Ellenden klacht* again provides valuable insight. In panel six, he depicts the "restoration and hope of peace" signaling the return of civic order and promise of victory under the leadership of William III. As head of the House of Orange and grandson of William

---

and *Sociology*, eds. Martin Endreß, Lukas Clemens, and Benjamin Rampp (Wiesbaden: Springer VS, 2020), 253–74.

[43] Michel Reinders, "Burghers, Orangists and 'Good Government': Popular Political Opposition during the 'Year of Disaster' 1672 in Dutch Pamphlets," *Seventeenth Century* 23, no. 2 (2008): 315–46. Reinders cites: *Dortse en Haagse woonsdag en saturdag, of nader opening van de bibliotheecq van mr. Jan de Witth, zijnde een samenspraak tusschen een Hagenaar en Dortenaar* (s.n. 1672).

the Silent, William III took advantage of the power vacuum created upon the death of Johan de Witt. By early July 1672, the powerful maritime provinces of Zeeland and Holland installed William as head of the military and stadhouder. This changing of the guard was indicative of a larger shift in political power away from the regents of Holland to the supporters of the House of Orange that would last until William's death in 1702. For De Hooghe, who was himself an "Orangist" (supporter of William III), this panel conveyed a significant, albeit visually subordinate, message. The print is an apologist political statement favoring the return of the House of Orange to a position of political primacy and the end of the first stadhouder-less period. Although the dominant visual motif for the print is tragedy, the "restoration and hope of peace" reminds the viewer that disasters are always in the eye of the beholder. A catastrophe for some may yield opportunities for others.[44]

From a military perspective, glimmers of hope appeared by the late summer of 1672. Dutch armies followed naval victories against the British with the lifting of the siege of Groningen and the recapturing of large parts of the eastern Republic. Even the weather seemed to shift in favor of Dutch defenders. The drought conditions that prevailed in the early months of 1672 had been atypical of the Maunder Minimum, which tended to feature wet, cold springs. The arrival of rainy weather in June and August more closely aligned with the climatic norm and meant that the army could fully implement the waterline to protect the core of Holland by intentionally inundating surrounding landscapes. This effectively halted the enemy advance. By 1674, under the leadership of William III, the Dutch Republic concluded treaties with England and the German bishoprics, only remaining at war with France. These strategic victories signaled the turning point of this particular conflict, but they also laid the groundwork for decades of costly, destructive war with France. The *Rampjaar* proved a watershed moment in Dutch political history both bemoaned and welcomed, but its consequences continued to haunt the Republic well into the eighteenth century.

The immediate impact of the *Rampjaar* substantially affected Dutch society. The war traumatized the Dutch population, especially those that

---

[44] Hanneke van Asperen offers a compelling, more pessimistic reading of the print's allegorical meaning, which warns against growing discord fostered by wealthy elites against poorer members of the Dutch society. Hanneke van Asperen, "Disaster and Discord: Romeyn de Hooghe and the Dutch State of Ruination in 1675," *Dutch Crossing* (2020): 1–22.

endured years of foreign occupation. Although the worst political and military effects of the *Rampjaar* ended by 1674, Dutch characterizations of the era in subsequent years remained inflexibly dismal. The Utrecht regent Bernard Costerus reflected that the *Rampjaar* reduced many "to poverty and distress and had lapsed into a wretched state, so that during their remaining years they were not able to repair their ruined livelihoods."[45] The Dutch considered the *Rampjaar* a totalizing disaster, the consequences of which touched every segment of society. Perhaps no single group suffered more than rural communities living in occupied lands. In early 1672, when French soldiers marched across the eastern Dutch borderlands, through Gelderland, Overijssel, and the Generality Lands (now North Brabant and Zeelandic Flanders), the countryside bore the brunt of the violence. Pamphlets described the pillaging of cities, and publishers distributed sensationalist prints depicting the "French Savagery" that took place in towns like Zwammerdam and Bodegraven in South Holland.[46] Local economies in these inland regions suffered during this onslaught as well as years of subsequent occupation.[47]

Even the waterline, heralded as a miracle of Dutch ingenuity and savior of the Republic, proved a disaster in the countryside. Rural communities actively resisted these military inundations because they protected the urban core of Holland at the expense of rural sacrifice zones. Flooded fields meant that farmers lost harvests and fodder for their livestock, and despite this loss of income, they remained responsible for repairing broken dikes after the war.[48] Faced with this onerous burden, rural villagers and citizens of border towns like Gouda, Gorkum, and Schoonhoven refused to comply with military plans for inundation and only acquiesced after Dutch soldiers arrived to quell resistance.[49] The

---

[45] Bernard Costerus, *Historisch verhaal ... raakende het formeren van de Republique van Holland ende West-Vriesland* (Leiden: by Coenraad Wishoff, 1736), 382.

[46] Donald Haks, *Vaderland en vrede, 1672–1713: Publiciteit over de Nederlandse Republiek in oorlog* (Hilversum: Verloren, 2013), 55.

[47] Israel, *The Dutch Republic*, 808. Piet van Cruyningen, "Behoudend maar buigzaam: boeren in West-Zeeuws-Vlaanderen, 1650–1850" (PhD diss., Wageningen University, 2000), 86–9.

[48] In Zeelandic Flanders, military inundations during the sixteenth century used seawater, and the ensuing tidal action resulted in massive land loss. Tidal impacts amplified the destructive consequences of inundations in 1672–3 as well. Adriaan M. J. de Kraker, "Flood Events in the Southwestern Netherlands and Coastal Belgium, 1400–1953," *Hydrological Sciences Journal* 51, no. 5 (2006): 918–19.

[49] Robert Fruin, *De oorlog van 1672* (Groningen: Wolters-Noordhoff, 1972), 174–85; Adriaan M. J. de Kraker, "Flooding in River Mouths: Human Caused or Natural Events? Five Centuries of Flooding Events in the SW Netherlands, 1500–2000,"

French added to these self-inflicted assaults by breaching dikes to slow the Dutch defenders. Dutch leaders understood the economic consequences of weaponizing water. When William III ordered the construction of a new dike to protect the fertile polder lands of Rijnland in 1672, for instance, he found the rural population unable to shoulder the immense financial burden. Loss of revenue in agricultural regions prevented timely and effective maintenance of dikes for years to come. Many of the same communities affected by military inundations would later experience dike breaches during the coastal floods of 1675 and 1702.[50] The environmental and economic consequences of the *Rampjaar* for rural populations proved both devastating and long lasting.

Beyond its destabilizing effects in the countryside, the invasion prompted a financial panic that resulted in one of the most dramatic stock market crashes in Dutch history. Shares in the *VOC* collapsed, and the value of *WIC* stocks became "virtually worthless" overnight. Although the market would rebound, other economic indicators like public construction, property rents, and the art market experienced slumps that would take decades to recover.[51] War forced the Dutch to recall warships protecting overseas commerce to defend domestic ports. As a result, fishing and key sectors of foreign trade ground to a halt.[52] These economic disasters compounded the shock of the military defeats and political upheaval.

On a cultural level, the most telling indication of the unsettled mentality of Dutch society was the explosion of pamphlet literature. The Dutch Republic was likely the most literate society in Europe, and its printing industry fostered diverse opportunities to read and disseminate information through newspapers, books, sermons, and state resolutions. Pamphlets were among the most popular media for news and polemic. They were relatively cheap, usually vernacular, and broadly accessible. Pamphlets were thus popular tools to broadcast commentary about

---

Hydrology & Earth System Sciences 19, no. 6 (2015): 9–10; Gottschalk, *Stormvloeden*, 239–40.

[50] Van Tielhof and van Dam, *Waterstaat in stedenland*, 215–16. Alfons Fransen, *Dijk onder spanning: De ecologische, politieke en fianciële geschiedenis van de Diemerdijk bij Amsterdam, 1591–1864* (Hilversum: Verloren, 2011), 130–2.
[51] Jonathan Israel, *Conflicts of Empires: Spain, the Low Countries and the Struggle for World Supremacy, 1585–1713* (London: The Hambledon Press, 1997), 327–8.
[52] Annette Munt, "The Impact of the Rampjaar on Dutch Golden Age Culture," *Dutch Crossing* 21, no. 1 (2016): 4.

current events, including disasters. Dutch printers published over 1,000 pamphlets in 1672.[53]

It is difficult to generalize about the authorship of pamphlets, but they certainly encompassed a wide spectrum of Dutch society and reflected its most pressing concerns and deepest divisions.[54] The documents from 1672 showcased the political conflicts that led to the revolt and coup, as well as moral, religious, and cultural tensions, such as the contested role of the Reformed Church in state governance and Dutch society's perceived depravity. A common complaint about Dutch culture was its decay from pious modesty to extravagance and greed. "In truth," one anonymous pamphleteer argued, "this our century has, through prosperity and good fortune, degenerated very far from the old simple manners and sobriety."[55] The *Rampjaar* was evidence of God's displeasure with this moral decline. Pamphlet literature spanned the gamut between sensationalist journalism intended to convey the horror of invasion to its readers, to social and political critique, to inventories of moral decay.

The spectacular growth and prosperity of the Dutch Golden Age had dazzled European onlookers during the seventeenth century, and Dutch citizens were justifiably optimistic about its future. To all onlookers, its near collapse in 1672 was shocking. One of the best-known foreign witnesses of the Dutch Golden Age, the English ambassador William Temple, saw 1672 as a break in Dutch history. "It must be avowed, That as This State, in the Course and Progress of its Greatness for so many Years past, has shined like a Comet; so in the Revolutions of this last Summer, It seem'd to fall like a Meteor, and has equally amazed the World by the one and the other."[56] Combined with the cost of the war and trauma of occupation, the *Rampjaar* profoundly transformed Dutch politics, society, and culture, testing Dutch confidence and resilience.

---

[53] Frijhoff and Spies, *Hard-Won Unity*, 221–3. Margaret Spufford, "Literacy, Trade and Religion in the Commercial Centres of Europe," in *A Miracle Mirrored: The Dutch Republic in European Perspective*, eds. Karel Davids and Jan Lucassen (New York: Cambridge University Press, 1995), 229–84. Michel Reinders, *Gedrukte chaos: populisme en moord in het Rampjaar 1672* (Amsterdam: Balans, 2010).

[54] Femke Deen, David Onnekink, and Michel Reinders, "Pamphlets and Politics: Introduction," in *Pamphlets and Politics in the Dutch Republic*, eds. Femke Deen, David Onnekink, and Michel Reinders (Leiden: Brill, 2010), 9–13, 22–3.

[55] Munt, "The Impact," 15. Quoted from *Huysmans-praetje, Voorgestelt tot onderrechtingh, Hoe men sich in desen verwerden en murmurerige toestandt des tijdts behoorden te dragen* (Amsterdam, 1672).

[56] William Temple, *Observations Upon the United Provinces of the Netherlands*, 2nd ed. (London: A. Maxwell, 1673), 259.

1.3 THE UNCERTAIN AFTERMATH AND DIVERGING DECLINE NARRATIVES

The ordeals of the *Rampjaar* notwithstanding, long-term decline was far from a foregone conclusion in 1672. To most contemporaries, it was impossible to ignore the fact that the Dutch Republic remained the wealthiest state in Europe and retained much of its previous prestige. The state had weathered numerous trials in the past, some of which seemed eerily similar in character to the *Rampjaar*. The disaster year of 1570, for instance, featured the combined impacts of military invasion during the Dutch Revolt from Spain and the All Saints' Day Flood (*Allerheiligenvloed*), which inundated large portions of the western Netherlands. That the Dutch Republic emerged victorious and stronger than ever following these collective calamities seemed evidence that a similar recovery was likely in 1673. Dutch statesmen in the 1670s certainly understood the gravity of their predicament. They perceived themselves beholden to an unwieldy public debt and the revenue of an overwhelmed tax base. Despite this, at the end of the seventeenth century, few thought these conditions would translate to long-lasting decline.[57]

The *Rampjaar*, however, was merely the first in a long series of mercantile and military conflicts with an expansionary France that lasted until 1713. This period, which some historians refer to as the "Forty Years' War," included the Franco-Dutch War (1672–78), the Nine Years' War (1688–97), and the War of the Spanish Succession (1701–13). It also featured numerous trade wars that periodically heightened political tensions.[58] These conflicts directly affected foreign and domestic commerce, and wartime expenses required loans that saddled provinces with enormous debt.[59] Reflecting on the consequences of this period of Franco-Dutch conflict, the Dutch-born agent of France Adrianus Engelhard Helvetius reported to the French state in 1705 that

---

[57] O'Brien, "Mercantilism and Imperialism," 489. Charles Wilson, "Taxation and the Decline of Empires, an Unfashionable Theme," in *Economic History and the Historian*, ed. Charles Wilson (London: Weidenfeld, 1969), 114–27.

[58] David Onnekink, *Reinterpreting the Dutch Forty Years War, 1672–1713* (London: Springer, 2017), 4, 339–46.

[59] Israel, *Dutch Primacy*, 339–46, 359–76. Wantje Fritschy, "The Poor, the Rich, and the Taxes in Heinsius' Times," in *Anthonie Heinsius and the Dutch Republic, 1688–1720. Politics, War, and Finance*, eds. J. de Jongste and A. J. Veenendaal (The Hague: Institute of Netherlands History, 2002), 242–58. Oscar Gelderblom and Joost Jonker, "Public Finance and Economic Growth: The Case of Holland in the Seventeenth Century," *The Journal of Economic History* 71, no. 1 (2011): 18–25.

"the commerce of the United Provinces in Europe has never been in a worse condition than today," which he attributed primarily to the financial burden of prolonged warfare.[60] His report revealed more than the *schadenfreude* expected of a geopolitical rival. It was a critique grounded in a reality whose significance slowly gained acceptance in the Republic itself.

Foreign observers were also among the first to examine the close connection many saw between the Republic's economic and environmental challenges. The English engineer Andrew Yarranton noted the relationship between taxation, decline, and the combined weight of military and environmental assaults. As early as 1677, he argued that "all people that know any thing of Holland, know that the people there pay great Taxes, and eat dearly, maintain many Souldiers [sic] both by Sea and Land and in the Maritime Provinces have neither good Water nor good Air ... and are many times subject to be destroyed by the devouring waves of the Sea's overflowing their banks."[61] Holland's economic and environmental vulnerabilities were mutually reinforcing. Unlike Dutch observers of this relationship like De Hooghe, Yarranton was far less willing to apply an optimistic lens to their implications.

Anxieties about decline gained increasing purchase in the Republic during the first decades of the eighteenth century. Rather than financial or environmental concerns, however, they pointed to moral and cultural degeneracy. Early eighteenth-century moralists argued that the Dutch Republic had lost sight of the core values of its prior Golden Age. Honesty, industriousness, thrift, and, above all, piety had been the bedrock upon which seventeenth-century success was built, and Dutch critics bemoaned its deterioration. Reformed clergymen published widely on these subjects, particularly those associated with Dutch Pietism. This movement advocated purification of personal and civic morality and saw decline as manifestation of divine judgment. Moralistic interpretations of decline were not restricted to Pietist ministers, however, nor even the clergy. Less puritanical interpretations appeared in poetry, the visual arts, state documentation, farmers' journals, and early-modern

---

[60] Translation taken from H. H. Rowen, *Low Countries in Early Modern Times* (London: Palgrave Macmillan, 1972), 226–7. The original French Mémoire Sur L'état Présent Du Gouvernement Des Provinces Unies can be found in M. van der Bijl, "De Franse politieke agent Helvetius over de situatie in de Nederlandse Republiek in het jaar 1706," *Bijdragen en Mededelingen van het Historisch Genootschap* 80 (1966): 159–94.

[61] Andrew Yarranton, *England's Improvement by Sea and Land: To Out-Do the Dutch without Fighting, to Pay Debts without Moneys* (London: R. Everingham, 1677), 6–7.

journalism. Despite their wide-ranging backgrounds and interests, these authors shared a dual conviction that at once asserted moral and cultural loss and advocated improvement.

No writer was as influential a critic as Justus van Effen (1684–1735), the early Enlightenment publisher of the literary magazine *De Hollandsche Spectator*. Van Effen used his platform to feed declensionist anxiety. He and many later spectatorial writers championed ideals that were inherently backward looking – hearkening to the republican virtue of the Golden Age. They were deeply critical of a decadent eighteenth century, which, in their view, appeared increasingly polluted by French manners and fashion. On the surface, little of this critique appeared new. Moralists had long warned that sin threatened Dutch prosperity, even during the seventeenth century.[62] More novel were the connections moralists found between sin and broader anxieties about economic, political, cultural, and environmental decline.[63] These critiques grew louder and more common over the course of the early eighteenth century. By the second half of the eighteenth century, decline had assumed a near universal character.

Cultural and moral anxieties about decline during the first half of the eighteenth century also informed Dutch responses to natural disaster. In the context of the early Dutch Enlightenment, these responses increasingly emphasized novelty, innovation, and perfectibility. Natural scientists, philosophers, merchants, and state officials employed these ideas to temper fears of decline, promote greater confidence in their ability to control nature, and check longstanding anxieties about reoccurring disasters like plagues, wars, and famines. According to Peter Gay, this optimism starkly contrasted with a general fear of stagnation that reigned in Europe before the late seventeenth century. Gay famously termed this transition the "recovery of nerve."[64] Early Enlightenment thinkers, state officials, and hydraulic experts in the Netherlands were active purveyors of these optimistic assessments. Disasters showcased how the rhetoric of

---

[62] Benjamin B. Roberts and Leendert F. Groenendijk, "Moral Panic and Holland's Libertine Youth of the 1650s and 1660s." *Journal of Family History* 30, no. 4 (2005): 327–46.
[63] Joris van Eijnatten, *God, Nederland en Oranje: Dutch Calvinism and the Search for the Social Centre* (Kampen: Kok, 1993), 64.
[64] Peter Gay, *Enlightenment*, Vol. 2 (Amsterdam: Knopf Doubleday, 2013), 3–55. Others term this shift an "optimistic feasibility philosophy," Bernd Rieken, "Learning from Disasters in an Unsafe World: Considerations from a Psychoanalytical Ethnological Perspective," in *Learning and Calamities: Practices, Interpretations, Patterns*, eds. Heike Egner, Marén Schorch, and Martin Voss (New York: Routledge, 2014), 27–41.

innovation and improvement gradually increased over the course of the eighteenth century.[65]

Natural disasters tested the Dutch mastery of nature and presented opportunities to develop new strategies to control and mitigate their effects. In contrast to bleak historical assessments of scientific stagnation, the international reputation of Dutch scientists and engineers actually increased in stature during the eighteenth century.[66] Luminaries like Willem Jacob 's Gravesande and Herman Boerhaave popularized Newtonian science on the continent and became the great teachers of Europe, while Nicholas Cruquius and Cornelis Velsen enjoyed international renown for their work in hydraulics and hydroengineering. Historians are no longer comfortable assigning declensionist labels to Dutch science in this era. In the words of one historian, "[W]hile in one field we can speak of decline, perhaps we can see progress in another, and what is decline from one perspective can be quite the opposite from another."[67] The significance of these Enlightenment-era figures reflected the growing confidence about the Dutch ability to understand and manipulate bodies and environments and respond to natural disasters.

Environmental challenges shaped the emerging perceptions of decline. On the one hand, natural disasters seemed to confirm the most pessimistic assessments of the Dutch Republic's changing fortunes. The repeated catastrophes of the early eighteenth century proved deadly and expensive, and each new calamitous event reinforced anxieties about the Netherland's flagging economy. In the wake of these disasters, community response fractured in ways that seemed to confirm the breakdown of Dutch society. Catastrophes laid bare a public moral degeneration that emphasized the Dutch Republic's diminishing providential favor. To make matters worse, the environmental liabilities already apparent in the mid-seventeenth century appeared to be worsening. The early eighteenth century seemed to present greater and deadlier disasters than ever before, or hazards of such uniqueness and novelty that they exposed the limits of Dutch political and technological capacity to respond.

---

[65] Davids, *Rise and Decline*, 53.
[66] Margaret Jacob has argued that Dutch failure to keep pace with industrializing rivals in the mid-eighteenth century reflects the failure of Dutch science. Margaret C. Jacob, *Scientific Culture and the Making of the Industrial West* (Oxford: Oxford University Press, 1997). Davids disputes this argument. Davids, *Rise and Decline*, 22.
[67] Klaas van Berkel, "Science in the Service of Enlightenment, 1700–1795," in *A History of Science in the Netherlands: Survey, Themes and Reference*, eds. K. van Berkel, A. Van Helden, and L. C. Palm (Leiden: Brill, 1999), 93.

On the other hand, disasters presented unequalled opportunities for adaptation and improvement. As trials of faith and reason, disasters encouraged moral self-examination and reconsideration of the distinction between their divine and "natural" origins. Natural disasters destroyed infrastructure, killed thousands, and incited confusion and violence, but they also catalyzed social solidarity, the development of new water management technologies, medical treatments to combat disease, as well as empirical and scientific examinations of the Dutch environment. Disasters revealed that decline was real, but they also showed the contours and contradictions of its development. In contrast to an eighteenth century characterized by economic torpor or what some historians have described as "a stagnation of spirit, a sapping of creative power, an end of greatness and a slide into social artificiality," Dutch response to catastrophe paints a more resilient picture.[68] These varied responses to disaster explain the broad trajectory of decline amidst profound changes in Dutch society and its relationship with its environment.

These welcome revisions to a uniformly dreary view of Dutch society privilege scientific and technological adaptation, yet these must be balanced against resistance to change and the continuing (in some cases increasing) exposure to adversity. New plans to regulate rivers and control floods did little to lessen the very real desperation of communities rebuilding in the wake of inundations, nor did they necessarily temper anxieties about decline. Physicians, enlightened citizen-scientists, and ministers experimented with novel tools like inoculation to combat cattle plague in the 1750s, yet most state responses remained wedded to centuries-old disease management strategies. This does not diminish the value or significance of either approach. Dutch communities coped with adversity in remarkable and creative ways. Both change and continuity characterize the early eighteenth-century experience – the first privileging optimism and innovation, the other memory and tradition. Both anxiously reached back to the golden achievements of the past.

---

[68] Charles Wilson, *The Dutch Republic and the Civilisation of the Seventeenth Century* (New York: McGraw-Hill, 1968), 230.

2

"Disasters in the Year of Peace"
The First Cattle Plague, 1713–1720

On July 14, 1713, the center of The Hague erupted into a fantastic spectacle of light and smoke. Fireworks burst from a grandiose platform outside the assembly hall of the Estates General of the Dutch Republic. Their light illuminated a "castle" bedecked with neoclassical sculptures, banners, and a large painting hung beneath a 50-foot-high triumphal arch. The entire construction rose out of an artificial lake called the *Hofvijver* which reflected the pyrotechnic display and dazzled onlookers. It was the centerpiece of a public event celebrating a general day of thanksgiving and prayer. The Estates General dedicated the festivities "to offer our hearts and to thank God Almighty who through his grace and goodness saved Our State and blessed us with many great successes in this recent War" as well as to "enjoy the desirous fruits of this good and holy peace ... and stave off all other well-deserved plagues"[1] (Figure 2.1). It celebrated the signing of the Treaty of Utrecht, which ended Dutch involvement in the War of the Spanish Succession and overt hostilities with France.

Since Louis XIV's invasion of the Netherlands during the *Rampjaar* of 1672, the Dutch Republic had endured four decades of near continuous conflict. War had not been kind to the Dutch Republic's finances or international prestige. The United Provinces fielded its largest army ever during the War of the Spanish Succession and paid for it largely through

---

[1] Nicholaas Christiaan Kist, *Neêrland's bededagen en biddagsbrieven: eene bijdrage ter opbouwing der geschiedenis van staat en kerk in Nederland. Deel 2* (Leiden: Luchtmans, 1849), 301–2.

FIGURE 2.1. The fireworks castle erected by the Estates General in 1713 celebrated the end of the war and promoted the renewal of Golden Age prosperity.
Source: Daniël Stopendaal, *Vuurwerk bij de viering van de Vrede van Utrecht, 1713*, Print, 1713, Rijksmuseum, Amsterdam, http://hdl.handle.net/10934/RM0001.COLLECT.474765.

credit. At its conclusion, the Republic bore an unprecedented debt, and high taxes burdened its population.[2] Natural disasters during this era exacerbated the country's struggles. A massive coastal flood struck the northern province of Groningen in 1686, killing almost 1,600 inhabitants, "bringing many thousands of people into misery."[3] Another large

---

[2] Israel, *Dutch Republic*, 968–90.
[3] H. Schenckel, L. Smids, and T. Brongersma, *Kort en bondige beschrijvinge van de schrickelijcke water-vloedt den 13 November 1686 over de provincie van Stadt en Lande ontstaen, behelsende een waerachtigh verhael van ... voorvallen ... alles door eygene ondervindinge ofte ... oor- en oogh-getuygen bevestight* (Groningen: Carel

coastal flood hit the Netherlands in 1702 that affected the same territories in Holland that defenders had intentionally inundated during the *Rampjaar*.[4] The winter of 1708–09 had been desperately cold, even by the standards of the frigid Maunder Minimum. It was likely one of the most severe winters in Europe since 1500. According to Dutch chroniclers, "postal riders died frozen on their horses," and the price of staple foods spiked due to harvest failures. This bitter winter contributed to subsistence crises and disease outbreaks across Europe.[5] Natural disasters exacerbated wartime privation. Even the Dutch Republic, which had remained relatively insulated from the "fatal synergy" of war, climate extremes, and disaster during the seventeenth century, could not completely escape these effects.[6]

The Peace of Utrecht was thus met with relief and jubilation. In a pamphlet published on a day of organized thanksgiving in the northern province of Friesland, the bookseller and publisher François Halma described the sense of release he and others felt after emerging from these trials. Halma had lived through the French occupation of Utrecht in 1672 and, along with his countrymen, endured the subsequent decades of dearth. "Now we thought we could take a breath after being so long

---

Pieman, 1687), preface. Jan Buisman, *Duizend jaar weer, wind en water in de Lage Landen. Vol 5, 1675–1750* (Franeker: Van Wijnen, 2006), 137.

[4] Fransen, *Dijk onder spanning*, 184–8.

[5] It was the fifth coldest winter in the Netherlands during the eighteenth century. Royal Netherlands Meteorological Institute (KNMI), "Monthly, Seasonal, and Annual Means of the Air Temperature in Tenths of Centigrade in De Bilt, Netherlands, 1706" (De Bilt: KNMI, 2007). http://climexp.knmi.nl/data/ilabrijn.dat. Jürg Luterbacher, Daniel Dietrich, Elena Xoplaki et al., "European Seasonal and Annual Temperature Variability, Trends, and Extremes since 1500," *Science* 303, no. 5663 (2004): 1500. Buisman, *Duizend jaar*, Vol. 5, 363–5. Simon Jacobszoon Kraamer, *Eenijge Merckwaerdige Gebeurttenisse, Voorgevalle Bij Mijn Tijdt.* c. 1752. GaZ. Doopsgezinde Gemeente Zaandam-West. KA 0012. 39. IV. anon., *Opregt dog droevig verhael, van de groote elende en droefheyd, veroorzaekt door de felle kou en sterk vriesend weer van deeze winter* (Jan Henriksz., 1709). Eleni Xoplaki, Panagiotis Maheras, and Jürg Luterbacher, "Variability of Climate in Meridional Balkans during the Periods 1675–1715 and 1780–1830 and Its Impact on Human Life," *Climatic Change* 48, no. 4 (2001): 597–8. Guido Alfani and Cormac Ó Gráda, "Famines in Europe: An Overview," in *Famine in European History*, eds. Guido Alfani and Cormac Ó Gráda (Cambridge: Cambridge University Press, 2017), 9–16.

[6] Daniel Curtis, Jessica Dijkman, Thijs Lambrecht, and Eric Vanhaute, "Low Countries," in *Famine in European History*, eds. Guido Alfani and Cormac Ó Gráda (Cambridge: Cambridge University Press, 2017), 119–40. The clearest elaboration on this "fatal synergy" from a global perspective is Parker, *Global Crisis*.

tortured and overburdened," he noted with relief.[7] The Treaty of Utrecht signaled a new beginning and the fireworks celebration represented that transition. State and public media went to great lengths to promote this view. For those unable to visit The Hague or other major cities hosting celebrations, they could purchase broadsides with images that conveyed their size and splendor. Newspapers and pamphlets interpreted the castles' classical iconography. Poems, plays, sermons, and books promoted the joyful atmosphere and optimistic assessments on the future of the Dutch Republic.[8] The fireworks castle in The Hague was triumphal and aspirational and everything about its appearance evoked themes of abundance, wealth, unity, and peace. The celebration ceremonially ended an era haunted by calamity in anticipation of a return to Golden Age prosperity.[9]

Harsh reality dampened the jubilant atmosphere. The financial strain of war forced the Republic to disband much of its military almost immediately, yet public debt remained enormous. Holland's liability alone doubled between 1688 and 1713, and less densely populated provinces like Groningen and Overijssel suffered as well.[10] Critics bemoaned the outcome of the treaty negotiations, which included trade agreements between Britain and France that seemed to disadvantage the Dutch.[11] The outcome of the treaty and its shrunken military meant that the Netherlands lost its position as a major power in Europe after 1713. Cultural historian Lotte Jensen has argued that the peace celebrations of 1713 inaugurated a new cynicism about the future of the Republic.[12] Up

---

[7] François Halma, *Godts wraakzwaardt over Nederlandt, vertoont in de zwaare sterfte onder 't rundtvee* (Leeuwarden: François Halma, 1714), 8.

[8] *Nederlansch Gedenkboek: Of, Europische Mercurius* (Amsterdam: A. van Damme, 1713), 301–9.

[9] The Estates General fireworks display was only one among several in the Netherlands, including a separate fireworks castle in The Hague sponsored by the province of Holland. W. Frijhoff, "Fiery Metaphors in the Public Space: Celebratory Culture and Political Consciousness around the Peace of Utrecht," in *Performances of Peace: Utrecht 1713*, eds. Renger Evert Bruin, Lotte Jensen, and David Onnekink (Leiden: Brill, 2015), 223–48.

[10] Fritschy, "The Poor, the Rich, and the Taxes," 242–58.

[11] *Korte schets van 'slands welwezen door de laatste vrede* (Voor den Autheur, 1714). Doohwan Ahn, "The Anglo-French Treaty of Commerce of 1713: Tory Trade Politics and the Question of Dutch Decline," *History of European Ideas* 36, no. 2 (2010): 167–80. Israel, *Dutch Republic*, 971–5. Aalbers, "Holland's Financial Problems," 79–81. G. N. Clark, "War Trade and Trade War, 1701–1713," *The Economic History Review* 1, no. 2 (1928): 278–9.

[12] Lotte Jensen, *Vieren van vrede: Het ontstaan van de Nederlandse identiteit, 1648–1815* (Nijmegen: Uitgeverij Vantilt, 2016), 107.

until this point, peace celebrations had largely supported the notion that "from that point everything would get better. A new Golden Age awaited where trade would again grow, agriculture would flourish, and general prosperity would increase."[13] The Treaty of Utrecht painted a cloudier, more uncertain picture.[14]

Making matters worse, nature seemed to have once again turned against them. A cool, wet summer, typical of conditions during the late Maunder Minimum, descended upon the Netherlands in 1713.[15] Aleida Leurink, a minister's wife from Overijssel, lamented that rains lasted through the summer of 1713. Wet midsummers presented numerous difficulties for both grain agriculture and grazing. Excessive or untimely rains affected plant growth, decreased the nutrient content of cereals, and, in the worst situations, could lead to harvest failure.[16] "On the 24th of June it began to rain," Leurink noted in her diary, "and rained very long every day, so that the rye could not be ripened, that people seemed to perish with hunger because all the old rye was exhausted by everyone."[17] Wet midsummers also presented challenges for pasturage. Grass was the agroecological basis for rearing, fattening, or dairying cattle. Wet-spring and midsummers depressed grass growth and undermined the hay harvest necessary for overwintering.[18] During sodden springs and summers, farmers delayed moving overwintered cattle from stalls to pasture. Lacking adequate fodder, cattle produced less milk and may have grown

---

[13] Ibid., 10.
[14] Willem Frijhoff questions whether the fireworks masked a general apprehension about decline. W. Frijhoff, "Utrechts vreugdevuur, Masker voor 's lands neergang?," *De Achttiende Eeuw* 40 (2008): 5–20.
[15] Jürg Luterbacher, Ralph Rickli, Elena Xoplaki et al., "The Late Maunder Minimum (1675–1715) – A Key Period for Studying Decadal Scale Climatic Change in Europe," *Climatic Change* 49, no. 4 (2001): 454.
[16] B. H. Slicher van Bath, "Agriculture in the Vital Revolution," in *The Cambridge Economic History of Europe*, eds. E. Rich and C. Wilson (Cambridge: Cambridge University Press, 1977), 57–9. Christian Pfister, "Weeping in the Snow: The Second Period of Little Ice Age-Type Impacts, 1570–1630," in *Kulturelle Konsequenzen der Kleinen Eiszeit*, eds. Wolfgang Behringer, Hartmut Lehmann, and Christian Pfister (Göttingen: Vandenhoeck & Ruprecht, 2005), 62–6.
[17] Jacobus Joannes van Deinse, ed., "Uit het dagboek van Aleida Leurink te Losser, 1698–1754," in *Uit het land van katoen en heide: Oudheidkundige en folkloristische schetsen uit Twente* (Enschede: van der Loeff, 1975), 538.
[18] J. M. G. van der Poel, "Het Noord Hollandse weidebedrijf om de 19ᵉ eeuw," *Holland: Regionaal-historische Tijdschrift* 3/4 (1986): 148–9. Bieleman, *Geschiedenis*, 162.

more susceptible to disease.[19] According to Leurink and other observers, these rainy conditions extended through August.[20]

In Holland, the dissonance provoked by damp summer weather and harvest shortages amidst an atmosphere of jubilation following the Peace of Utrecht was pronounced enough to be memorialized in poetry. The farmer-poet Hubert Korneliszoon Poot described his despair at having survived war only to face the consequences of poor weather and further hardship. In his poem entitled "Disasters in the Year of Peace," he contrasted the relief he felt upon the conclusion of the war with the "heavy rains [that] clapped and burst and quickly arrived, in the heart of the harvest and sweet summer, both fields of grain and lush meadows sodden."[21] More ominously, Poot described the recent arrival of a new disaster that spread across the Republic like a "whip of dark red blood and fire."[22] A cattle panzootic that had decimated herds in southern and central Europe for the past three years finally appeared in the Dutch Republic.[23]

Between the first reported outbreak in Holland in May 1713 and the last in 1720, cattle plague devastated herds across the Netherlands and across Europe. The disease, likely the morbillivirus rinderpest, was invasive and deadly.[24] Beginning in 1709, it swept across pastures from

---

[19] Sam White, "A Model Disaster: From the Great Ottoman Panzootic to the Cattle Plagues of Early Modern Europe," in *Plague and Contagion in the Islamic Mediterranean*, ed. N. Varlik (Kalamazoo: Arc Humanities Press, 2017), 91–116. Richard Oram, "'The Worst Disaster Suffered by the People of Scotland in Recorded History': Climate Change, Dearth and Pathogens in the Long 14th Century," *Proceedings of the Society of Antiquaries of Scotland* 144 (2014): 229. Slavin, "The Great Bovine Pestilence," 1245–7. Timothy P. Newfield, "A Cattle Panzootic in Early Fourteenth–Century Europe," *Agricultural History Review* 57, no. 2 (2009): 178. The high mortality of healthy cattle to cattle plague complicates this assessment. Timothy P. Newfield, "Domesticates, Disease and Climate in Early Post-classical Europe: The Cattle Plague of c. 940 and Its Environmental Context," *Post-Classical Archaeologies* 5 (2015): 102–3, 8.

[20] Van Deinse, "Uit Het Dagboek," 538. Buisman, *Duizend jaar*, Vol. 5, 399–400.

[21] H. K. Poot, *Rampen van het vredejaer* (1713), 4.

[22] Buisman, *Duizend jaar*, Vol. 5, 400. Poot, *Rampen*, 5. Poot was an active purveyor of the Arcadian "idyll" in the eighteenth-century literature. Leemans and Johannes, *Worm en Donder*, 414–43.

[23] Clive A. Spinage, *Cattle Plague: A History* (New York: Kluwer Academic/Plenum, 2003), 104.

[24] This diagnosis reflects a general, albeit speculative, consensus based on documented characteristics of illness and outbreaks. "Cattle plagues" also resulted from bovine pleuropneumonia, foot-and-mouth disease, and anthrax. Paleomicrobiological evidence confirming the disease was rinderpest is unavailable. Timothy P. Newfield, "Early Medieval Epizootics and Landscapes of Disease: The Origins and Triggers of European

eastern Europe and the Balkans north and westward. By 1713, the disease had already moved along a circuitous route through Italy and southern Europe before it emerged in the Baltic and North Sea basins. Cattle plagues during the early modern period spread through networks of trade or warfare. These networks satisfied the appetites of increasingly large armies and urban populations and connected environments and disease ecologies across two continents. It was likely no coincidence that the virus had emerged in 1709 following one of the harshest European winters in the past millennium and spread westward amidst the War of the Spanish Succession. The synergistic interplay of climatic extremes and conflict may have triggered the initial outbreak in eastern Europe and influenced the spread of infection westward.[25] Networks of trade, meanwhile, introduced the disease into the Republic. The transnational oxen trade between Denmark and Holland that supplied its burgeoning cities with meat almost certainly served as the conduit that brought this cattle plague into the Republic.[26]

The outbreak of 1713 was the first of three deadly cattle plague episodes in the Republic during the eighteenth century and one of the most severe in early-modern history. Livestock diseases were relatively common during this era, but the outbreak of 1713 seemed different.[27] Rinderpest infection rates could reach as high as 100 percent and mortality ranged from 60 to 90 percent. Farmers looked upon its arrival with understandable dread. The plague's symptoms horrified them. Cattle afflicted with the disease suffered fevers, discharges from their eyes and

---

Livestock Pestilences," in *Landscapes and Societies in Medieval Europe East of the Elbe: Interactions between Environmental Settings and Cultural Transformations*, eds. Sunhild Kleingärtner, Timothy P. Newfield, Sébastien Rossignol, and Donat Wehner (Toronto: Pontifical Institute of Mediaeval Studies, 2013), 95–6.

[25] Luterbacher et al., "European Seasonal and Annual Temperature Variability," 1499–1503. White, "A Model Disaster," 107–9.

[26] P. A. Koolmees, "Epizootic Diseases in the Netherlands, 1713–2002: Veterinary Science, Agricultural Policy, and Public Response," in *Healing the Herds: Disease, Livestock Economies, and the Globalization of Veterinary Medicine*, eds. Karen Brown and Daniel Gilfoyle (Athens: Ohio University Press, 2010), 19–41. The disease chronology is outlined in Franciscus Cornelius Hekmeijer, *Korte geschiedenis der runderpest: benevens eene opgave van al de over deze ziekte handelende geschriften, die van de vroegste tijden tot op heden zijn uitgekomen* (Amersfoort: Jacobs en Meijers, 1845). The authoritative account is Spinage, *Cattle Plague*.

[27] An epidemic of what was likely hoof and mouth disease struck the Dutch Republic in 1682. Ronald Rommes, "'Geen vrolyk geloei der melkzwaare koeijen': Runderpest in Utrecht in de achttiende eeuw," *Jaarboek Oud Utrecht* (2001): 89.

noses, blisters on their tongues, diarrhea, and finally death.[28] The outbreak in 1713 struck with shocking intensity. Tens of thousands of cattle died in the first two years of the plague. The province of Friesland alone lost 66,000 cattle between 1713 and 1715. Although records of cattle mortality during this deadliest wave are incomplete, estimates place the number lost in the Republic between 120,000 and 300,000 animals.[29] Mortality varied by province and region, but to many anxious onlookers, the epizootic was a disaster that affected the entire Republic. It seemed to spread like wildfire, indiscriminately wiping out herds of animals. Cattle plague, in this view, cruelly inverted the cornucopian optimism promoted by the peace celebrations. Every new corpse added to the struggles of rural populations still burdened by decades of warfare, extraordinary taxation, and other natural disasters. Instead of Arcadian fertility and a return to Golden Age prosperity, the Dutch Republic's fertile meadows had become burial grounds. Cattle plague reified fears of decline.

Contemporaries linked cattle plague to decline for two reasons. First, because cattle occupied a unique, symbolic space in Dutch culture. Cattle represented rural fertility, economic prosperity, and providential favor. Each were cultural artifacts of the early-modern management of Dutch landscapes. Cattle plague inverted these associations, imparting new cultural and moral meanings. Diseased cattle bodies and depleted pastures reflected the decay of Dutch morality, economic vitality, and the loss of their favored providential status. Second, the virulence and timing of the plague seemed extraordinary. It killed so quickly and arrived so soon after the peace negotiations that it seemed only explicable through divine providence. The disease amplified moral anxieties about the future of the Republic that had festered since the *Rampjaar*. Cattle plague in this view was a problem that affected the Netherlands as a whole, indicative of a broader decline in Dutch society.

---

[28] Spinage, *Cattle Plague*, 5.
[29] Koolmees, "Epizootic Diseases in the Netherlands," 23. The first estimate appears low because the province of Holland lost 160,000 cattle during the less severe outbreak in the 1770s. A. M. van der Woude, *Noorderkwartier*, 585–93. The higher estimate from a nineteenth-century source has not been confirmed by modern scholars. Hekmeijer, *Korte Geschiedenis*, 6. This same source estimates that 1.5 million cattle died throughout Europe. J. A. Faber, "Cattle Plague in the Netherlands during the Eighteenth Century," *Mededelingen van de Landbouwhogeschool te Wageningen* 62, no. 11 (1962): 2. It is difficult to estimate mortality without figures of the total number of cattle. Faber estimated that the province of Friesland suffered a 41 percent loss, although intensive cattle-raising areas lost as much as 66 percent. J. A. Faber, *Drie eeuwen Friesland: Economische en sociale ontwikkelingen van 1500 tot 1800* (Leeuwarden: De Tille, 1973), 466.

Disaster response was not limited to religious and moral appraisals. Dutch towns and provinces enacted a variety of strategies to counter the epizootic and restore the cattle economy. The effectiveness of these regulations depended upon the participation of farmers on a local scale, who also relied on a host of additional strategies to protect their animals. Historians have tended to criticize the decentralized approach to disease management in the Republic. They argue it was unsystematic and ineffective, pointing to high mortality and the extended duration of the plague. Others condemn the perceived overreliance of farmers on popular remedies and spiritual solutions during an era when other European states had begun experimenting with innovative containment strategies. These assessments reinforce negative assumptions about the capacity of the Dutch state to cope with crisis during the early eighteenth century.[30] The regressive influence of its religious culture, static medical marketplace, and archaic state structure resulted in the Republic's stagnation, or even backwardness, relative to other European powers.

The proactive response of Dutch state and religious authorities and the public belie the declensionist narratives of contemporaries and later negative assessments of response by scholars. Moralists crafted causal stories that transformed divine providence into politically expedient coping strategies. Provincial efforts worked to contain the disease and alleviate its financial burden. Although certainly decentralized, it promoted a diversity of responses across multiple scales of governance. In keeping with prevailing environmentalist and contagionist theories of disease transmission, most provinces opted for a universalist strategy to cleanse polluted environments and limit animal movement into the Republic and within its borders. Few promoted innovative medical or management strategies, preferring to rely on tried-and-true methods of disease management. Cattle plague nevertheless prompted careful evaluation of the origin, meaning, and management of the disease. Dutch response was indeed backward looking, but it was not backward.

2.1 DUTCH CATTLE AND DANISH OXEN IN THE GOLDEN AGE

Cattle held significant meaning in eighteenth-century Dutch society and culture. Like many early-modern societies, the Dutch read their

---

[30] Koolmees, "Epizootic Diseases," 23–5. Jan Bieleman, *Boeren in Nederland: Geschiedenis van de landbouw 1500–2000* (Amsterdam: Boom, 2008), 214. Faber, "Cattle Plague," 4–5. Rommes, "Geen vrolyk," 123.

landscapes metaphorically.[31] Cattle were both symbols and material embodiments of wealth, fertility, and prosperity. These associations emerged out of centuries-long environmental and agricultural transformations of the Dutch landscape, which crystallized during the seventeenth century. Cattle came to represent both agricultural productivity and an optimistic, providential relationship between God and the Republic that was characteristic of the Golden Age idyll. In visual arts and in literature, cattle scenes, cattle portraits, and pastoral imagery promoted these associations in popular culture, and the success of rural themes in the art market, in theater, and poetry illustrated their broad, enduring acceptance by the public.[32]

The symbolic associations between cattle and the Republic could be even more overt. Along with the Dutch "maid" and "lion," cattle were the most frequently employed symbols of the Dutch "fatherland" by Golden Age artists and writers.[33] In his *Allegorical Print on Dutch Prosperity*, Hendrick Hondius depicted muscular cattle bodies as part of a series of political prints in 1644 (Figure 2.2). Their physical strength reflected the Republic's power at the height of the Golden Age. Cattle (like the Dutch maid depicted in *Ellenden klacht*) also symbolized the tenuous balancing act required to maintain that power. "Honorable guardians, be vigilant in your care that our cattle are not stolen," Hondius's text reminded viewers.[34] The fireworks castle sponsored by Estates General in 1713 also featured pastoral imagery beneath its large, central arch. A cow and farmer lounge contentedly at the bottom of the frame, with Ceres (goddess of the harvest) and Mercury (god of commerce) looking down from above. The scene reflected the idealized Arcadian relationship between agricultural fertility and Dutch wealth that their recent treaty ensured, but it also reminded viewers how attentive they would need to be to preserve it.

Cattle's symbolic significance reflected their perceived material importance in the Golden Age economy. Seventeenth-century historian Wouter

---

[31] Alexandra Walsham, *The Reformation of the Landscape: Religion, Identity, and Memory in Early Modern Britain and Ireland* (Oxford: Oxford University Press, 2011), 328.

[32] Johan Koppenol, "Noah's Ark Disembarked in Holland: Animals in Dutch Poetry, 1500–1700," in *Early Modern Zoology: The Construction of Animals in Science, Literature and the Visual Arts* (Leiden: Brill, 2007), 467–72. C. Boschma, ed., *Meesterlijk vee: Nederlandse veeschilders 1600–1900* (Zwolle: Waanders, 1988).

[33] Marijke Meijer Drees, "'Vechten voor het vaderland' in de literatuur, 1650–1750," in *Vaderland: Een Geschiedenis van de Vijftiende Eeuw tot 1940*, ed. N. C. F. van Sas (Amsterdam: Amsterdam University Press, 1999), 130–4.

[34] Hendrick Hondius, "Koeien in een landschap," *Rijksmuseum*, 1644, http://hdl.handle.net/10934/RM0001.COLLECT.491844.

The First Cattle Plague, 1713–1720 61

FIGURE 2.2. Cattle symbolized the Dutch state in Hondius's print, as well as Golden Age fertility and strength. *Source*: Hendrick Hondius, *Koeien in een landschap*, Print, 1644, Rijksmuseum, Amsterdam. http://hdl.handle.net/10934/RM0001.COLLECT.491844.

van Gouthoeven attributed "the richness of the land of Holland" to three "mines or mountains of wealth," two of which related to livestock. The "horses, oxen, cattle and sheep hauled to markets from Holland" and the "innumerable quantities of butter, cheese and other dairy found in

Holland and traded throughout the world" underpinned the Dutch economy.[35] Although later writings on the Republic's political economy often pointed to industry and overseas commerce as the motor of Dutch prosperity, agriculture remained the largest single sector of the economy well into the eighteenth century, and cattle stood at the center of Dutch rural life.[36] Dairy cattle produced staple foods such as milk, butter, and cheese while both cattle and oxen supplied products such as meat, tallow, hides, and horn. Oxen provided protein and labor. The specialized nature of Dutch cattle production necessitated a wide variety of professions, from cattle breeders, grazers, milkmaids, and dairy farmers to cattle traders, shoemakers, and tanners. Textile bleachers used lactic acid from buttermilk to whiten their cloth. Farmers depended on oxen as draft animals, and farmers growing tobacco and other nutrient-hungry crops needed cattle manure to fertilize fields.[37]

The character of cattle holding varied widely by region and landscape type. In the drained peat landscapes of southern Holland, herds were relatively large, production was capital intensive, and farmers produced dairy products to trade in local, regional, and international markets. In the sandy soils of rural Drenthe in the eastern Republic, by contrast, herds were smaller, and farmers historically focused on cattle husbandry. Even in the reclaimed polders of Zeeland, where cattle scarcely produced enough butter and cheese to serve local markets, farmers still depended upon their manure for arable production.[38] Everywhere they appeared,

---

[35] The third pillar was the herring industry. Wouter van Gouthoeven, *D'oude chronijcke ende historien van Holland (met West-Vrieslandt) van Zeeland ende van Utrecht* (Dordrecht: Peeter Verhaghen, 1620), 6.

[36] De Vries and van der Woude, *First Modern*, 195. Ida J. A. Nijenhuis, "Shining Comet, Falling Meteor: Contemporary Reflections on the Dutch Republic as a Commercial Power during the Second Stadholderless Era," in *Anthonie Heinsius and the Dutch Republic 1688–1720. Politics, War, and Finance*, eds. J. A. F. de Jongste and Augustus J. Veenendaal Jr. (The Hague: Institute of Netherlands History, 2002), 115–31.

[37] Ronald Rommes, "Twee eeuwen runderpest in Nederland (1700–1900)," *Argos* 31 (2004): 35.

[38] Filip van Roosbroeck and Adam Sundberg, "Culling the Herds? Regional Divergences in Rinderpest Mortality in Flanders and South Holland, 1769–1785," *TSEG/Low Countries Journal of Social and Economic History* 14, no. 3 (2018): 48. Bas van Bavel, "The Economic Origins of Cleanliness in the Dutch Golden Age," *Past & Present* 205, no. 1 (2009): 55–8. Jan Bieleman, "Rural Change in the Dutch Province of Drenthe in the Seventeenth and Eighteenth Centuries," *The Agricultural History Review* 33, no. 2 (1985): 110. Peter R. Priester, *Geschiedenis van de Zeeuwse landbouw circa 1600–1910* (Utrecht: HES, 1998), 244–5.

cattle performed important economic functions. Cattle were a keystone species in the rural economy of the Dutch Republic.

Although cattle appeared timeless in seventeenth-century text and imagery, their significance in society and culture resulted from the historical conjuncture of long-term environmental and economic transitions. The intensification of Dutch merchant involvement in the Baltic grain trade during the sixteenth and seventeenth centuries freed farmers to pursue specialized agriculture, including dairy farming and the production of industrial and horticultural crops. Dutch cattle breeding produced varieties renowned for their size and the quality of their milk. This transition was particularly pronounced in the peat and clay landscapes of Holland and Friesland.[39] By the early seventeenth century, high agricultural prices incentivized wealthy urban investors to reclaim bogs and drain large inland lakes in Holland to create polders for agricultural use. These lakes were the legacy of centuries of soil subsidence brought on by intensive reclamation for grain cropping and peat extraction for fuel. This had provoked an ecological crisis by the mid-fourteenth century when grains proved unsuited to these new conditions, and small peat mires expanded into sizable lakes.[40] By reclaiming these landscapes, Dutch investors not only created ideal conditions for intensive dairying and fattening, but they did so by self-consciously fashioning new land. The glorification of the Dutch pastoral in art, literature, and cartography reflected pride in this cultural landscape.[41] It also memorialized the close ties between urban investment and rural productivity. Cattle, especially dairy cattle, occupied a central space in both narratives.

Oxen (castrated bulls) likewise bore significant symbolic and economic heft during the seventeenth century. The oxen trade demonstrated the

---

[39] Bieleman, *Five Centuries*, 49–55. Peter Sutton, "The Noblest of Livestock," *J. Paul Getty Museum Journal* 15 (1987): 107. Jan de Vries, *The Dutch Rural Economy in the Golden Age, 1500–1700* (New Haven, CT: Yale University Press, 1974), 142–3.

[40] Bieleman, *Geschiedenis*, 40–1. Petra J. E. M. van Dam, "Sinking Peat Bogs: Environmental Change in Holland, 1350–1550," *Environmental History* 6, no. 1 (2001): 32–45. Jan Luiten van Zanden, *The Rise and Decline of Holland's Economy: Merchant Capitalism and the Labour Market* (Manchester: Manchester University Press, 1993), 30. Wilhelmina Maria Gijsbers, "Danish Oxen in Dutch Meadows: Beef Cattle Trading and Graziery in the Netherlands between 1580 and 1750," in *Facing the North Sea: West Jutland and the World; Proceeding of the Ribe Conference*, eds. Mette Guldberg, Poul Holm, and Per Kristian Madsen (Esbjerg: Fiskeri-og Søfartsmuseet, 1993), 129–48. Van Bavel and van Zanden, "Jump-Start," 522–33.

[41] Alette Fleischer, "The Garden Behind the Dyke: Land Reclamation and Dutch Culture in the 17th Century," *Icon* 11 (2005): 16–32.

expansive reach of the Republic's commercial influence. Dutch cattle markets had already tapped an expansive international oxen trade to provide beef for growing urban centers by the first half of the fourteenth century. These trade routes connected Dutch consumers to suppliers in Denmark and Schleswig-Holstein. Trade along this northwest European network peaked in the early seventeenth century, at which point the Dutch imported 40,000–50,000 heads annually. Dutch protectionist trade policies after 1686 limited cattle imports in an attempt to stabilize fluctuating prices amidst agricultural recession, stimulate domestic breeding and fattening, and increase taxable revenue. Nevertheless, Dutch merchants continued to ship Danish cattle across the North Sea or drive them along the coast by the thousands into the eighteenth century where they would fatten themselves after their journey in rich Dutch pastures and polder lands before merchants sold them at market.[42]

"Danish" oxen driven or shipped to Dutch cattle markets became powerful symbols of economic power, agricultural fertility, and the interdependencies of urban wealth and the countryside (Figure 2.3). Prosperous urban merchants in Holland not only invested in reclamation projects that expanded pasturage but also many became active agents in the Danish cattle trade. To be an *ossenweider* (grazier) was a sign of wealth and class privilege. Common *koe-boeren* (cattle farmers) enjoyed much lower status but consistently featured in representations of rustic fertility.[43] Jan van der Velde's print *Terra* from 1622 centers both Dutch wealth and agricultural productivity on a bustling market. Richly clad buyers surrounded by cattle and other agricultural products haggle with "bristly rural farmers" over this "show of wealth of their land."[44] The hustle and bustle of the city whose buildings crowd the left of the print pleasantly contrast with the pastoral landscape in the background. The market stood as an emblematic meeting place between rural fertility and urban prosperity.

Just as with dairy production, economic and environmental changes underpinned the rise and development of the early-modern oxen trade. The spectacular growth of Dutch cities during much of the sixteenth and seventeenth centuries spurred demand for meat. The high price of land, population density, and specialized agriculture in the coastal provinces

---

[42] Wilhelmina Maria Gijsbers, *Kapitale ossen: De internationale handel in slachtvee in Noordwest-Europa (1300–1750)* (Hilversum: Verloren, 1999), 15, 81–3.
[43] Alison M. Kettering, "After Life: Rembrandt's Slaughtered Ox," *Artibus et Historiae* 79 (2019): 272–5. Gijsbers, *Kapitale ossen*, 239–40.
[44] From the Latin inscription to Jan van de Velde, "Terra (De Vier Elementen)," *Rijksmuseum*, 1603–41. http://hdl.handle.net/10934/RM0001.COLLECT.333114.

FIGURE 2.3. Van de Velde's print depicts the meeting ground between rural fertility and urban wealth. *Source*: Jan van de Velde after Willem Buytewech, *Terra* (The Elements), Print, 1622, Rijksmuseum, Amsterdam. http://hdl.handle.net/10934/RM0001.COLLECT.333119.

incentivized farmers to outsource land-extensive pasturage to regions like Denmark where land and labor were cheaper. This was part of a broader pattern of core Western European economies displacing their ecological pressures to the periphery, which by the early eighteenth century extended deep into eastern Europe.[45] Relatively expensive beef also presented an opportunity for wealthy Dutch burghers' consumption to mirror their status. The expansion of transoceanic trade in the seventeenth century necessitated a simultaneous increase in salted meat for ships' crews.[46] Dutch demand for Danish beef connected environments separated by hundreds of kilometers and created pathways for disease along the North Sea coast and into central Europe. Ecological transformations lay the foundation for the cattle economy and its symbolic significance in Dutch culture. They also set the stage for disaster.

---

[45] Karl Appuhn, "Ecologies of Beef: Eighteenth-Century Epizootics and the Environmental History of Early Modern Europe," *Environmental History* 15, no. 2 (2010): 268–87. Gijsbers, "Danish Oxen in Dutch Meadows," 133–7. White, "A Model Disaster," 105–8.
[46] Priester, *Geschiedenis*, 244–54. Bieleman, *Geschiedenis*, 63.

When the plague arrived in June 1713, it was no surprise that it first appeared in Amsterdam, site of the largest market for Danish cattle. Pamphleteers immediately implicated the Danish oxen trade. It was "through Danish oxen that this evil was introduced to the beasts in our Fatherland," one pamphlet noted in 1714. Networks of trade that moved cattle from urban markets to stalls and pastures, then back to markets for slaughter in the fall, served equally well as pathways for disease. Poor grass growth during this first summer of the cattle plague may have increased susceptibility of some animals to disease, but conditions would improve in subsequent years.[47] Cattle mobility proved more crucial to epizootic spread, including during these early months. "So hastily were the cattle driven to market in November," an early report noted, "that the infection spread and the plague intensified so astonishingly quick that scarcely any survived."[48] The same cattle markets that symbolically bridged urban capital with rural abundance and demonstrated the power and reach of Dutch trade now opened the door to an outbreak.

2.2 CATTLE PLAGUE AND DECLENSIONIST INVERSIONS OF PROSPERITY

To many nervous citizens, the arrival of cattle plague in the spring of 1713 brought a profound unease about the future. Newspapers and pamphlets reported the progress of the disease and readers warily monitored its progress. The Frisian farmer Jan Wopkes first described cattle plague in his journal on August 20, months before it arrived in Friesland. "Many cattle are dying in Holland in the jurisdiction of Amsterdam," he noted, "already more than 8,000 cattle, both oxen and cows."[49] The disease spread unevenly across the Netherlands, manifesting at different times in different provinces. It appeared in Utrecht in July 1713 and reached Friesland by the end of the year. Regional markets likely served as channels for transmission. Wopkes noted in 1713 that "very many cattle sent to Wommels [for the annual market] suffered from the terrible sickness, so that it is hard to find one house that is now free of this evil

---

[47] Buisman, *Duizend jaar*, Vol. 5, 935.
[48] *Bedenkingen, en raad, noopende de tegenwoordige stervte onder het rundvee* ('s Gravenhage: Pieter van Thol, 1714), 5–6.
[49] Jan Wopkes, "Aantekeningen van Jan Wopkes, Boer of Koster Te Wommels, over Geboorten, Huwelijken, Overlijden En Andere Gebeurtenissen," TRE. Bib 1244 (1720): 17.

sickness."⁵⁰ By the end of 1714, pamphleteers widely accepted that disease had become a disaster of Republic-wide proportions affecting Danish oxen and Dutch cattle alike.

Almost immediately, writers framed cattle plague as an inversion of cattle's Golden Age association with wealth. The earliest accounts came from moralizing poets and ministers who produced dramatic descriptions of the misery inflicted upon Dutch families. "The farmer sees shortly his cattle, his wealth, entirely robbed ... and looks with jaded sadness, with wife and tender offspring on the ability to preserve their lives," wrote the moralist Is Centen.⁵¹ Centen also acknowledged its unequal consequences for the lower classes, declaring that the "miserly market in dairy, meat, and butter only makes this disaster all the worse for the poor."⁵² Others described the threat of cattle plague in broader terms, affecting entire provinces. Addressing the provincial assembly of Friesland, François Halma pronounced cattle the "silver mine of Friesland's fertile land, and treasure and lifeblood of the land's commonwealth." That wealth was under threat by "the flaming sword and burned with the fire of plague."⁵³ The writer and historian Abraham Moubach interpreted the plague in even more catastrophic, broader terms. "During all calamitous and dangerous disasters," he began, "namely those that affect an entire country, and threaten equally the destruction of all, the most miserable were those for which no help or solace could be found. Can there be a greater misery than the cattle plague?"⁵⁴ As plague spread from province to province, the inversion of Golden Age wealth seemed an increasingly Dutch phenomenon.

Despite growing fears of a broadly shared disaster, the plague's impacts were in fact highly variable. Many of the largest losses occurred in the regions that had undergone significant ecological transformations in previous centuries making possible intensive pasturing. These areas were most dependent on specialized dairy or beef production. Holland,

---

⁵⁰ J. J. Spahr van der Hoek, *Geschiedenis van de Friese landbouw* (Drachten: Friesche Maatschappij van Landbouw, 1952), 218.
⁵¹ Is. Centen, *Gods oordelen over Nederland, in de sterfte van 't rundvee, den zwaren storm, en hogen watervloed* (Amsterdam: Johannes Oosterwijk, 1718), 4.
⁵² Ibid., A4. The class dimensions of cattle plague are little understood, though it likely elevated food prices by freezing the international oxen trade, reducing stock available in urban areas. Rommes, "Geen vrolyk," 89.
⁵³ Halma, *Godts Wraakzwaardt over Nederlandt*, A2.
⁵⁴ Abraham Moubach, *Schaadelijke Veepest, Ontstaan sedert den Jare 1711 in Italie, Duitsland, en Zwitserland; en die in deeze Nederlanden tot heden zoo jammerlyk onder 't Hoornvee heeft gewoed* (Amsterdam: Joannes Oosterwyk, 1719), 37.

Friesland, Utrecht, and Groningen had the largest herds and, as a result, likely suffered the greatest losses. The most complete records cover the province of Friesland, where up to 66 percent of total stock succumbed in the hardest hit regions.[55] In the *Landschap* of Drenthe, by contrast, cattle were fewer in number and cattle plague was less virulent. By 1716, the income from taxation on horned animals in the region had decreased by 21 percent.[56] Even within provinces, mortality differed by region and locality. Pasturelands experienced greater rates of infection than areas specializing in arable production.[57] This variability likely resulted from different land-use patterns, the size of herds, and the proximity of those herds to each other.[58] Large herds of oxen or dairy cattle grazing in open polder lands had far greater potential contact with diseased animals than cattle employed in systems of mixed husbandry where animals were fewer and less mobile. Once infected, however, cattle in arable areas experienced higher mortality than pasturelands, likely because farmers were less financially dependent upon them, and they received less care.[59]

Losses in tax revenue reflected these disparities. Most provinces relied on cattle production for taxable income. Plague destabilized their tax farming systems because they reduced the total number of assessable animals and undermined farmers' ability to afford other forms of taxation. Provinces imposed multiple taxes on farmers and many directly taxed animals (called *hoorngeld*).[60] The cattleholders in the provinces and regions most dependent upon large herds stood to lose the most income. In Utrecht, for instance, total cattle losses approached 35–40 percent during the outbreak, but here again the total amount of lost tax revenue varied by region. Herds were larger in the western districts of the province that specialized in pasturage than the sandier regions to the east and losses correspondingly more severe.[61] Farmers engaged in specialized beef or

---

[55] Faber, *Drie Eeuwen*, 466.
[56] Jan Bieleman, *Boeren op het Drentse zand, 1600–1910: Een nieuwe visie op de "oude" landbouw* (Wageningen: Landbouwuniversiteit, 1987), 320.
[57] In the Hasselt Parish of Overijssel, only 10.9 percent of cattle remained in 1714, whereas 2.7 percent survived in the municipality of Staphorst. B. H. Slicher van Bath, *Een samenleving onder spanning: geschiedenis van het platteland in Overijssel* (Utrecht: HES Publishers, 1957), 513. Bieleman, *Boeren in Nederland*, 150.
[58] Filip van Roosbroeck demonstrates this phenomenon in the Austrian Netherlands during the third outbreak of cattle plague in "To Cure Is to Kill?: State Intervention, Cattle Plague and Veterinary Knowledge in the Austrian Netherlands, 1769–1785" (PhD diss., University of Antwerp, 2016).
[59] Faber, *Drie Eeuwen*, 174–5. Bieleman, *Geschiedenis*, 110.
[60] De Vries and van der Woude, *First Modern*, 213.   [61] Rommes, "Geen vrolyk," 105.

dairy production were also more vulnerable because they often had fewer options to spread financial risk. Farmers less dependent upon cattle might temporarily tolerate their loss by shifting emphasis to arable production, although this was more difficult for pastoral farmers, particularly in western polder landscapes unsuited to grain cropping. They might also shift to other agricultural activities such as sheep holding or hay production. Others consolidated or enlarged their farms or shifted from labor-intensive dairying to land-extensive fattening.[62] Farmers were certainly not without options, but cattle plague nevertheless presented unequal financial challenges depending upon region and land use.

Little of this nuance filtered into early state consideration of the disease, however. Provinces treated the plague as a single, undifferentiated threat. Even in Friesland where the provincial governing body (Estates of Friesland) ordered municipalities to provide precise numbers of dead cattle so that the province could determine "in what manner this defect in the lands finances would result," the language of vulnerability remained inflexibly general.[63] A resolution on 10 March 1714 stated that "the greater parts of these good inhabitants have lost all their animals and many others are not able to pay the *floreen* rent or other aforementioned taxes."[64] Lacking a complete understanding of the differential implications of the disease, provinces interpreted its affects in the direst possible terms.

As the disease spread more widely, moralizing literature took center stage. Reformed ministers published books and sermons that framed cattle plague as a disaster that affected the whole of the Netherlands. The Holland minister Isbrandus Fabricius lamented in his book *Dutch Judgments and Disasters (Nederlantse oordelen en rampen)* that the disease "rages not only here and there ... in one or another corner of the country, no! but that appears to exterminate everything like an avenging angel in the entire country. Gelderland, Holland, Utrecht, Friesland, Overijssel, Groningen; the whole country trembles and wails in these times."[65] By 1715, the significance of cattle plague extended

---

[62] Many of these changes were already under way during the ongoing agricultural recession. Bieleman, *Geschiedenis*, 128–9, 166–71.
[63] Resolution (10 March) 1714. TRE. Gewestelijke bestuurinstellingen van Friesland 1580–1795. 5.253, 62–3.
[64] Ibid.
[65] Isbrandus Fabricius, *Nederlantse oordelen en rampen: na een schriftmatige verklaringe van den twist godts met Israël* (Alkmaar: Nikolaas Mol, 1718), 69.

beyond its consequences for rural families or even the provincial economies most dependent upon beef or dairy production. "Without a doubt," wrote the minister Johannes Mobachius, "many people's wealth existed primarily in cattle, and they must have been brought into poverty; but the land's prosperity was also ruined. The power and sinews of the country, in particular that which also existed in cattle were broken." Mobachius went further, bringing his argument full circle back to the Peace of 1713. Cattle plague "is as destructive for the provinces as this most recent, terrible burdensome war that exhausted our country."[66]

Ministers framed the disaster as an inversion of cattle's association with an idealized Golden Age. Referring to cattle as "Lovely Arcadia-Nature's valued gift," the Groningen minister Henricus Carolinus van Byler proudly reminisced that his home was once "a dream of fattened cattle filled as far as the eye can see, as plentifully as sand in the sea." The memory of this pastorale sharply contrasted with a new reality. "What does one find now," he asked, "your meadows lay wasted, wherever one looks one finds new reason for sadness and misery."[67] Johannes Sluiter, a minister in Overijssel, contrasted morbid imagery of fields filled with corpses with scenes from the recent past when "this land, rich with grass was adorned and covered with herds of cattle."[68] Cattle plague created a rift with this Golden Age past. The promises of the treaty celebrations of 1713 forgotten, cattle now reminded people how illusory those hopes now appeared.

2.3 CATTLE PLAGUE AND PROVIDENCE

Cattle plague promoted declensionist anxieties among moralists and created significant economic challenges in the countryside, but only the most pessimistic of observers resigned themselves to fatalistic acceptance of the status quo. As dire as the disease outlook appeared, nearly all agreed

---

[66] Johannes A. Mobachius, *De waare oorzaken van Neerlands plage, Wegens de groote Sterfte onder het Rundvee ... vertoond ... uit de klaagliederen van Jer. Cap. 3 vs. 38, 39, 40. Op den voorgaanden Bid-dag ... in 't Jaar 1715 ... Alwaar met een ingevoegt is, een Wederlegging van het gevoelen van F. Leenhof, aangaande den Hemel op Aarden* (Groningen: Jurjen Spandaw, 1716), 122.

[67] Hendrik Carel van Byler, *Historis-verhaal van de sterfte die in vorige eeuwen onder het rundvee, in deze en andere landen geweest is, en nog duurt* (Groningen: Jurjen Spandaw, 1719), 59.

[68] Johannes Sluiter, *Vernedert Nederlant, Of Klagte over Nederlants sonden, gestraft door de sterfte van het Rundervee* (Steenwyk: Pieter and Hendrik Stuyfzant, 1715), 4.

that a correspondingly dramatic response was needed. What that response should be was far less clear. Cattle plague interpretations diverged almost as soon as the disease entered Dutch territory. Medical authorities, moralists, and the broader public debated the cause and meaning of cattle plague and crafted causal stories that translated the disaster into defined, actionable positions. These interpretive dialogues illustrated causation, assigned responsibility, and promoted a set of responses for the disaster.[69]

The most common causal stories that appeared in print, by far, were providential. Cattle plague, according to this view, was the direct result of human sin. "War, famine, and disease are three general heavy punishments and scourges of God that have frequently assaulted countries and kingdoms," explained the historian and translator Abraham Moubach in 1719. To Moubach, "the present terrible cattle plague that has for many years now insidiously spread everywhere" revealed God's righteous judgment.[70] In a 1714 broadside entitled *The Plaintive Netherlands*, the Amsterdam writer, satirist, and publisher Jan van Gyzen warned of "God's striking hand now sent among us." The disease impoverished farmers, he argued. It threatened Dutch prosperity and seemed an indication that God had abandoned them.[71] The disease was the direct, causal result of human malfeasance.

Providential interpretations like those of Moubach and Van Gyzen developed out of a long tradition of gleaning moral meaning from environmental change, including natural disasters. "Providence" functioned as a widely held and commonly understood conceptual tool to interpret the divine presence in creation. The doctrine of "general providence" framed the normal functioning of nature as a function of God's grace. One could read God's intentions from environmental phenomena, using either scripture (revealed theology) or the Book of Nature (natural theology). This certainly applied to instances of dramatic environmental change such as natural disasters. Earthquakes, floods, comets, and plagues were opportunities for early-modern observers to interpret God's will.[72]

Providential narratives were ubiquitous in the wake of calamity. They dominated print media in the Dutch Republic. Books, pamphlets, and

---

[69] Stone, "Causal Stories."   [70] Moubach, *Schaadelijke Veepest*, 1.
[71] Jan van Gyzen, *Klaagend Neederland, bezogt met sterfte onder het rund-vee en swaare stormwinden* (Amsterdam: Jacobus van Egmont, 1714).
[72] Alexandra Walsham, *Providence in Early Modern England* (Oxford: Oxford University Press, 1999), 2–3. Morgan, "Understanding Flooding," 39–40. Eric Jorink, *Reading the Book of Nature in the Dutch Golden Age, 1575–1715*, trans. Peter Mason (Leiden: Brill, 2010), 18, 30.

sermons were the most common pedagogical vehicles to transmit and interpret this causal story. Although we know less about the ways that listeners absorbed their lessons, historians have argued that these documents performed a variety of tasks, including religious instruction, fostering emotional connections with victims, developing communal identities, and encouraging social solidarity.[73] The pervasiveness of providential interpretations across confession and their frequent appearance in other formats, including prints, state publications, and manuscript sources, speaks to the likely penetration of this causal story into society at large. Providential narratives were in general quite durable as well. Many accounts adhered to genre-specific sets of formulas, biblical motifs, images, and expressions that demonstrated remarkable coherence across medieval and early-modern Europe.[74]

Emphasizing these surface similarities, however, obscures real differences of meanings and purpose and elides their changing use over time.[75] Although many promoted a peccatogenic worldview that attributed disasters to human sin in a direct causal relationship, the definitions of moral misconduct and utility of these characterizations were flexible.[76] They frequently reflected the specific character of disaster outcomes and the larger social and environmental contexts in which they occurred. In the context of cattle plague, providential thinking performed multiple tasks. Some employed providence as a coping strategy and explanatory model to understand otherwise inexplicable tragedy.[77] Others wielded

---

[73] Joris van Eijnatten, "Getting the Message: Towards a Cultural History of the Sermon," in *Preaching, Sermon and Cultural Change in the Long Eighteenth Century*, ed. Joris van Eijnatten (Leiden: Brill, 2009): 343–88. Marijke Meijer Drees, "'Providential Discourse' Reconsidered: The Case of the Delft Thunderclap (1654)," *Dutch Crossing* 40, no. 2 (2016): 108–21. Adriaan Duiveman, "Praying for (the) Community: Disasters, Ritual and Solidarity in the Eighteenth-Century Dutch Republic," *Cultural and Social History* 16, no. 5 (2019): 543–60. Marie Luisa Allemeyer, "Profane Hazard or Divine Judgement? Coping with Urban Fire in the 17th Century," *Historical Social Research/Historische Sozialforschung* 32, no. 3 (2007): 148–52.

[74] Morgan, "Understanding Flooding," 40. Christian Rohr, "Writing a Catastrophe. Describing and Constructing Disaster Perception in Narrative Sources from the Late Middle Ages," *Historical Social Research/Historische Sozialforschung* 32, no. 3 (2007): 88–102.

[75] Raingard Esser, "'Ofter gheen water op en hadde gheweest' – Narratives of Resilience on the Dutch Coast in the Seventeenth Century," *Dutch Crossing* 40, no. 2 (2016): 99. Joseph Hardwick and Randall J. Stephens, "Acts of God: Continuities and Change in Christian Responses to Extreme Weather Events from Early Modernity to the Present," *Wiley Interdisciplinary Reviews: Climate Change* 11, no. 2 (2020): 1–16.

[76] Wolfgang Behringer, *A Cultural History of Climate* (Cambridge: Polity, 2010), 133–5.

[77] Jussi Hanska, *Strategies of Sanity and Survival: Religious Responses to Natural Disasters in the Middle Ages* (Helsinki: Finnish Literature Society, 2002). J. Spinks and C. Zika,

providence as a weapon to condemn or police the immorality of individuals, social groups, or Dutch society as a whole. Finally, authors promoted providential rituals as prophylaxes or treatments for the disease. Moralists fashioned providential causal stories to suit their needs and the needs of their communities.

Cattle plague presented several possibilities for providential interpretation. Nearly all agreed that the unique characteristics of the plague justified a peccatogenic reading. The disease behaved so brutally and spread so widely that it seemed almost a new type of disaster. "Where has there ever been so terrible a plague as God's justice has now brought upon these Lands?" asked François Halma in 1714.[78] Disease symptoms horrified onlookers and seemed intentionally designed to appeal to their moral sympathies. Lambert Rijckszoon Lustigh, a farmer and alderman from Holland, described his impressions of the plague in a handwritten chronicle in May 1713: "It is hardly possible to speak of the desperate pain that most of these animals suffer in their last few days." He went on: "[O]ne must have a bitter and disdainful conscience if he does not regard it with compassion."[79] Lustigh believed cattle plague was divine judgment, and the death and suffering of animals compounded God's rebuke because the animals, rather than their owners, suffered due to human sin.

The behavior of the disease made it particularly well suited to providential interpretation. Rinderpest was highly contagious and spread via direct contact between infected animals. It could not propagate via vectors and was not airborne. As a result, the period of greatest susceptibility to infection occurred when farmers stalled their animals during the fall and winter. This behavior perplexed observers, particularly those like Mobachius convinced that disease resulted from divine disfavor as well as environmental sources of pollution. Why was it that animals remained healthy while grazing in herds during the summers, "but thereafter in the barns or the stalls, they become sick and die?" he asked.[80] The disease also affected individual animals differently. Those that survived infection gained lifelong immunity, which slowed the rate of infection and reduced the susceptibility of herds during later outbreaks. Outbreaks thus waxed

Disaster, Death and the Emotions in the Shadow of the Apocalypse, 1400–1700 (London: Palgrave Macmillan UK, 2016), 4.
[78] Halma, *Godts Wraakzwaardt over Nederlandt*, 7.
[79] Lambert Rijckxzoon Lustigh, "Kroniek van Lambert Rijckxz. Lustigh te Huizen voornamelijk betreffende Gooiland, 1713–1722," NHA Collectie van Losse Aanwinsten 176.1527 (1713), fol. 5.
[80] Mobachius, *Waare Oorzaken*, 111.

and waned in severity over months and years depending on changing herd immunity.[81] This dynamic confounded explanations grounded in direct environmental or moral causation.

Even for those who expected the disease to operate according to the logic of contagion, the plague seemed too unpredictable. Why else but by divine will would provinces like Holland, Utrecht, and Overijssel fall victim to the disease within weeks of the first outbreak, whereas Groningen remained disease-free for over a year? For those who likened plague to a fire indiscriminately sweeping across the landscape, the selective geography of outbreaks defied expectations and confirmed its providentialist origins.

Historical context and the recent experience of other calamities likewise confirmed its divine provenance. Moralists, especially Pietist ministers, saw the outbreak as merely the latest episode in a prolonged series of economic, social, and natural disasters. In his sermon on Revelations 3:20, the Maassluis minister Aegidius Francken summed them up:

What rods has God not already used to beat upon the backs of the Netherlands' peoples? How many times has he beaten us to death with storms winds, whereby many were swallowed alive in the angry waves, then by bitter frosts, whereby the winter grains in the fields froze ... [then] through the terrifying breach of our dikes and dams with floods so that many lands seemed more sea than country. Has the Lord now knocked on the door of the Dutch people with a general decay of commerce? walk through the cities and villages of our country, ah! how many stores are not closed? How many trades do not stand silent? Now has the Lord not knocked on our door with long and devastating wars, which emptied our treasury, shed much heroic blood, and made so many widows ... And now that the war is changed to peace, the Lord has slaughtered our Cattle.[82]

Francken's juxtaposition of the peace and disease underscored the importance of timing. Cattle plague too neatly counterbalanced the ostentatious displays of optimism and confidence in 1713 to be coincidence. "The year of the Peace had not yet come to an end before it became a year of terror," said the Rotterdam minister Alardus Tiele. Yet this time "death did not climb through our windows, but slept in our fertile fields

---

[81] P. B. Rossiter and A. D. James, "An Epidemiological Model of Rinderpest. II. Simulations of the Behavior of Rinderpest Virus in Populations," *Tropical Animal Health and Production* 21 (1989): 73–4.

[82] Aegidius Francken, "Geestelyk houwelyk, dat is, een Verhandeling van de Ondertrouw der gelovigen met Kristus, door AEgidius Franken, Bedienaar van 't H. Evangely te Maas sluys. Te Dordregt ...," in *Boekzaal der Geleerde Wereld* (Amsterdam: Gerard onder de Linden, 1715).

and in our cattle stalls."[83] Everything about the disease, from its horrific symptoms to its unprecedented scale and severity to its timing, confirmed providential interpretations of the disaster. Cattle plague was "God's striking hand" sent as punishment for Dutch sin. This providential reasoning operated as a useful coping strategy that imposed clarity and meaning upon a disorienting disaster.

Providence also justified moral responses to disaster. If cattle plague was punishment for human sin, then spiritual therapies and prophylaxes deserved privileged consideration. Dutch ministers needed only to identify the culpable sins, a task they proved eminently capable of handling. In his poem *Humbled Netherlands, or Lamentations over Dutch Sins* (*Vernedert Nederlant, of Klagte over Nederlants Sonden*), Johannes Sluiter attributed the cattle plague to "pomp and splendor, drunkenness, and foul language."[84] The East Frisian minister Jacob Harkenroht added "idolatry, forgetting the word of God, insensitivity, excess and waste" to the list of responsible sins.[85] Pietist ministers promoted programs of personal repentance to purify their congregations, stem the statewide outbreak, and prevent its reappearance. Other authors connected the events to the plagues of Egypt in Exodus or Psalms.[86] Most identified specific sins in their congregations that remained generalizable. This allowed them to at once assert the general character of the affliction and exert local control over the problem.

Ministers also wielded providence as a weapon to externalize blame. Catholic rituals in rural Holland, for instance, posed a threat to strict Reformed sensibilities. In 1713, newspapers published reports of the miraculous healing waters of the *Runxputte* (fountain) near the Catholic community of Heiloo in North Holland. Rumors spread quickly, and farmers from "far and wide came thither with jugs and vats to get the

---

[83] Alardus Tiele, *Schouwburg der oordeelen en gerigten Gods, geoeffend en uitgevoerd door watervloeden, stormwinden, oorlogen, aardbevingen, pestilentien, siektens over menschen en vee ...: in drie deelen* (Rotterdam: Nikolaas Topyn, 1736), 649.

[84] Sluiter, *Vernedert Nederlant*, 5.

[85] Jacobus Isebrandi Harkenroht, *Kerkreede over Oostfrieslands rundvees pest, gevreest, en helaas gekoomen, aangetoont uit Jes: VII. vs. 21. ... op een maandelijke bededag aan de gemeente te Larrelt in Oostfriesland* (Embden: Enoch Brantgum, 1716), 25–7.

[86] Taken from Exodus 9:3 of the state translation of the Bible. With the sixth plague, Moses promised that God "shall bring a terrible pestilence of your livestock in the fields, or the cattle and the smaller animals," *Biblia, dat is: de gantsche H. Schrifture, vervattende alle de Canonycke Boecken des Oude en des Niewen Testaments (Statenvertaling 1637)* (Leiden: Paulus Aertsz van Ravensteyn, 1637), pg. 4v.

holy water as a remedy for their sickened cattle."[87] Protestant ministers condemned the event and ridiculed the "incredibly large influx of foolish people" seeking cures for their cattle.[88] Henricus Carolinus van Byler reported an historical account of a Catholic farmer near Leiden that hung an image of St. Anthony of Padua around his cow's neck to prevent the spread of disease. "The superstition is, from time to time, growing today," he noted, and he condemned these "nefarious beliefs of fools."[89] These reports highlighted the deep divisions between Catholic and Protestant interpretations of spiritual remedies. Rituals interpreted as therapeutic from the Catholic perspective appeared symptomatic of the diseased moral state of the Republic to others.

For the Protestant majority, prayer, fasting, and thanksgiving proved popular strategies of treating and mitigating disaster. The Estates General ordinarily restricted its concerns to foreign affairs, yet it promoted public participation in these spiritual therapies in their official days of "Thanksgiving, Fasting, and Prayer." Provincial governments and the Estates General produced a steady stream of these official proclamations of prayer throughout the seventeenth century, oftentimes amidst wars, natural disasters, or notable political events.[90] Beginning in 1713, national prayer days became annual events (*vaste bededagen*). A wide variety of faiths were invited by local authorities to observe the ritual, including dissenters from the Reformed Church, such as Lutherans, Mennonites, and Arminians. Invitations were occasionally extended to Jews and, by the end of the eighteenth century, Roman Catholics as

---

[87] Ottie Thiers, '*T Putje van Heiloo: Bedevaarten naar O. L. Vrouw ter Nood* (Hilversum: Verloren, 2005), 30. Livestock owners across Catholic Europe turned to pilgrimages, processions, and the curative power of the saints as spiritual therapies during the cattle plagues of the eighteenth century. Carsten Stühring, "Managing Epizootic Diseases in 18th Century Bavaria," in *Economic and Biological Interactions in Pre-Industrial Europe from the 13th to the 18th Century*, 1000–1008, ed. Simonetta Cavaciocchi (Florence: Firenze University Press, 2010), 473–80.

[88] Ibid. Reformed ministers in neighboring cities punished Catholic pilgrims making the journey and applied intense legal pressure to stop it. P. Leendertz et al., *De Navorscher: een middel tot gedachtenwisseling en letterkundig verkeer tuschen allen, die iets weten, iets te vragen hebben of iets kunnen oplossen* (Amsterdam: Frederick Muller, 1862), 294–5.

[89] Van Byler, *Historis-verhaal*, 49.

[90] Prayer days were officially sanctioned at the Synod of Dordrecht in 1618. See article 66 of the Church Ordinance of the Services approved at the Synod of Dordrecht. P. Biesterveld and H. H. Kuyper, *Kerkelijk Handboekje: bevattende de bepalingen der Nederlandsche synoden en andere stukken van beteekenis voor de regeering der Kerk* (Kampen: J.H. Bos, 1905).

well.[91] Clergy and their parishioners performed these rituals at their local churches, and provincial governments established fines for noncompliance. These penalties may be evidence of public skepticism, particularly since prayer days forced parishioners to give up valuable time for work or leisure.[92] At the same time, they demonstrated the power of providential reasoning. Declarations of thanksgiving and prayer represented the official sanction of providential remedies, and their punishments for noncompliance policed them.

Prayer days served as disaster prophylaxes and insurance, as coping mechanisms following disasters, or as civic rituals of provincial or interprovincial solidarity.[93] Provinces declared extraordinary prayer days before the arrival of cattle plague, during the outbreak, and upon their conclusion. On December 15, 1714, for instance, provincial officials first acknowledged cattle plague in Groningen and declared an "extraordinary day of fasting and prayer" for the following January.[94] A little over eight years later, Groningen marked the end of the plague with a new day of thanksgiving, crediting the "infinite mercy of God" for ridding them of the plague.[95] Friesland went further. They declared a "mandatory prayer day" on the first Wednesday of every month.[96] These official documents demonstrated the success of providence as a causal story. States called for prayer days because they distilled the disruptive chaos of the epizootic into an understandable causal relationship (sin-and-punishment), because they promoted and policed unity and because the sermons at the center of these rituals rarely implicated the state. State censorship of published Prayer Day sermons in the Netherlands was rare compared to other European states, yet secular authorities still exerted a degree of control over what they contained and, in the case of disasters, even outlined a

---

[91] O. C. Edwards Jr., "Varieties of the Sermon: A Survey of Preaching in the Long Eighteenth Century," in *Preaching, Sermon and Cultural Change in the Long Eighteenth Century*, ed. Joris van Eijnatten (Leiden: Brill, 2009), 39.

[92] In the late eighteenth century, Dutch clergymen complained about lack of attendance and interest in prayer days. Buisman, *Tussen Vroomheid*, 201.

[93] Leo Noordegraaf, "Of bidden helpt? Bededagen als reactie op rampen in de Republiek," in *Of bidden helpt? Tegenslag en cultuur in Europa circa 1500–2000*, eds. Marijke Gijswijt-Hofstra and Florike Egmond (Amsterdam: Amsterdam University Press, 1997). Peter van Rooden, "Dissenters en bededagen. Civil Religion ten tijde van De Republiek," *BMGN: Low Countries Historical Review* 107, no. 4 (1992): 703–12. Duiveman, "Praying for (the) Community."

[94] Proclamation (15 December) 1714. GrA. Staten van Stad en Lande, 1594–1798. 1.477.

[95] Proclamation (20 February) 1723. GrA. Staten van Stad en Lande, 1594–1798. 1.477.

[96] Resolution (8 March) 1714. TRE. Gewestelijke bestuursinstellingen van Friesland 1580–1795. 5.253, fol. 24.

general set of causal stories in their calls for prayer.[97] To ministers, lay moralists, and a public certain of its divine origins, cattle plague may have reflected the broader decline of Dutch morality and inversion of Golden Age prosperity, but few overtly connected that decline to Dutch authorities. Providence, as a result, enjoyed widespread appeal in state publications about the disease.

2.4 DISEASE MANAGEMENT AND CATTLE MOVEMENT

The value of providential interpretations notwithstanding, few promoted these causal stories exclusively. "God's Striking Hand" was the ultimate or "first" cause of the disease, but most authors (including moralists) sought secondary, "natural" causes as well. Whereas religious authorities and moralists acknowledged the natural behavior of the disease but emphasized their first causes, secular authorities tended to acknowledge providence and act upon natural causes. Identifying the proximate material origins of disease was not simple, however. Was the disease contagious? If so, how was it transmitted? Did it emerge out of infected pastures or an unhealthy climate? Or perhaps some combination of these conditions? Provincial governments staggered through the initial weeks of the outbreak, uncertain what to emphasize.

Early disease management strategies reflected this disequilibrium. Utrecht was among the first provinces to enact new regulations in response to the plague. On July 13, 1713, the Estates of Utrecht published their first of a series of official notices warning of the "great sickness and death amongst the oxen, cattle, and other livestock [that] reigns in the lands surrounding this province."[98] Officials had anxiously monitored the arrival and spread of the disease since it appeared in Holland and they worried that it might cross into Utrecht. Utrecht quickly banned the import of sick animals as well as infected meat "whether raw, cooked, roasted salted, or smoked."[99] The state provided no guidance about how to interpret this order, however, much less how they would enforce it.

---

[97] Pasi Ihalainen, *Protestant Nations Redefined: Changing Perceptions of National Identity in the Rhetoric of the English, Dutch and Swedish Public Churches, 1685–1772* (Leiden: Brill, 2005), 50–1.

[98] J. van de Water, *Groot Proclamationboek Vervattende Alle de Placaten, Ordonantien En Edicten, Der Edele Mogende Heeren Staten 's Lands Van Utrecht* (Utrecht: Van Poolsum, 1729), 642.

[99] Claes Arisz. Caescooper describes the "deaths of cattle and oxen" first on May 10, 1713. Lambert Rijksz Lustigh noted disease "already in the months of April, May, June, and

What constituted a "sick" animal? Must it exhibit symptoms? Was prior contact with disease required, or should animals arriving from polluted environments or regions where sickness had been observed be banned as well? The lack of consensus about its origin and possible transmission straightjacketed nascent containment efforts.

Lacking a coherent causal story, subsequent provincial orders vacillated between environmentalist and contagionist interpretations of the disease. Publications from July and August (both at the behest of farmers living in municipalities along the Holland–Utrecht border) forbade the pasturing and movement of animals out of polders that seemed to harbor the plague.[100] These early local restrictions failed to appear in the next provincial notification published on August 12. Instead, it laid out six new regulations intended to halt the spread of the disease. It prohibited the import of cattle and cattle products like hides and meat into the province, barred the dumping of cattle bodies into rivers or drainage ditches, ordered farmers to separate sick from healthy animals in their herds, required they report the presence of disease to local authorities, and finally "in order to better prevent, as much as possible the further spread of the infection throughout the province," banned the movement of animals across municipal borders.[101] By early August, in other words, Utrecht treated the disease as both contagious and the result of environmental factors like polluted water.

Utrecht's early containment strategies reveal the interplay of two interpretations of disease propagation in the early-modern period. The contagionist model privileged contact between infected bodies and/or materials. According to this view, states could control the spread of the outbreak by managing the movement of cattle, usually via import or export restrictions, quarantine, and by surveilling cattle as they moved into and out of pastures. The environmentalist (or miasma) model identified polluted environments with unhealthy air as the source of disease. Disease might arise from corrupt conditions inherent to environments such as bogs, or they might emanate from diseased or decaying bodies. Provinces employed a variety of environmentalist strategies to limit exposure, including draining wetlands, cleansing and ventilating stalls, and interring corpses.[102] Environmentalist strategies enjoyed ascendancy

---

July of 1713," Buisman, *Duizend jaar*, Vol. 5, 400. Lustigh, "Kroniek van Lambert Rijckxz. Lustigh te Huizen," fol. 176.
[100] Van de Water, *Proclamationboek*, 642.   [101] Ibid., 643.
[102] I use the term "environmentalist" here in the manner described in J. C. Riley, *The Eighteenth-Century Campaign to Avoid Disease* (London: Macmillan, 1987).

in the early eighteenth century, and their implementation during this outbreak of cattle plague reflected the increasing influence of neo-Hippocratic theories of disease transmission.[103]

These two explanations for disease were not mutually exclusive, however. Diseases might arise from miasmatic airs, contagious particles, or some combination of the two. As a result, states often chose to quarantine cattle and institute import restrictions while at the same time regulating methods of corpse disposal. Management never took the effectiveness of these strategies for granted, which partly explains the tendency of administrators to employ multiple approaches simultaneously. Disease, they believed, manifested along a spectrum of environmental and contagionist influences, and effective response depended upon negotiation and debate between medical practitioners, laypeople, and states as to its precise character.[104]

The early response of Utrecht, as well as those of the Estates of Holland, Friesland, and Groningen reflected this negotiation. Initially, provincial governments emulated Utrecht's efforts to contain the disease by restricting international and interprovincial movement of cattle. Friesland enacted its first import ban from Holland in November 1713. Holland waited to impose limits until January 1714 but added restrictions on the movement of dogs because they saw them as possible vectors for the disease.[105] Groningen banned import of cattle, hides, and hay from Holland, Overijssel, and Utrecht in December 1713.[106] In their case, it was a preventative strategy the Estates enacted a full year before the disease crossed its borders in December 1714. Recognizing that imported cattle from Denmark posed a risk to their domestic herds as well, the Estates of Groningen placed a near complete ban on their import in March 1714.[107] The provinces based their early attempts to control the disease on its behavior. It appeared mobile and likely contagious. The descriptive language they employed, describing the "sickness of cattle" as

---

[103] Ibid.
[104] Dominik Hünniger, "Umweltgeschichte kulturhistorisch: Tierseuchen in den Lebenswelten des 18. Jahrhunderts," in *Natur und Gesellschaft: Perspektiven der interdisziplinären Umweltgeschichte*, eds. Manfred Jakubowski-Tiessen and Jana Sprenger (Göttingen: Universitätsverlag Göttingen, 2014), 179–81.
[105] Proclamation (January) 1714. SA. Archief van de Burgemeesters: Publicaties van de Staten-Generaal en van de Staten van Holland en West-Friesland. 5022.10.
[106] Resolution (21 December) 1713. *GrA*. Staten van Stad en Lande, 1594–1798. 1.477.
[107] Resolution (5 March) 1714. *GrA*. Staten van Stad en Lande, 1594–1798. 1.477.

either *contagieuse* or *besmettelijk*, denoted contagion. Despite this, the precise mechanism of its transmission remained ill defined.

Provincial orders likewise regulated environmental determinants of disease. Authorities considered cattle corpses as public health hazards when dumped in bodies of water or left to rot, because emanations from their decomposition polluted environments. Cattle stalls and other cloistered, poorly ventilated environments promoted illness. Farmers therefore managed disease by aerating their barns. Farmers participated in this evolving dialogue of disease causation as well. Lambert Rijckzoon Lustigh, for instance, determined that disease likely spread via infected dung based on his observations of when farmers mucked out their stalls and when their cattle died.[108] Fears of contamination understandably increased as the cattle plague intensified. The speed with which provinces instituted environmental management strategies demonstrated its virulence. The first regulations requiring interment of dead cattle appeared as early as December 1713 and continued to gain popularity the following year.[109]

Medical pamphlets and books promoting these strategies demonstrate the fluidity across which contagionist and environmentalist interpretations ranged. One pamphlet from 1714 described how breath and sweat transmitted the disease, but also argued that cattle plague manifested differently depending on climate and season. It proved most virulent when farmers stalled cattle, because disease propagated via polluted manure and straw. Farmers should, therefore, clean the stalls with vinegar and gunpowder.[110] Others argued that poisoned dew in pastures spread the illness. "The more cattle that continue to eat such infected grasses and drink from such polluted waters, the more will become infected."[111] In response, farmers burned hay from infected stalls, and water authorities drained excess water from polders to "purify" their fields.[112] Authorities,

---

[108] Lustigh, "Kroniek van Lambert Rijckxz. Lustigh te Huizen," fol. 27–9. Thérèse Peeters, "'Sweet Milk-Cows' in Huizen and 'Memorable Incidents' in Oost Zaandam: Identity and Responsibility in Two Eighteenth-Century Rural Chronicles," *Volkskunde* 2 (2014): 163–79.

[109] Van de Water, *Proclamationboek*, 648.

[110] *Bedenkingen, En Raad*, 12–13. This preventative makes an appearance in many Dutch publications as well as John Mills, *A Treatise on Cattle: Shewing the Most Approved Methods of Breeding, Rearing, and Fitting for Use, Horses, Asses, Mules, Horned Cattle, Sheep, Goats, and Swine* (Dublin: W. Whitestone, 1776), 454.

[111] Mobachius, *Waare Oorzaken*, 110. Lustigh, "Kroniek van Lambert Rijckxz. Lustigh te Huizen," fol. 227.

[112] Lustigh, Kroniek van Lambert Rijckxz. Lustigh te Huizen, fol. 16.

in other words, promoted a diverse array of responses based on two distinct, though frequently intersecting interpretations of disease. They worked across multiple scales, from international regulation of cattle movement down to the level of individual stalls.

Both regimes of disease management relied upon compliance from farmers, merchants, and other participants in the cattle economy. Considering the serious economic and social consequences of trade restrictions and interment, they did not always offer it willingly. Jan Wopkes noted that thousands of diseased cattle in Friesland were slaughtered during the outbreak and their "flesh cooked for tallow ... [and also were] salted and eaten by many people."[113] This was in direct opposition to a decree from the Estates of Friesland the previous year that ordered the burial of diseased corpses.[114] Provinces reprinted their proclamations every few months and called out "selfish men who want to make some foul profit" by shirking the law. Provincial courts sometimes prosecuted offenders. A lower court in Groningen, for instance, charged Harm Faasterman and Claas Goutier with "importing infected cattle meat" in October 1715. Aelbert de Cruiff, a farmer in Utrecht, received a fine of 250 guilders for transporting cattle without a certificate of health from Gelderland into Utrecht.[115] Punishments for noncompliance like these appear rare, however. Criminalizing noncompliance presented financial consequences – not only for individuals but also for the provinces that depended on their taxes.

From the perspective of state authorities, regulating the cattle trade seemed almost as fraught as the disease itself. The illness claimed large numbers of cattle, more than could be reared in Dutch territories. If farmers turned to imported cattle to restock their herds, the risk of new outbreaks increased. Early in 1714, the Estates General attempted to solve this problem by turning to Danish suppliers. They lifted protectionist tariffs that had been in place since 1686. From the Estates perspective, this was "necessary to facilitate ... the trade of lean oxen that are annually brought from Jutland [Denmark]." The economic risks facing the cattle economy were greater than the risk of increased exposure. "One

---

[113] Spahr van der Hoek, *Geschiedenis*, 218.
[114] Resolution (21 December) 1713. TRE. Gewestelijke bestuursinstellingen van Friesland 1580–1795. 5.253.
[115] "Harm Faasterman En Claas Goutier Wegens De Invoer Van Bedorven Rundvless," (1715). GrA. Volle Gerecht van de stad Groningen, 1475–1811. 1534.1447. K. van Schaik, *Overlangbroek op de kaart gezet: Dorp, landschap en bewoners, waaronder een familie De Cruijff* (Hilversum: Uitgeverij Verloren BV, 2008), 15.

the other hand," they conceded, "care must be taken to that no oxen may be imported from places where there is also cattle plague." They set up a health certification system and printed forms that state officials could use to identify the Danish suppliers, the number of cattle and location of cattle stalls, and parties responsible for confirming this information.[116] Other regulations attempted to preserve existing stocks by banning cattle exports, especially the sale of calves. State regulations were the first line of defense against the disease, yet they inevitably added burdens to the cattle economy. Dutch provinces therefore issued, revoked, and reissued regulations based on perceived risk of infection versus economic necessity. Historian Dominik Hünniger noted similar policies at work in Northern Germany and termed this strategy "continuous crisis management."[117] It accommodated not only the fiscal demands of authorities and the citizenry but also multiple interpretations of disease causation. Provincial governments employed a flexible, universalist approach to prevent the spread and mitigate the impact of cattle plague.

This flexibility had limits, however. A third realm of response, termed "stamping out" or the "Lancisi system," enjoyed no support. This strategy mandated the culling of both infected animals and those suspected of harboring disease. The pope's personal physician, Giovanni Lancisi, popularized preventative slaughter around 1714 in Italy.[118] Italian states had struggled to contain the disease since it began sweeping across southern Europe in 1710 and preventative slaughter proved an effective, albeit expensive, management strategy. A version of this system successfully eradicated cattle plague in England in a span of six months.[119] In all cases where the state imposed preventative slaughter, however, it met strong resistance from cattleholders.[120] Its chief proponent in England,

---

[116] Proclamation (24 February) 1714. *Groot placaet-boeck, vervattende de placaten, ordonnantien ende edicten van de ... Staten Generael der Vereenighde Nederlanden, ende vande ... Staten van Hollandt en West-Vrieslandt, mitsgaders vande ... Staten van Zeelandt* ('s Gravenhage: Isaac en Jacobus Scheltus, 1725), 1628.

[117] Dominik Hünniger, "Policing Epizootics Legislation and Administration during Outbreaks of Cattle Plague in Eighteenth-Century Northern Germany as Continuous Crisis Management," in *Healing the Herds: Disease, Livestock Economies, and the Globalization of Veterinary Medicine*, eds. K. Brown and D. Gilfoyle (Athens: Ohio University Press, 2010).

[118] Jean Blancou, *History of the Surveillance and Control of Transmissible Animal Diseases* (Paris: Office international des épizooties, 2003), 172.

[119] John Broad, "Cattle Plague in Eighteenth-Century England," *The Agricultural History Review* 31, no. 2 (1983): 104–15.

[120] Madeleine Ferrières, *Sacred Cow, Mad Cow: A History of Food Fears* (New York: Columbia University Press, 2006).

Thomas Bates, complained that farmers skirted regulations by failing to notify authorities of the presence of disease and continuing to sell diseased animals to distant markets.[121] Historians have thus tended to credit strong centralized governments for effective implementation of the stamping-out policy because they provided the necessary coordination and policing power to counter this resistance. They simultaneously criticize "the lack of a strong central government" in the Dutch Republic for failing to do so.[122] The Dutch, these critics argue, ignored the benefits of "radical slaughter and isolation policies" and relied instead upon spurious "cures" and "medicines."[123]

This position discounts the extensive efforts of provincial governments to monitor and control the movement of cattle through certification, import and export restrictions, and later quarantine. Most regulation occurred on the provincial scale with smaller administrative districts assuming responsibility for implementation and enforcement. This division of responsibility allowed administrators to begin managing the disease quickly. Utrecht required certification that proved cattle originated in "such lands, or from meadows, where not infection has yet been seen" as early as August 1713. They implemented and enforced this policy through existing hierarchies of municipal and provincial governance.[124] When other provinces and localities imposed restrictions, they learned from prior experience. That same month in North Holland, for instance, the village of Uitgeest forbade cattle imported from neighboring Assendelft "where the terrible sickness among the cattle is reigning."[125] Gathering and tabulating data about cattle deaths also occurred on a local level. In Friesland, municipalities reported the number of cattle killed to the province to assess tax remissions. Local deputies posted announcements in public places, oftentimes in churches. A January 1714 proclamation in Groningen, for instance, ordered clergymen to read their new restrictions aloud during religious services and place the

---

[121] Lise Wilkinson, *Animals and Disease: An Introduction to the History of Comparative Medicine* (Cambridge: Cambridge University Press, 1992), 53.
[122] Cornelis Offringa, *Van Gildestein naar Uithof: 150 jaar diergeneeskundig onderwijs in Utrecht. 1. 's Rijksveeartsenischool (1821–1918): Veeartsenijschool Hoogeschool (1918–1925)*. (Utrecht: Rijksuniv. te Utrecht, Fac. der diergeneeskunde, 1971), 15. C. Huygelen, "The Immunization of Cattle against Rinderpest in Eighteenth-Century Europe," *Medical History* 41, no. 2 (1997): 183.
[123] Wilkinson, *Animals and Disease*, 53.   [124] Van de Water, *Proclamationboek*, 648.
[125] "Verbod op de invoer van vee uit het met veepest besmette Assendelft," *NHA* (August 16, 1713) Ambachts- en Gemeentebestuur van Uitgeest en Markenbinnen. 1052.942.

documents in public view so that "nobody may claim any ignorance."[126] Working through existing institutions and permitting localities to spearhead initial responses allowed provincial governments to mobilize response quickly.[127]

The critique of Dutch response also ignores the very real limitations facing early-modern states intent on imposing the stamping-out policy, especially on the continent. The Lancisi system made little sense if neighboring states failed to impose similar restrictions. The Dutch Republic, with its porous borders and numerous ports, faced substantial challenges for policing. The size and structure of Dutch commercial cattle holding ensured that once the disease appeared, it would spread rapidly, particularly in the areas with larger herds such as Holland and Friesland. The cattle economy and the plague depended on the stabling of breeding stock through the winter, which encouraged close, frequent contact between animals. The incorporation of new lean animals into overwintered herds in the spring, potentially sourced from regions where illness raged, encouraged the introduction of the virus. On top of this, the open polder landscapes where commercial cattle holding flourished virtually guaranteed interchange between herds.[128] The close commercial connections between the Dutch Republic and its neighbors, the nature of landscape use, and the dependence of commercial cattle-holding regions on large herds limited the attractiveness of stamping out.

Dutch states also rejected the Lancisi system because it was prohibitively expensive. Culling herds of diseased cattle was costly enough, but slaughtering all suspected cattle produced much more severe financial consequences. It met fierce resistance, even when imposed on an ad hoc basis. Lambert Rijckzoon Lustigh, for instance, described his dismay upon hearing of a planned slaughter in northern Holland in 1713. Having discovered that two of his neighbor's seventeen cattle had fallen ill, local authorities in Purmerend and the Beemster offered to purchase his herd to slaughter them, a proposition Lustigh condemned as "terrible" and "abhorrent."[129] Preventative slaughter proved very unpopular, even in regions of Europe where the state indemnified farmers. The Lancisi

---

[126] *Resolution* (4 January) 1714. GrA. Staten van Stad en Lande, 1594–1798. 1.477.
[127] The benefits of the Dutch Republic's decentralized state structure were evident in water management as well. Piet van Cruyningen, "Dealing with Drainage: State Regulation of Drainage Projects in the Dutch Republic, France, and England during the Sixteenth and Seventeenth Centuries," *The Economic History Review* 68, no. 2 (2015): 420–40.
[128] Van Roosbroeck and Sundberg, "Culling the Herds?," 31–55.
[129] Lustigh, "Kroniek van Lambert Rijckxz. Lustigh te Huizen," fol. 236.

system enjoyed little support from medical professionals in the Netherlands, who preferred to emphasize the economical character of their alternative remedies.[130] Finally, accepting culling as a viable strategy presupposed widespread acceptance of contagion as the sole causal story of the plague. Few were willing to support this in the face of evidence supporting environmental and providential causes.[131] The perceived failure of the state to adopt this policy is largely responsible for later scholarly criticisms of Dutch response. Faced with dramatic financial consequences, little reason to expect success, and with few, if any, proponents among state authorities or the public, Dutch provinces understandably rejected stamping out.

Although the Dutch Republic produced no unified response to the cattle plague, the similarities between provincial strategies overshadowed their differences. As the first and deadliest wave of the epizootic passed over the Republic by the end of 1714, Dutch provinces settled on a similar suite of preventative strategies drawing on both contagionist and environmentalist interpretations of disease, import and export restrictions, certification of healthy animals and pastures, and increased surveillance on the movement of cattle between localities. These strategies revealed the power of contagionist interpretations. Meanwhile, burial regulations and orders to ventilate and clean cattle stalls drew upon environmentalist interpretations. This universalist strategy demonstrated the compatibility of these interpretations, rather than indecision on the part of provincial administrators. In selecting this suite of responses and trusting in the added value of providential remedies to address the "first causes" of the disease, Dutch authorities capitalized on multiple causal stories of disease.

2.5 CONCLUSION

Centuries of landscape manipulation and economic development in the Netherlands produced a pastoral environment that was at once highly diverse, culturally resonant, and uniquely susceptible to epizootic outbreaks. The high degree of commercialization in the countryside, particularly in Holland, resulted in large herds, and networks of trade connected Dutch farmers to landscapes across the Republic and across Europe. These commercial connections, in combination with warfare and environmental triggers, likely transported the epizootic into the Republic in

---

[130] *Bedenkingen, en Raad*, 2.   [131] Van Roosbroeck, "To Cure Is to Kill," chapter 3.

1713 and encouraged its spread. The most vulnerable cattleholders engaged in specialized production and managed the largest, more interconnected herds. Even if those losses varied sharply by region and locality, however, the pervasiveness of cattle in agriculture across the Republic ensured widespread mortality.

Considering the uneven consequences of cattle plague and the decentralized nature of Dutch response, it is perhaps unsurprising that state authorities remained focused on local and provincial affairs. They did not link cattle plague to the broader phenomenon of decline. Dutch provincial governments focused their efforts on disease management and implemented a variety of strategies to prevent and contain its spread between 1713 and 1715. Providential interpretations of disease provided useful coping strategies for the public, which religious and state authorities could weaponize or use to promote solidarity and deflect blame. Dutch response was decentralized, but the provinces nevertheless adhered to a common set of containment strategies. Although not backward, it was backward looking. Provinces eschewed innovation, opting instead for tried-and-true strategies that adhered to both environmentalist and contagionist interpretations of disease.

Despite the proactive efforts of the provinces to contain the plague, they remained ineffective at eradicating it completely. Cattle plague lingered in the Republic for seven years. The feeling of powerlessness, more than anything, promoted declensionist interpretations that linked cattle plague to other social, economic, and environmental problems. These arguments largely centered on the experiences of the farmers who suffered the greatest cattle losses, who were simultaneously the most exposed to the financial consequences of the ongoing secular decline in agricultural prices and swelling tax burden left as legacy of the War of the Spanish Succession.[132] To make matters worse, those same communities coped with additional, concurrent natural disasters. Severe floods hit coastal communities in Holland and Friesland in 1714 and 1715, and plagues of mice ravaged harvests in Groningen in 1716.[133] The conjuncture of decreasing agricultural prices and increasing costs resulting from

---

[132] Bieleman, *Geschiedenis*, 107–9.
[133] Arnoldus Rotterdam, *Gods weg met Nederland, of vervolg op Blomherts geschiedenissen van het Vereenigde Nederland; ... onder 't verstandig beleid van Willem den IV* (Amsterdam: S. v. Esveldt, 1753), 18. Buisman, *Duizend jaar*, Vol. 5, 405, 414–15. H. Blink, *Geschiedenis van den boerenstand en den landbouw in Nederland* (Groningen: J. B. Wolters, 1904), 285.

war and natural disasters created conditions increasingly difficult to manage into the second decade of the eighteenth century.

Declensionist rhetoric and imagery in the second decade of the eighteenth century remained confined to a vocal group of moralists who grew increasingly convinced that the persistence of cattle plague and its arrival amidst other disasters confirmed dire providential interpretations about the future of the Republic. They linked disasters together in an unbroken causal chain that painted a picture of inversion and decline. The simultaneous appearance of coastal flooding and cattle plague in 1714 prompted the writer Jan van Gyzen to ask why God "who did everything before for the well-being of his people, now gives us no peace." The cattle now die "in the fields as well as the stalls," and those "cattle that had until then survived the plague, they [farmers] saw miserably drift in the flood waters." He was convinced that these disasters portended greater troubles ahead. He added a prayer – a prophylactic against future disaster.

O God of Gods, look at this people kneeling before your throne, and graciously free them from new wars, pestilence, and loss of commerce; tame the winds, silence the cattle plague, confirm those that through you we have peace, and make my grass-rich land again full of cows, and oxen, so that the milk and honey may again flow, repair the commerce upon which I depend, hasten the peace between me and the constable; give abundance of grain, and oil for my wounds; for already I'm warned of troublesome neighbors, and give us healthy, fat, fruitful, pious years, that I can again see the smoke from the Netherlands' altar of thanks.[134]

Although hopeful, little in this prayer bespoke confidence. It certainly showed none of the conviction observed during the peace celebrations that a new Golden Age awaited. The minister Gerardus Outhof published a similar pamphlet on the cattle plague. This one featured a fictional dialogue between two farmers named Alkon and Damon. "I'm afraid, neighbor Damon," Alkon stated, "that these unfortunate circumstances only foretell of future disasters, even more severe."[135] These fears were disastrously realized on Christmas Eve 1717 amidst what was likely the deadliest coastal flood in North Sea history.

---

[134] Jan van Gyzen, *Klaagend Neederland, bezogt met sterfte onder het rund-vee*.
[135] Gerardus Outhof, *Damon. Ofte herdersklagte over de sterfte onder 't rundvee ... waar mede Godt Oostvrieslandt ... bezoekt* (Emden: E. Brantgum, 1716), 23.

3

"The Fattened Land Turned to Salted Ground"

The Christmas Flood of 1717 in Groningen

The weather on Christmas Eve night gave coastal villagers in Groningen little reason to worry. The late December of 1717 had been mild and somewhat stormy, but a "strong and persistent southwest wind" had blown across the region for days.[1] Coastal communities knew winter often brought dangerous storm surges, but they rarely occurred when wind moved water away from shore. Storm surges manifested when north or northwesterly winds pushed water over the shallow Wadden Sea toward the coast. If combined with a spring tide, waves could crest several meters higher than normal. Gales produced by westerlies were common during the winter storm season, but winds moving from the southwest rarely produced dangerous floods because northern provinces such as Groningen lay protected in the lee of the storm.[2] Christmas Eve did not bring a spring tide, and the persistent southwesterly winds may have assured coastal residents they could sleep soundly.[3]

Some noted apprehension despite these benign conditions. In the town of Leens, a minister recorded the weather worsening during the night as a

---

[1] Buisman, *Duizend jaar*, Vol. 5, 439. Gerhardus Outhof, *Gerhardus Outhofs Verhaal van alle hooge watervloeden, In meest alle plaatsen van Europa, van Noachs tydt af, tot op den tegenwoordigen tydt toe: Met een nieuw Kaertje van 't verdronken Landt in den Dollaart, afbeeldinge van Kosmas en Damianus en van den steenenman in Vrieslandt, Met eene breede beschryvinge van den zwaaren Kersvloedt van 1717, En bygevoegde Tafels der watervloeden...* (Embden: H. van Senden, 1720), 645.

[2] Wheeler et al., "Atmospheric Circulation and Storminess." Luterbacher et al., "The Late Maunder Minimum." Ton Lindemann, "De stormvloed van 25 December 1717: Astronomische, hydrologische en meteorologische achtergronden," *Meteo Maarssen*, Report MM-17.3 (2017), 11.

[3] Lamb and Frydendahl, *Historic Storms*, 72.

storm brought "thunder, lightning, and hail."[4] A major cold front had been moving quickly southeast across the North Sea at the same time that warm, wet air moved northwards, creating conditions typical of severe storms.[5] The provincial official tasked with water management, Thomas van Seeratt, also noted unease in his journal. During his dike inspection, he talked with "an old man of eighty years" who said he "had never seen the water so low," indicating a turn to strong southeasterly winds. The sky appeared ominous. The air grew "heavier like an incoming hurricane," perhaps reminding Van Seeratt of his past life as a sea captain in the West Indies.[6] Around ten o'clock in the evening, the wind reversed direction and blew from the north/northwest, creating conditions ideal for a storm surge.

Eyewitness accounts of the flood vary, but all agreed the initial moments were terrifying. Victims likened its sound to an animal roar. The incoming water poured into houses, burying sleeping inhabitants. The minister Gerhardus Outhof despaired that "thousands of people of every age, men and women, drown[ed] in the salty seawater, many of whom were overtaken in their beds by the rushing waters."[7] The nearer a house or village was to a breach, the greater its exposure. One disaster chronicle described the horror of initial impact in verse:

> How terrible, the water's power that pierced the dike and brought itself unto the land;
> How the unmastered sea swelled and rushed through the breaches in the dikes;
> How indescribable, the fury of the waves, the land buried further and further;
> The pounding waves break the houses and of those, most are ground to dust;
> How many people sought to flee this dangerous and deadly invader;
> On top of houses, roofs, beams, and trees, all stranded by salty streams;
> This, children, servants, men, and women were all forced to watch ....[8]

The following hours and days tested the limits of peoples' endurance. Willems van Pieterburen managed to save his wife and six children, including a fourteen-day-old infant, by strapping them to a hay bale, only

---

[4] 25 December (1717) GrA. Doop, Trouw en Begraaf (DTB), Kerkelijke Gemeente Leens Doop en Trouwboek 1680–1748. 124.248. fol. 231.

[5] Lamb and Frydendahl, *Historic Storms*, 74. Lindemann, "De Stormvloed," 41–7.

[6] Thomas van Seeratt, "Journaal Van De Commies Provinciaal Thomas Van Seeratt Betref De Dijken over De Jaren 1716–1721," GrA, Staten van Stad en Lande, 1594–1798, 1.818, fol. 15.

[7] Outhof, *Verhaal*, 656.

[8] A. E. Crous, *Opregt en nauwkeurig historis-verhaal van de verwonderenswaardige, droevige, schrikkelike en seer schaadelike waaters-vloed, voorgevallen in de provincie van Groningen en Ommelanden, op Kersdag den 25. december ao. 1717* (Groningen: Seerp Bandsma, 1719), 1.

to see them drown as the night wore on.[9] A man from the town of Vierhuizen lost his wife, five children, and sister-in-law to the flood and drifted to the nearby town of Ulrum.[10] By Christmas morning, *Stad Groningen* (the city of Groningen) mobilized and dispatched relief. They commandeered every available ship, including its peat barges.[11] The city suffered little flooding, but water inundated the *Ommelanden* (the countryside surrounding the city of Groningen). Parts of the *Ommelanden* were submerged up to ten feet. By midday, relief boats returned to *Stad Groningen*, with people and livestock rescued from surrounding areas.[12] The city provided shelter, food, and raised collections to assist refugees. Survivors occupied the squares, churches, and state buildings. In neighboring East Friesland, the minister Johann Christian Hekelius noted the perversity of the disaster hitting on a day associated with joy and festivity. "In place of celebrating the happy event of the birth of Jesus Christ, there was nothing to be heard aside from screaming, lamenting, and crying."[13]

The Christmas Flood of 1717 was astonishingly deadly and remarkably expansive. Seawater breached dikes and inundated fields and cities from England to Denmark. The extent and severity of the extreme event easily placed it in the top tier of North Sea storms since 1500, but climate and weather alone cannot explain the severity of the Christmas Flood. The long history of human transformation of the coastal environment had produced communities extremely vulnerable to the storm.[14] Coastal Germany was the hardest hit, but the Dutch coast also suffered widespread devastation, especially the northeastern-most province of Groningen (Figure 3.1). Estimates vary, but as many as 13,352 people may have died in the flood, of which almost 2,300 lived in Groningen.

---

[9] Ibid.   [10] Ibid., 16.

[11] David Hartsema, "De Kerstvloed van 1717," *Vroeger en later* 1 (1989): 31.

[12] Van Seeratt, "Journaal," 20. Ten feet equaled approximately 3 meters. J. M. Verhoeff, *De oude Nederlandse maten en gewichten databank* (1983). www.meertens.knaw.nl/mgw/maat/70.

[13] I. H. Zijlma, *De Kerstvloed van 1717 in de Marne*. GrA. Bibliotheek RAG (Losse Boeken). 350.1760. Johann Christian Hekelius, *Ausführliche und ordentliche Beschreibung Derer beyden erschrecklichen und fast nie erhörten Wasserfluthen in Ost-Frießland* (Neue Buchhandlung: Halle im Magdeb, 1719), 39.

[14] Tim Soens, "Resilient Societies, Vulnerable People: Coping with North Sea Floods before 1800," *Past & Present* 241, no. 1 (2018): 143–77. Richard Paping, "De Kerstvloed van 1717 in Vliedorp en Wierhuizen," *Stad en Lande: Cultuurhistorisch Tijdschrift voor Groningen* 26, no. 4 (2017): 10–13. Adam Sundberg, "Gemeenschappelijke verantwoordelijkheid en weerstand: De Kerstvloed van 1717 in Groningen," *Historisch Jaarboek Groningen* (2018): 32–49.

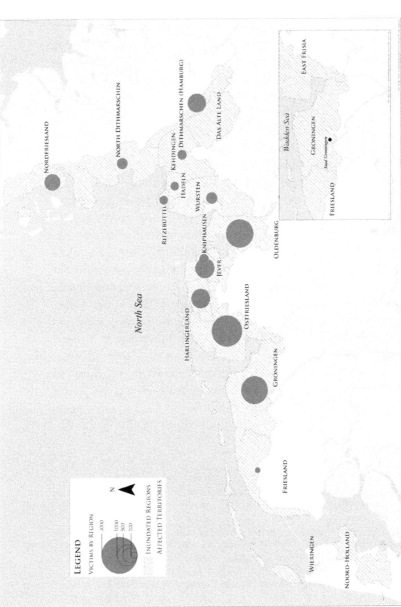

FIGURE 3.1. Christmas Flood mortality across the North Sea. Grey circles represent total mortality during the Christmas Flood as reported in Buisman, *Duizend jaar*, Deel 5. Sources: Inundated regions derived from Johann Baptist Homann, *Geographische Vorstellung der jämmerlichen Wasser-Flutt in Nieder-Teutschland*, Map, 1718, GrA, NL-GnGRA_817_2454. Boundaries modifed from Boonstra, NLGis shapefiles. MPIDR [Max Planck Institute for Demographic Research] and CGG [Chair for Geodesy and Geoinformatics, University of Rostock] 2011: MPIDR Population History GIS Collection (partly based on Hubatsch and Klein 1975 ff.) – Rostock. W.

The Christmas Flood of 1717 in Groningen

According to historian Tim Soens, it was likely the deadliest flood disaster in the human history of the North Sea.[15]

In the wake of the Christmas Flood, coastal communities in Groningen and across the North Sea struggled to make sense of the disaster and recover from its devastating impacts. Floods were certainly not rare along the North Sea coast. Groninger victims of the Christmas Flood might even remember the last great inundation. The St. Martin's Flood of 1686 had killed almost 1,600 people and flooded many of the same territories. Coastal communities had spent centuries adapting to inundations. Flood cultures emerged that relied upon complex systems of water defense governed by powerful management institutions but also required the skills, labor, and capital of at-risk communities. Flood disasters often revealed conflicts internal to these arrangements, but the cultural memory of prior inundation, legal precedent, and longstanding traditions of dike building and maintenance lent some clarity to post-disaster response. Cultures of coping depended on these interpretations of the past. They provided malleable templates to guide recovery during even the worst disasters.[16]

The Christmas Flood precluded easy comparisons, however. Everything about the disaster – the behavior of the weather, the height and extent of the flood, and the severity of its consequences – seemed unprecedented. Rather than presenting a template for response, the cultural memory of flooding demonstrated the flood's exceptionality. The Christmas Flood occurred amidst the cattle plague and during an era when anxieties about the future of the Republic were building. Unlike the plague, however, the Christmas Flood was localized. Large portions of Groningen experienced flooding, but the worst impacts were confined to communities along the Wadden Sea coast. The severity of the flood prompted moralists to expand the scale of moral responsibility from localities to the province, or even the Republic at large. Dikes ordinarily

---

[15] Soens, "Resilient Societies," 162. Tim Soens, "Waddenzee wordt moordzee: De Kerstvloed van 1717 en de kwetsbaarheid voor stormvloeden," *Tijdschrift voor Geschiedenis* 131, no. 4 (2018): 605–30. Most contemporary accounts of the Christmas Flood focus on German areas. The most complete authoritative account remains Manfred Jakubowski-Tiessen's *Sturmflut 1717: Die Bewältigung einer Naturkatastrophe in der Frühen Neuzeit* (Oldenbourg: Oldenbourg Wissenschaftsverlag, 1992). Buisman, *Duizend jaar*, Vol. 5, 451–3. Lamb and Frydendahl, *Historic Storms*, 9.

[16] Adam Sundberg, "Claiming the Past: History, Memory, and Innovation Following the Christmas Flood of 1717," *Environmental History* 20, no. 2 (2015): 238–61.

represented ingenuity, prosperity, and providential favor. The exceptional severity of the flood upended these meanings. The failure of Groningen's dikes implicated broad social and moral failings.

At the same time, the flood presented an unprecedented opportunity for the province to improve water infrastructure. The flood washed away entire stretches of coastal dikes. Provincial technocrats promised higher and stronger replacements that would improve upon older, inferior designs. Numerous challenges constrained these ambitions, however. The most significant problem was the issue of financial responsibility for reconstruction. From the perspective of the rural *Ommelanden*, the Christmas Flood was an exceptional disaster that warranted scaling up financial responsibility from the local to the provincial level. *Stad Groningen* resisted this interpretation and the *Ommelanden*'s appeals. Technocratic improvements depended upon the timely resolution of these conflicts.

Groningen's dikes stood at the center of these moral and financial negotiations. They were palimpsests that materialized the fluidity of coastal culture. Moralists, provincial authorities, and technocrats all navigated the uncertainties presented by the Christmas Flood by inscribing new meanings and purpose. Cultural memory, precedent, and history proved powerful sources of legitimacy, which they employed in unique and sometimes contradictory ways to promote their interests. They drew on the past to maintain the status quo and promote improvement; to evaluate risk and celebrate resilience; to welcome the disaster as opportunity or interpret it as a sign of decline. The Christmas Flood connected these expressions of coastal culture together.

3.1 FLOOD CULTURE AND FLOOD MEMORY IN THE DUTCH REPUBLIC

"Of all of the righteous plagues and punishments which the Lord visits upon a Land due to their sins, none is greater nor more oppressive, than when He ... sends floods with furious and swelling waves."[17] This opening passage to Groningen poet and author Albert Ebbo Crous's chronicle of the Christmas Flood emphasized two important features of Dutch flood culture. First, floods resulted from human sin. This causal story was the central pillar of premodern disaster interpretation. In this

---

[17] Crous, *Opregt*, 1.

respect, floods differed little from other divine judgments, whether warfare, cattle plague, or the decline of commerce. Second, Crous stressed the distinctiveness of flooding. Floods occupied a unique place in Dutch history. They had tested coastal communities along the North Sea since earliest settlement. They ranged from mundane occurrences to among the deadliest and most destructive disasters in Dutch history. This long and sometimes traumatic history provoked distinctive social and technological adaptations to control water. By the eighteenth century, water management had evolved into an important component of Dutch society, culture, and identity. Floods thus triggered intense scrutiny of water infrastructure and coastal mentalities. Flood *"memoria"* like Crous's chronicle reflected this cultural investment.

Like many regions along the North Sea, Dutch flood culture centered on water management. Building, maintaining, and repairing dikes were among its most important tasks.[18] Dikes facilitated reclamation and drainage and protected vulnerable coastal territory from inundation. In what would become the Dutch Republic, coastal dike building was an integral part of the large-scale transformation of the coastal environment beginning during the medieval era. The reclamation of peat bogs for agriculture resulted in subsidence when the organic material in drained peat soils oxidized and shrank. Peat extraction exacerbated this issue in the early-modern period. Daily tides scoured pathways for seawater into these sinking landscapes.[19] Beginning around 1000 AD, people began protecting their settlements with sea dikes. These low, small dikes were built and maintained by local landholders, but they slowly expanded in size, extent, and expense. By 1200, dikes protected most coastal areas and, by 1300, many riverine communities as well. Dikes grew larger and higher during the medieval era, requiring additional construction elements such as wood and organic material.[20] These large, complex dikes

---

[18] Maria Louisa Allemeyer, "Nature in Conflict. Disputes Surrounding the Dike in 17th Century Northern Frisia as a Window into an Early Modern Coastal Society," in *Historians and Nature. Comparative Approaches to Environmental History*, eds. Ursula Lehmkuhl and Hermann Wellenreuther (New York: Oxford University Press, 2007), 92–109. Maria Louisa Allemeyer, *"Kein Land ohne Deich – !": Lebenswelten einer Küstengesellschaft in der Frühen Neuzeit* (Göttingen: Vandenhoeck & Ruprecht, 2006).

[19] Van de Ven, *Man-made Lowlands*. For detailed studies on this phenomenon in Holland, see Van Dam, "Sinking Peat Bogs," 34. William H. TeBrake, *Medieval Frontier: Culture and Ecology in Rijnland* (College Station: Texas A&M University Press, 1985), 185–220.

[20] W. Barendsen, ""De zeedijk van zijn ontstaan tot het jaar 1730," *OTAR* 45, no. 9 (1961): 196–205.

hardened expansive stretches of the coastline, facilitating and accelerating further transformation of the environment.

Water management influenced the economic and political landscape as well. Dikes imposed stricter discipline upon littoral space, which often reduced the frequency of small floods and fostered intensive land use and settlement. Between 1250 and 1600, regional administrative bodies called water boards (*dijkrechten* and *zijlvesten* in Groningen, *hoogheemraadschappen* in Holland) emerged along the North Sea coast and rivers to manage increasingly complex and expensive systems of dikes, dams, and sluices. These institutions were regionally diverse, but many enacted legal requirements for settlement requiring labor or investment in dike infrastructure as a condition of occupancy.[21] If landholders failed to maintain their dikes (or pay dike taxes), they lost legal rights to the property.[22] Dike laws, institutions, and technologies were cultural manifestations of the social and economic investment in dry land.[23]

Dikes also served an important offensive function against water. Winning dry, productive territory from marshes, bogs, lakes, and seas had been a centuries-long, often pyrrhic struggle. Into the sixteenth century, large storms coupled with subsiding peatlands claimed at least as much Dutch land as was gained from the sea. By the mid-sixteenth century, however, the tide turned and increasing prosperity fostered a spectacular increase in reclamation projects. They ranged from the peat lake reclamations of Holland (*droogmakerijen*) to coastal reclamations

---

[21] Erik Thoen, "Clio Defeating Neptune: A Pyrrhic Victory? Men and Their Influence on the Evolution of Coastal Landscapes in the North Sea Area," in *Landscapes or Seascapes? The History of the Coastal Environment in the North Sea Area Reconsidered*, eds. Erik Thoen, Guus J. Borger, Adriaan M. J. de Kraker et al. (Turnhout: Brepols, 2013), 417–20. Petra J. E. M van Dam, "Ecological Challenges, Technological Innovations: The Modernization of Sluice Building in Holland, 1300–1600," *Technology and Culture* 43, no. 3 (2002): 500–20.

[22] In coastal Germany, the expression *"Kein land ohne dike, kein Deich ohne Land"* (No land without dikes, no dikes without land) referred to customary law adopted across the North Sea that required landowners' active participation in dike building. Allemeyer, "Kein Land Ohne Deich." Tim Soens, *De spade in de dijk?: waterbeheer en rurale samenleving in de Vlaamse kustvlakte (1280–1580)* (Ghent: Academia Press, 2009).

[23] A. A. Beekman, *Het dijk-en waterschapsrecht in Nederland vóór 1795*, Vol. 2 ('s Gravenhage: M. Nijhoff, 1907), 1502. Franz Mauelshagen, "Disaster and Political Culture in Germany since 1500," in *Natural Disaster, Cultural Responses: Case Studies toward a Global Environmental History*, eds. Christof Mauch and Christian Pfister (New York: Lexington Books, 2009), 53. Milja van Tielhof, "Regional Planning in a Decentralized State: How Administrative Practices Contributed to Consensus-Building in Sixteenth-Century Holland," *Environment and History* 23, no. 3 (2017): 431–53.

(*landaanwinning*) in Zeeland and the Dollard sea arm in Groningen.[24] The Dutch maxim that "God shaped the world, but the Dutch created the Netherlands" is almost cliché in the environmental historiography of the Netherlands, yet it clearly illustrated the Golden Age pride in this cultural landscape.[25]

Dikes became powerful symbols of coastal culture. They guarded the coast, ordered a dynamic landscape, and represented the best and worst of Dutch industry and solidarity. Eighteenth-century authors described dike building in duty-laden terms. "The dike must ever preserve this land," wrote the Mennonite minister Johannes Deknatel in 1722, "like the rudder to a ship in angry waves."[26] Dikes stabilized coastal society like a rudder oriented a ship. They promoted dryness and productivity. Foreign visitors to the Republic in the seventeenth century marveled at the seeming preternatural ability of the Dutch to reclaim coastal territory and convert coastal marshes to productive polder landscapes. "They plough the very bowels of the deep," proclaimed the Anglo-Welsh historian James Howell. "In the wrinkled forehead of Neptune being the furrows that yield them increase."[27] It was precisely this capacity to transform wasted marine and aquatic environments into Arcadian plenty that fostered the Golden Age reputation for water wizardry. Dutch success in water management encouraged an optimism and self-confidence that privileged the mastery of nature.[28]

Floods were also formative elements of coastal culture. The material risk posed by frequent small floods and the dramatic consequences of the largest inundations ensured that coping strategies assumed an important place in coastal life. Greg Bankoff argues that shared flood risk led to broad similarities between cultures of disaster across the North Sea

---

[24] Van de Ven, *Man-Made Lowlands*, 151–4, 159–65.
[25] This maxim is ascribed to René Descartes. "Dieu créa le monde, mais les Néerlandais créèrent la Hollande," Thimo de Nijs and Eelco Beukers, eds., *Geschiedenis van Holland, Deel I* (Hilversum: Verloren, 2002), 12.
[26] J. Deknatel, *Klaag en troost-dicht over den tegenwoordigen staat van Oost-vriesland, door zwaare watervloeden van vyf achter een volgende jaaren in de uyterste ellende gebracht* (Amsterdam: s.n., 1722).
[27] C. D. van Strien, *British Travellers in Holland during the Stuart Period: Edward Browne and John Locke As Tourists in the United Provinces* (Leiden: Brill, 1993), 192.
[28] Clarence J. Glacken, *Traces on the Rhodian Shore: Nature and Culture in Western Thought from Ancient Times to the End of the Eighteenth Century* (Berkeley: University of California Press, 1967), 476.

basin.²⁹ Similar environments – and in the case of Frisian lands, a shared cultural patrimony – encouraged comparable coping strategies, many of which worked through water management institutions and technologies.³⁰ Petra van Dam has identified an "amphibious culture" in the low-lying regions of the Netherlands that relied upon a shared suite of flood adaptations, such as the compartmentalization of the landscape, building settlements at elevation, and relying upon water-based transportation.³¹ Although her case studies focus on the coasts of the Southern Sea, much of her model applies elsewhere, including Groningen. The fixed nature of this enviro-cultural relationship should not be overstated. Floods tested the limits of these adaptations and evolving property relations, legal entitlements to flood protections, dike financing, religious confession, and other social factors influenced flood cultures and determined access and entitlement to safety.³² Most work on flood cultures, as a result, focuses on the dynamic response of regional or local actors.³³

---

[29] Greg Bankoff, "The 'English Lowlands' and the North Sea Basin System: A History of Shared Risk," *Environment and History* 19, no. 1 (2013): 3–37.

[30] Franz Mauelshagen, "Flood Disasters and Political Cultura at the German North Sea Coast: A Long-Term Historical Perspective," *Historical Social Research/Historische Sozialforschung* 32, no. 3 (2007): 133–44. Otto S. Knottnerus, "Angst voor de zee: Veranderende culturele patronen langs de Nederlandse en Duitse waddenkust (1500–1800)," in *De Republiek tussen zee en vasteland: Buitenlandse invloeden op cultuur, economie en politiek in Nederland*, eds. Karel Davids, Marjolein 't Hart, Henk Kleijer, and Jan Luccassen (Leuven: Garant, 1995), 57–81. Rieken, *Nordsee*. Michael Kempe, "'Mind the Next Flood!' Memories of Natural Disasters in Northern Germany from the Sixteenth Century to the Present," *The Medieval History Journal* 10, no. 1/2 (2007): 327–54.

[31] Petra J. E. M. van Dam, "An Amphibious Culture: Coping with Floods in the Netherlands," in *Local Places, Global Processes: Histories of Environmental Change in Britain and Beyond*, eds. David Moon, Paul Warde, and Peter Coates (Oxford: Windgather Press, 2016), 78–93.

[32] Soens, "Flood Security." Otto S. Knottnerus, "Vroegmoderne cultuurgebieden in Nederland en Noordwest-Duitsland gedachten over behoudzucht en dynamiek," *De Zeventiende Eeuw* 16 (2000): 14–28. Milja van Tielhof, "After the Flood. Mobilising Money in Order to Limit Economic Loss (the Netherlands, Sixteenth-Eighteenth Centuries)," in *Atti delle "Settimane di Studi" e altri Convegni*, Vol. 49 (Florence: Firenze University Press, 2018), 393–411.

[33] Maria Louisa Allemeyer, "The World According to Harro: Mentalities, Politics, and Social Relations in an Early Modern Coastal Society," *Bulletin of the German Historical Institute* 3 (2006): 53–76. Norbert Fischer, *Wassersnot und Marschengesellschaft: zur Geschichte der Deiche in Kehdingen* (Stade: Schriftenreihe des Landschaftsverbandes der ehemaligen Herzogtümer Bremen und Verden, 2003). Soens, "Waddenzee wordt Moordzee," 606.

Despite these variations, the Dutch Republic produced the closest approximation of a unified flood culture of any state in the North Sea region. This was less the result of a unique or cohesive set of flood adaptations or social practices than a shared appreciation of the rich moral meaning of inundation. The control of water loomed large in this shared identity. To win and defend land was both divine imperative and evidence of exceptional, providential favor. The sixteenth-century dike builder Andries Vierlingh famously declared, "The making of new land belongs to God alone, which some people are graced with the intelligence and power to act upon. It takes love and great labor, and not everybody can play that game."[34] Dutch industriousness was an expression of piety, their ingenuity and prosperity a product of providential favor. Simon Schama has argued that this "diluvian" moral geography produced a "hydrographic society" during its Golden Age. When storm surges destroyed dikes or inundated reclaimed landscapes, they inverted this cultural meaning.[35]

The past supported these providential interpretations and lent authority to Dutch claims of a special relationship with God. Floods appeared in the Republic's origin story and they featured prominently in histories and chronicles of cities and provinces. Some stories offered clear lessons. The arrival of timely rains during the Siege of Leiden in 1574 signaled divine sanction for the Dutch Revolt from Habsburg Spain. Rain allowed Dutch leaders to employ military inundations, flooding the landscape and lifting the siege.[36] The All Saints' Day Flood of 1570, which intensified tensions between Dutch regents and Spain, left greater room for interpretations, and providential responses broke down along confessional lines.[37] Some interpreted the flood as evidence of God's anger over the recent Iconoclasm. Others saw it as an omen of dramatic changes yet to come. Histories and chronicles recounted these stories because they were rich in interpretive possibilities. Some floods were mundane and expected – by-

---

[34] Andries Vierlingh, *Tractaet van dijckagie* (The Hague: Martinus Nijhoff, 1920), 18. http://resources.huygens.knaw.nl/retroboeken/vierlingh/.
[35] Schama, *Embarrassment*, 44.
[36] Robert Tiegs, "Hidden Beneath the Waves: Commemorating and Forgetting the Military Inundations during the Siege of Leiden," *Canadian Journal of Netherlandic Studies* 35 (2014): 1–27.
[37] Robert Tiegs, "Wrestling with Neptune: The Political Consequences of the Military Inundations during the Dutch Revolt" (PhD diss., Louisiana State University, 2016).

products of peoples' proximity to dynamic bodies of water.[38] Others were trials by adversity, and authors used them as heuristic tools to explore sin and redemption. Whether enemy or ally, floods strengthened the Dutch covenant with God and confirmed their chosen status as the new Israel (*Neerlands Israël*).[39]

A genre of chronicles dedicated to historical flooding gained popularity at the end of the seventeenth century. Prior to this era, accounts of floods appeared in city and provincial chronicles, news media, and pamphlets. The polymath Martin Schoock published the first flood chronicle in 1652 – a Latin account that listed every inundation between 860 and 1650.[40] The Frisian historiographer Simon Gabbema wrote a chronicle in Dutch that listed every major flood in the Netherlands between 760 and 1686. Not coincidentally, these works focused on the same period that birthed Dutch flood culture.[41] The Dutch minister in Emden, Gerhardus Outhof, produced an even more ambitious chronicle in 1718. It reached even deeper into the past to the biblical deluge and carried the story forward to the Christmas Flood.[42] The deluge and Christmas Flood appeared as logical correlates in a moral geography that saw flooding as a mechanism of cleansing and redemption.

Disaster *memoria* preserved the material and moral meaning of inundation. They could be practical or symbolic and often both. Memory stones embedded in buildings (*hoogwatermerken*) recorded high water levels that served as benchmarks for later adaptations just as they presented a daily reminder of physical and spiritual vulnerability. State resolutions and memoranda preserved the institutional memory of flood responses. These documents invoked providential authority and established legal precedents. Dikes themselves stood as visual reminders of this

---

[38] Raingard Esser, "Fear of Water and Floods in the Low Countries," in *Fear in Early Modern Society*, eds. William G. Naphy and Penny Roberts (New York: Manchester University Press, 1997), 62–77.

[39] Schama, *Embarrassment*, 25–50.

[40] Martin Schoock, *Martini Schoockii Tractatus de Inundationibus, Iis maxime, quae Belgium concernunt. Quatuor Disputationibus propositus In Academia Groningae et Ommelandiae* (Groningen: Johannis CollenI, 1652). Harm Pieters, "Herinneringscultuur van overstromingsrampen, gedenkboeken van overstromingen van 1775, 1776 en 1825 in het Zuiderzeegebied," *Tijdschrift voor Waterstaatsgeschiedenis* 21 (2012): 48–57.

[41] Simon Abbes Gabbema, *Nederlandse watervloeden, of naukeurige beschrijvinge van alle watervloeden voorgevallen in Holland ... en de naabuirige landen* (Gouda: Lucas Cloppenburg, 1703).

[42] Outhof, *Verhaal*, 1720.

history of shared risk and responsibility.[43] Sermons, newspapers, pamphlets, and visual culture like maps and prints served similar purposes. They emphasized flood frequency and exceptionality, highlighted resilience, and commemorated the deep lineage of disaster in the Netherlands. Disaster *memoria* reminded readers that the Dutch may have created the Netherlands, but the ultimate credit for dry, fertile land rested with God.

Crous's chronicle of the Christmas Flood performed each of these roles. It outlined a provincial culture of coping by noting historic water levels, provincial relief efforts, and the size and location of dike breaches. He tallied the number of houses and livestock washed away and the number of human lives lost.[44] Crous packaged this practical information in providential language befitting a hydrographic society. He was first and foremost a moralist who wanted to ensure that the providential meaning of the flood "would forever be remembered."[45] Large floods often demanded this type of memorialization, but the exceptionality of the Christmas Flood challenged moralists to revisit core elements of this disaster culture.

3.2 SPECIAL PROVIDENCE, THE INVERSION OF DRY LAND, AND DECLINE

"Has not the Almighty over the last several years changed our times, both for the good, but also and not the least for the bad?" The Pietist minister Johannes Mobachius posed this question to his congregation in Groningen less than two weeks after the Christmas Flood. He reminded them that the flood was one of a series of dramatic changes in their recent past. In 1713, they had celebrated the end of the "terrible, burdensome, bloody and very destructive war." Shortly thereafter, they suffered a "great plague of such devastating sickness and death amongst our cattle in many provinces (including this one [Groningen])."[46] Mobachius had

---

[43] Dike bodies were visual examples of what historian Michael Kempe refers to "memory of things" (memories produced by objects in daily life that remind one of the past). Kempe, "Mind the Next Flood!," 329.
[44] Crous, *Opregt*, 14.   [45] Ibid., 2.
[46] Johannes A. Mobachius, *Groningerlands zeer Hooge en Schrikkelyke Watervloed ... op Kerst-tyd den 25 Decemb. 1717. Verhandeld ... Op den eersten Biddag, gehouden in Groningerland, den 5 Jan. 1718. En daar na uitgebreid, en ter Gedagtenis in 't ligt gegeven* (Groningen: Jurjen Spandaw, 1718), 4.

already published sermons covering both events.⁴⁷ The Christmas Flood added yet another link in this moral chain of causation. "Have not our times changed again, when the Lord ... brought a strong storm wind out of the northwest, brought such a high, terrible, yes! Unprecedented flood over our land, in particular over this province (but also other regions)?"⁴⁸ Cattle plague had already undermined his optimism about returning to Golden Age peace and prosperity. The Christmas Flood deepened this rift with the past.

But whose past? Christmas Flood impacts in the Dutch Republic were largely confined to one province. Groningen suffered 94 percent of Dutch casualties. The Christmas Flood appeared provincial compared to the widespread consequences of the War of the Spanish Succession and the cattle plague. Some of Groningen's ministers and chroniclers emphasized this provinciality. City and provincial identities often superseded identification with the Republic in the early eighteenth century.⁴⁹ Mobachius even entitled his flood sermon "*Groningerland's very high and terrible flood.*" Yet Mobachius carefully separated the flood's moral meaning from its spatial extent. Groningen showcased the sins responsible for the flood – "evil thoughts, murder, adultery, fornication, lying, and blasphemy" – but so did the Netherlands more generally. According to Mobachius, the "persistent sins and extreme iniquity of the Dutch people (*Neerlands Volk*)" were to blame.⁵⁰

Mobachius was not alone. Authors outside Groningen expanded upon this Dutch interpretation of the Christmas Flood. Some supported it by exploring the Christmas Flood's effects beyond Groningen. The flood seemed widespread enough to the Amsterdam translator and poet Adriaan Bogaert, who documented its impacts in Zeeland, Friesland, and Holland, as well as Groningen.⁵¹ Other Dutch writers described

---

⁴⁷ Johannes A. Mobachius, *Neerlands lang gewenschte Vreede, tot Roem van den Almagtigen ... Verkondigd en aangedrongen Uit Psalm CXLVII. 12, 13, 14. Op den ... 14. van de Maand Juny 1713* (Groningen: Seerp Bansma, 1713). Mobachius, *Waare oorzaken*.

⁴⁸ Mobachius, *Groningerlands*, 4.

⁴⁹ E. O. G. Haitsma Mulier, "Het begrip 'vaderland' in de Nederlandse geschiedschrijving van de late zestiende tot de eerste helft van de achttiende," in *Vaderland: Een geschiedenis van de vijftiende eeuw tot 1940*, ed. N. C. F. van Sas (Amsterdam: Amsterdam University Press, 1999), 179.

⁵⁰ Mobachius, *Groningerlands*, 166–71.

⁵¹ A. Bogaert, *De Kersvloedt van den jare mdccxvii, vermengt met de gedenkwaardigste vloeden sedert den algemeenen* (Amsterdam: Gerrit Bosch, 1719). Holland suffered no casualties, though the flood inundated areas surrounding the IJ river. Guus J. Borger,

its international impacts in Flanders, England, East Friesland, and elsewhere. Yet even these accounts invariably focused their moral attention upon the sins of the Netherlands.[52] The Christmas Flood may have been an international catastrophe, and Groningen may have suffered the lion's share of losses in the Republic, but this was a Dutch tragedy. "*Neêrlands Volk!*" were culpable for the "terrible and destructive flood" that impacted "Holland, Friesland, the Ommelanden and its neighbors."[53]

Two characteristics of the Christmas Flood encouraged this reinterpretation. First, the disaster was extraordinarily severe, even by the standards of large floods. Flood disasters rarely claimed more than 1,000 lives, and almost 2,400 died in Groningen alone.[54] Large floods received far greater attention than smaller, mundane floods, which by nature of their ubiquity were interpreted as the ordinary functioning of nature (general providence).[55] Truly exceptional floods highlighted God's "special providence." Special providence explained how and why God suspended natural law.[56] When the winds shifted to the northwest on Christmas Eve, Groningers interpreted this as an extra-natural, divine act.[57] Special providence also explained seemingly miraculous

---

"De betekenis van de Kerstvloed van 1717 voor de gebieden rond het IJ," *Geografisch Tijdschrift* 1 (1967): 97–103.

[52] *Omstandig verhaal van de schrikkelyke en nooid gehoorde watervloed, voorgevallen in Engeland, Zeeland, Holland, Frisland, Groeningen en Duytsland, &c. op den 25 December 1717* (Middelburg: Johannes op Somer, 1718). C. J., *Korte beschryvinge van de schrikkelyke watervloed, veroorsaekt door een sterke stormvint/voorgevallen tusschen den 24. en 25. december 1717: soo in Holland/Friesland/Groninger-Embder en Oost-Friesland* (Leeuwarden: Johannes Thijssens, 1718).

[53] H. F., *Treurdigt, Ter droeviger gedagtenisse Van de vreesselyken en verderffelyken Watervloed. Den 25sten van Wintermaand MDCCXVII en enige volgende Dagen, gebragt oover Holland, Friesland, Omlanden en derselver Nabuuren* (Amsterdam: Johannes Douci, 1718).

[54] Van Bavel et al., "Economic Inequality."

[55] Adriaan M. J. De Kraker argues that floods (especially smaller floods) were interpreted as mundane, "natural" events in "Two Floods Compared: Perception of and Response to the 1682 and 1715 Flooding Disasters in the Low Countries," in *Forces of Nature and Cultural Responses*, eds. Katrin Pfeifer and Niki Pfeifer (Dordrecht: Springer, 2012), 287–302.

[56] Gary B. Deason, "Reformation Theology and the Mechanistic Conception of Nature," in *God and Nature: Historical Essays on the Encounter between Christianity and Science*, eds. David C. Lindberg and Ronald L. Numbers (Berkeley: University of California Press, 1986), 167–91.

[57] Manfred Jakubowski-Tiessen identifies a similar trend in affected German regions. Jakubowski-Tiessen, "'Harte Exempel göttlicher Strafgerichte.' Kirche und Religion in Katastrophenzeiten: Die Weihnachtsflut von 1717," *Niedersächsisches Jahrbuch für Landesgeschichte* 73 (2001): 119–32.

phenomena like unlikely rescues. Sermons and chronicles were thus replete with rescue narratives. One story that appeared in multiple accounts described a large ship (*tjalk*) "launched by the waves, not through, but over the dike . . . and onto the house of a shoemaker."[58] Rather than destroying it, ten refugees boarded the unlikely ark, and special providence ensured they saved a further thirty people before morning. These extraordinary events positioned the flood outside the realm of ordinary causation and justified moral reconsideration.

Providential accounts wrestled with the exceptionality of floods by contextualizing them. Chronicles of the history of flooding enjoyed a small resurgence after the Christmas Flood. According to the Groninger antiquarian Ludolph Smids, the Christmas Flood had breathed "new life" into Simon Gabbema's chronicle, written decades earlier. Smids published an update that listed all the Dutch floods that occurred since its last edition.[59] The Christmas Flood prompted Gerhardus Outhof to write his *Account of All Large Floods* (*Verhaal Van alle Hooge Watervloeden*) and he included a special, detailed report on the Christmas Flood as a final chapter. Flood chroniclers rarely compared disasters. Their recurrence alone served their purposes. "Never forget this judgment of God" Outhof warned, "but consider carefully how the Lord's disasters affected us because of our sins, from which we still suffer today."[60] Deep historical perspectives cautioned against forgetfulness and focused a more vigilant eye to the future.

Other types of flood *memoria* explicitly emphasized the Christmas Flood's exceptionality. Pamphlets and sermons claimed that water levels shattered previous records. Most Groninger accounts compared the Christmas Flood to the St. Martin's Day Flood of 1686. "Yes!" proclaimed Mobachius. "It is true, it is more than two feet higher than the previous flood and in all the old histories."[61] Another anonymous pamphlet declared, "No heavier floods can be found in the histories of previous eras, not even the St. Elizabeth's Flood (1421) and the All Saints' Day Flood (1570) were nearly so high."[62] Pamphleteers compared water levels across the Netherlands, reinforcing its wide-ranging significance. According to one Frisian account, floodwaters in Amsterdam rose almost

---

[58] Crous, *Opregt*, 19.
[59] Ludolf Smids, *Diluviana of Daghwyser der Nederlandsche waternooden van het Jaar 793 Tot deesen laatsten van den 25 November des verledene Jaars 1717* (Amsteldam: Hendrik van de Gaete, 1718), 4.
[60] Outhof, *Verhaal*, 644.   [61] Mobachius, *Groningerlands*, 142.
[62] H. F., *Treurdigt*, 5.

a foot higher than the flood of 1682, waters rose nine "thumbs" over the flood level of 1715 in Friesland, and on the North Sea island of Ameland, "the elderly could not remember the waters so high."[63] *Memoria* even recorded high water in Holland's riverlands, inland from the coast. Memory stones commemorated elevated water levels near Gouda, Schiedam, and Ammerstol.[64] Flood *memoria* framed the disaster as a unique moment in Dutch history, subject to special providential causation. Although sometimes exaggerated, these accounts performed the rhetorical task of setting the Christmas Flood apart.

The Christmas Flood's timing during other calamitous events of consequence also warranted attention. Four years into the disaster, the cloud of cattle plague still hovered over farmers worried about their herds and ministers worried about their flocks. Isbrandus Fabricius, the Reformed minister from Northern Holland, described the combined impacts of disasters in his book *Dutch Judgements and Disasters (Nederlantse oordelen en rampen)* in 1718: "I am assaulted from every side and consumed by disaster; still not free from the first, already come more destructive plagues upon me! Desperately besieged; without breath, robbed of flesh; I see the dried bones of my misery ground to dust by plague after plague."[65] Fabricius recounted the trauma of war and his exhilaration following the Treaty of 1713. Peace now seemed a distant memory. "People expect after peace, after rest, healing and comfort, but no! there came a new, and terrible fear ... one evil after the other."[66] Hubert Korneliszoon Poot, the same poet who penned "Disasters in the Year of Peace" in 1713, described the Christmas Flood in a new poem, "On the High Water Flood." "What livestock have you left?" he asked. "[T]he wealth of your pastures. Only recently has the fiery pestilence left death everywhere, now the flood smothers their stalls."[67] It seemed extraordinarily cruel that floods drowned cattle that survived the plague. The Netherlands had once been a "second Canaan, richly overflowing with

---

[63] C. J., *Korte Beschryvinge*, 4, 7, 14–15. One Frisian "foot" was approximately 0.3 meters. Nine "thumbs" was roughly 22 cm. Verhoeff, *Maten en gewichten*.

[64] Buisman, *Duizend jaar*, Vol. 5, 893.

[65] Isbrandus Fabricius, *Nederlantse oordelen en rampen ... betooght in de weeklaaglijke en nog durende runt-vee sterfte, en verschrijkelijke storm-winden en water-vloeden* (Alkmaar: Nikolaas Mol, 1718), 1.

[66] Ibid., 74.

[67] Hubert Korneliszoon Poot, *Op den hoogen watervloet, omtrent het einde des Jaers MDCCXVII* (1717), 7.

milk and honey, with prosperity and God's blessing," another pamphleteer recalled. Now, "the country is full of disasters."[68]

The inversion of prosperity and optimism had been a common trope in cattle plague literature and it easily translated to the Christmas Flood as well. The upending of Arcadian fertility was a powerful metaphor of decline. Flood literature harnessed that discourse and expanded it. If anything, water and flooding occupied an even more central place in Dutch identity than cattle. Ministers employed biblical analogies to perform this inversion. Psalm 107:34 ("A fruitful land into salted waste, for the wickedness of them that dwell therein") simultaneously imparted a causal story for the flood and characterized judgment as inversion. "The power of the Netherlands is broken," the Groningen lawyer and poet Johannes Kemner proclaimed, "robbed of grain and wheat fields, the fattened land turned to salted ground."[69]

When floodwaters breached dikes and drowned landscapes, they transformed its moral geography. The cultural imperative to control water ensured that inundated landscapes became powerful metaphors of loss. The Christmas Flood's timing and extraordinary severity expanded its moral meaning beyond Groningen. Moralists writing in 1717 had lived through years of war and grown increasingly worried about the loss of international prestige and an uncertain economic future. If cattle plague confirmed these fears, the Christmas Flood magnified them. The language of inversion captured and conveyed growing moral anxieties about this prolonged reversal of fortune.

3.3 FLOOD SEVERITY, ENVIRONMENTAL CHANGE, AND INSTITUTIONAL CRISIS

Thomas van Seeratt embodied the cultural imperative to manage and control water. He served as *commies provinciaal*, a provincial bureaucrat and engineer tasked with maintaining Groningen's public works, including its water infrastructure. Between 1717 and 1721, this included the monumental task of rebuilding and improving Groningen's dike infrastructure. He documented this experience with pride in a handwritten journal. By his own reckoning, his work yielded significant profits for the

---

[68] Centen, *Gods oordelen*, 4–5.
[69] Johannes Kemner, *De suchtende landtman in de provincie van stadt Groningen en Ommelanden over de hooge water-vloedt op kers-tydt den 25. van winter-maandt in het jaar onses heeren 1717* (Groningen: Johannes van Velsen, 1718), 5.

province and ensured that the lives of "many thousands in [Groningen] were spared."[70] To Van Seeratt, Groningen's reconstructed dikes represented diligence and ingenuity – foremost his own. They encouraged productive reclamation projects and ensured those lands remained dry. He tempered his personal pride in these accomplishments with the providential language of humility. "I thank God for offering me the strength and power to execute the orders of the [province] with zeal and speed."[71] Van Seeratt's confidence that God granted him the ability to promote safety and prosperity exemplified the most optimistic reading of Dutch coastal culture.

Where moralists saw broken dikes as indicators of moral decline, Van Seeratt saw opportunity. The flood damaged or destroyed large sections of coastal dikes along Groningen's northern border. According to Van Seeratt, this was due to poor maintenance and faulty design. Rather than simply rebuild the dikes, he wanted to improve them. This ambition required cooperation from provincial authorities, local elites, and laborers.[72] Instead, his plans faced resistance at every stage. Provincial authorities questioned his designs, and his coerced labor force intentionally slowed construction. The most serious impediment centered on the question of financial responsibility. Dikes remained the responsibility of landholders in the *Ommelanden*, yet the flood had devastated those rural communities. Many victims were still recovering from the combined impact of a rural recession and other natural disasters. The past seemed to provide limited guidance for navigating this ruinous state of affairs.

Provincial disputes that hindered dike reconstruction arose out of a culture of coping that struggled to balance asynchronous changes to its physical and social environment. Eighteenth-century Groningen's flood identity was largely tied to the Wadden Sea. The Wadden Sea is an intertidal zone of the North Sea situated between the northern Dutch coast (as well as parts of Germany and Denmark) and a chain of barrier islands. As was the case across the coastal Netherlands, the risk of tidal inundation and storm surges affected Groningen's early settlement

---

[70] Van Seeratt, "Journaal," 251. [71] Ibid., 26–7.
[72] Archived copies of Van Seeratt's journal highlight the importance of navigating between power brokers in Groningen. Two copies went to governing bodies representing provincial authority in Groningen – the Deputy Estates of Groningen and the Estates of the *Ommelanden*. He sent an additional two copies to *jonkers* (rural elites) in the *Ommelanden*. A fifth copy went to the Burgermeesters in Stad Groningen. L. J. Noordhoff, "Thomas van Seeratt, zijn levensloop en zijn betekenis voor de provincie Groningen," *Groningse Volksalmanak* (1961): 56.

history. The ebb and flow of the Wadden Sea produced fertile coastal soils consisting of sand, sea clay, and peat, which early immigrants exploited for crops and pasturage. Cattle production was prominent by the eighteenth century.[73]

Frequent inundations prompted coastal inhabitants to develop and refine water management strategies that mirrored those across the North Sea. They elevated houses or entire villages on artificial dwelling mounds called *terpen* (or *wierden* in Groningen) to escape floodwaters and constructed continuous sea dikes by the thirteenth century. No strategies prevented flood disasters entirely. Dikes remained relatively low and storm surges easily overtopped them. Dikes built on sandier soil were prone to dike breaches and subsidence due to reclamation, and peat excavation depressed Groningen's already bowl-shaped topography.[74] In response to these evolving environmental challenges and expanding settlements, Groninger communities formed regional drainage boards called *zijlvesten* and dike boards called *dijkrechten* to advise water management.[75] The authority of these water boards would grow throughout the early-modern period, but dike maintenance remained the financial responsibility of individual landholders in a system called *verhoefslaging*.

The risk of coastal inundation increased during the fifteenth and sixteenth centuries. The erosion of barrier islands and salt marsh foreshores (*kwelders*) reduced natural buffers to storm surges. This resulted in numerous coastal flood disasters, especially in vulnerable peatlands. Regular inundation coupled with large storms prompted significant loss of land in the eastern portion of the province, which eventually culminated in the formation of the Dollard Sea arm after 1509 and the displacement of dozens of settlements. Renewed reclamation between the sixteenth and seventeenth centuries encouraged new settlement and

---

[73] Otto S. Knottnerus, "History of Human Settlement, Cultural Change and Interference with the Marine Environment," *Helgoland Marine Research* 59, no. 1 (2005): 2–8. R. F. J. Paping, "De ontwikkeling van de veehouderij in Groningen in de achttiende en negentiende eeuw; een grove schets," *Argos* 24 (2001): 175.

[74] L. Hacquebord and A. L. Hempenius, *Groninger dijken op deltahoogte* (Groningen: Wolters-Noordhoff, 1990), 44. Jeroen Zomer, "Middeleeuwse veenontginningen in het getijdenbekken van de Hunze: Een interdisciplinair landschapshistorisch onderzoek naar de paleogeografie, ontginning en waterhuishouding (ca 800–ca 1500)" (PhD diss., Rijksuniversiteit Groningen, 2016).

[75] B. W. Siemens, *Dijkrechten en zijlvesten* (Groningen: Tjeenk Willink, 1974). *Zijlvesten* managed drainage in the interior of the province, the construction and maintenance of canals and roads, and the governance of polders and sluice gates (*zijlen*). *Dijkrechten* managed coastal dikes.

cultivation in these lower lying coastal regions.[76] Flood infrastructure was capable of managing regular inundation, but dikes remained relatively low (most less than three meters in height) across the province.[77] In the context of growing rural populations and expanding settlement during this period, the risk of flood disasters slowly rose.

Flood risk may have also increased due to changing atmospheric and ocean circulation patterns, though the connections between changing climate, storminess, and flood disasters are difficult to parse. Paleoclimatic research based on proxies from peat bogs, ice cores, and shifting dune landscapes coupled with documentary sources indicate that the North Sea during the Little Ice Age was in general stormier than the present, but storminess varied intensely on decadal and annual time scales.[78] In Western and Central Europe, storminess is often associated with changes in the North Atlantic Oscillation (NAO), which is characterized by a low-pressure zone near Iceland and a high-pressure zone near the Azores. When the Azores High intensified and Iceland Low deepened, this "positive" setting of the NAO brought stronger westerlies and more intense storms over Western Europe. The negative setting of the NAO by contrast brought fewer storms, colder temperatures, and increasing precipitation.[79] The negative setting of the NAO dominated the coldest periods of the Little Ice Age, including the Maunder Minimum, yet

---

[76] Daniel R. Curtis, "Danger and Displacement in the Dollard: The 1509 Flooding of the Dollard Sea (Groningen) and Its Impact on Long-Term Inequality in the Distribution of Property," *Environment and History* 22, no. 1 (2016): 113–19.

[77] Johan Kooper, *Het waterstaatsverleden van de provincie Groningen* (Groningen: J. B. Wolters, 1939), 99–102.

[78] S. M. J. F. Jelgersma, M. J. F. Stive, and L. Van der Valk, "Holocene Storm-Surge Signatures in the Coastal Dunes of the Western Netherlands," *Marine Geology* 125, no. 1–2 (1995): 109–10. Michele L. Clarke and Helen M. Rendell, "The Impact of North Atlantic Storminess on Western European Coasts: A Review," *Quaternary International* 195, no. 1–2 (2009): 32. Rixt de Jong, Svante Björck, Leif Björkman, and Lars B. Clemmensen, "Storminess Variation during the Last 6500 Years as Reconstructed from an Ombrotrophic Peat Bog in Halland, Southwest Sweden," *Journal of Quaternary Science* 21, no. 8 (2006): 916. Lisa C. Orme, Liam Reinhardt, Richard T. Jones et al., "Aeolian Sediment Reconstructions from the Scottish Outer Hebrides: Late Holocene Storminess and the Role of the North Atlantic Oscillation," *Quaternary Science Reviews* 132 (2016): 21–2. Lamb and Frydendahl, *Historic Storms*, 22–32. De Kraker, "Reconstruction of Storm Frequency," 51–69.

[79] Joaquim G. Pinto and Christoph C. Raible, "Past and Recent Changes in the North Atlantic Oscillation," *Wiley Interdisciplinary Reviews: Climate Change* 3, no. 1 (2012): 80. Valérie Trouet, Jan Esper, Nicholas E. Graham et al., "Persistent Positive North Atlantic Oscillation Mode Dominated the Medieval Climate Anomaly," *Science* 324, no. 5923 (2009): 78–9.

counterintuitively featured an increase in powerful storms. The NAO was not the only atmospheric or oceanic circulation that affected Western European storminess, and paleoclimatic models point to multiple potential drivers of these changes.[80] Historical documentation confirms that the coldest periods of the Maunder Minimum (1690–1720) also featured an intensification of powerful storms, even as their frequency decreased.[81] The Christmas Flood of 1717, arriving in the winter at the close of the Late Maunder Minimum, corresponded to these prevailing trends.

Not every storm, however, was equally likely to produce a flood. Flood disasters in the coastal Netherlands generally occurred when storm surges propelled by gusty winds blowing from the North and West combined with high water levels (usually a spring tide) to overtop coastal barriers.[82] Ship logbook data indicates that winds generally intensified and that westerly winds grew more frequent after 1700, particularly during relatively warm winters such as 1717.[83] Transient weather events did not always correspond to prevailing climate, but in the case of the Christmas Flood, these conditions may have increased its likelihood.

Even the most intense north and northwesterly storms and associated surges, however, did not necessarily result in flood disasters. Storm surges emerged in the context of stormy weather potentially fueled by climatic

---

[80] Recent paleoclimatic modeling has attempted to reconcile this discrepancy by arguing that an increased meridional temperature gradient resulted in fewer, more intense storms. Valérie Trouet, J. D. Scourse, and C. C. Raible, "North Atlantic Storminess and Atlantic Meridional Overturning Circulation during the Last Millennium: Reconciling Contradictory Proxy Records of NAO Variability," *Global and Planetary Change* 84 (2012): 48–55. Orme et al., "Aeolian," 22–3.

[81] Wheeler et al., "Atmospheric Circulation and Storminess." R. C. Cornes, "Historic Storms of the Northeast Atlantic since circa 1700: A Brief Review of Recent Research," *Weather* 69, no. 5 (2014): 121. Relying on dike maintenance records, Adriaan de Kraker notes that while the region became in general less flood-prone after 1672, large storms struck in 1682 and 1715, and the period between 1714 and 1717 was particularly stormy. Adriaan M. J. de Kraker, "Storminess in the Low Countries, 1390–1725," *Environment and History* 19, no. 2 (2013): 166–70. He resisted connecting these events to climate changes. De Kraker, "Flood Events in the Southwestern Netherlands," 920. Paleoclimatic modeling supports these findings. C. C. Raible, Masakazu Yoshimori, T. F. Stocker, and Carlo Casty, "Extreme Midlatitude Cyclones and Their Implications for Precipitation and Wind Speed Extremes in Simulations of the Maunder Minimum versus Present Day Conditions," *Climate Dynamics* 28, no. 4 (2007): 409–23.

[82] De Kraker, "Storminess in the Low Countries," 13.

[83] Colder winters, however, often featured increasing easterlies, especially before 1700. Mellado-Cano et al., "Euro-Atlantic Atmospheric Circulation," 3851–55. Wheeler et al., "Atmospheric Circulation and Storminess," 266–70. Degroot, *Frigid Golden Age*, 44–6.

drivers, yet flood catastrophes also depended on social interference with the coastal landscape and associated changes in social vulnerability.[84] This was certainly the case in Groningen prior to the Christmas Flood. During the mid-seventeenth century, farmers aggressively transformed the *kwelders* outside the Wadden Sea dikes into productive agricultural lands.[85] The province encouraged this coastal reclamation in 1637 by offering to freeze taxes on new lands for twenty-five years.[86] These coastal reclamation projects opened a new frontier of exploitable lands, but they also removed an important environmental defense. *Kwelders* absorbed the kinetic energy of waves as they approached the shore and a recent analysis based on contemporary maps and documentation of the size and location of breaches has demonstrated that areas with wide salt marshes experienced fewer, less catastrophic dike breaches in 1717.[87] Anthropogenic changes to the coastline during this period of reclamation fashioned a new geography of risk.

Lacking *kwelders* as a natural buffer, dikes required expensive wooden breakers. Reclamation also demanded costly osier constructions and additional earthen dikes to protect and expand the newly won land. By the mid-seventeenth century, the financial burden to maintain these structures grew untenable for individual landholders, and, in 1650, the province grudgingly accepted partial responsibility for dike upkeep. Considering the "inability and negligence of the landholders and the long-feared inundation ... which would lead to the immense suffering and ruin of thousands of people and notable loss of the provincial resources," they agreed to pay two-thirds of the cost of wooden barriers (*post en paalwerk*) that protected the dikes.[88] This breach of tradition

---

[84] Soens, "Resilient Societies," 150.

[85] W. E. van Duin, H. Jongerius, A. Nicolai et al., "Friese en Groninger kwelderwerken: Monitoring en beheer 1960–2014," *Wettelijke onderzoekstaken Natuur & Milieu* 68 (2016): 17.

[86] Hacquebord and Hempenius, *Groninger dijken*, 47. State incentives were common across the eastern and southern edge of the North Sea. Piet van Cruyningen, "State, Property Rights and Sustainability of Drained Lands along the North Sea Coast, Sixteenth-Eighteenth Centuries," in *Rural Societies and Environments at Risk Ecology, Property Rights and Social Organisation in Fragile Areas*, eds. Bas van Bavel and Erik Thoen (Turnhout: Brepols, 2013), 181–207.

[87] Zhenchang Zhu, Vincent Vuik, Paul J. Visser et al., "Historic Storms and the Hidden Value of Coastal Wetlands for Nature-Based Flood Defence," *Nature Sustainability* 3, no. 10 (2020): 1–3.

[88] Resolution. 8 March (1650). *GrA*. Staten van Stad en Lande, 1594–1798. 1.11.

presented potentially worrisome implications for both urban and rural interests.

Historically, Groningen's deputy administrators (*Gedeputeerden*), which included urban and rural representatives, had not been heavily involved in dike maintenance and they rarely organized relief following floods. Long-standing antipathy and divergent interests between Groningen's urban and rural authorities undermined collective action. This situation evolved during the second half of the seventeenth century as the deputies began asserting greater influence in water management.[89] In addition to assuming partial upkeep of the *post en paalwerk*, they established a committee, called the *Gecommittteerden van Stad Groningen en Ommelanden Tot de Dijken cum plena*, to guide reconstruction after the St. Martin's Flood of 1686. This committee continued to provide oversight over dike maintenance into the eighteenth century. At the same time, the urban–rural rift deteriorated further. Some disputes even required outside arbitration from the Estates General of the Netherlands. In this context, *Stad Groningen*'s political influence relative to the *Ommelanden* grew considerably.[90]

Social changes in the countryside reflected the city's growing power as well. Wealthy, urban elites and landowners began displacing smaller rural landholders from positions in water management. Those most vulnerable to inundation, such as farmers in the reclaimed territories along the Wadden Sea, grew increasingly marginalized. The remaining landholders' inability to subsidize dike maintenance, coupled with the concentration of political power in the hands of unaccountable elites, and the growing power of the city relative to the countryside produced an unsustainable environment for effective water management.[91] Coupled with the loss of *kwelder* buffers due to reclamation, it was no coincidence that the municipalities that suffered

---

[89] This struggle against dike centralization or communalization (called *gemeenmaking*) proceeded at different rates and for different reasons across the North Sea. Milja van Tielhof, "Forced Solidarity: Maintenance of Coastal Defenses along the North Sea Coast in the Early Modern Period," *Environment and History* 21, no. 3 (2015): 319–50.

[90] G. A. Collenteur, G. A. C. Pim Kooij, and P. Kooij, eds., *Stad En Regio* (Assen: Van Gorcum, 2010). 't Hart, "Town and Country in the Netherlands," 86–7.

[91] Otto S. Knottnerus, "Yeomen and Farmers in the Wadden Sea Coastal Marshes, c. 1500–c. 1900," in *Landholding and Land Transfer in the North Sea Area (Late Middle Ages–19th Century)*, eds. P. Hoppenbrouwers and Bas van Bavel (Turnhout: Brepols, 2004), 149–86. R. F. J. Paping, "De zijlschotregisters, dijkrollen en registers van schouwbare objecten in Groningen tot circa 1800," in *Broncommentaren 4: Bronnen betreffende de registratie van onroerend goed in middeleeuwen en Ancien Régime*, ed. G. A. M. van Synghel (The Hague: Instituut voor Nederlandse Geschiedenis, 2001), 277–310. Soens, "Resilient Societies," 143–77. Milja van Tielhof, *Consensus en conflict: Waterbeheer in de Nederlanden, 1200–1800* (Hilversum: Verloren, 2020), 244–5. A similar process

The Christmas Flood of 1717 in Groningen 113

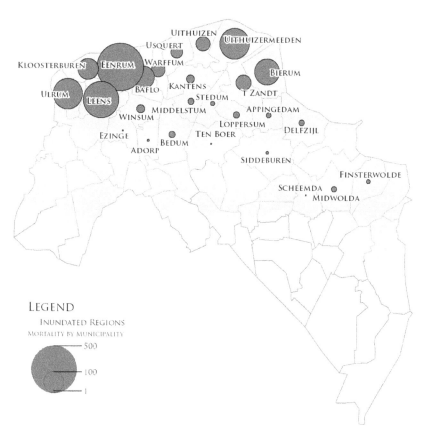

FIGURE 3.2. Christmas Flood fatalities by municipality in Groningen. Size of grey circles indicates relative mortality against total provincial losses.
Source: Register, van geleden verlies in de provincie van Stadt En Lande. Aan wechgespoelde huysen verdronken menschen hoornbeesten peerden verkens schapen in de jongste Watervloedt van Midwinter 1717. Groningen, 1718. Boundaries modified from Boonstra, NLGis shapefiles.

greatest mortality during the Christmas Flood were located in or near these regions along the Wadden Sea coast (Figure 3.2). Uithuizermeeden was one such community, which by Tim Soens's estimation may have lost one out of every five inhabitants to the flood.[92]

Disputes over water management in 1718 developed out of this toxic arrangement between rural and urban interests. Immediately following the Christmas Flood, however, it seemed as if conflict could be

---

played out after a flood in 1715 in the "calamitous polders" of Zeeland. Van Cruyningen, "Sharing the Cost," 375–9.

[92] Soens, "Resilient Societies," 165.

transcended. *Stad Groningen*'s relief efforts during the flood received high praise from *Ommelander* deputies.[93] The city's response was coordinated and effective. This brief period of amity ended a few weeks later when *Ommelanders* took issue with the provincial plan to finance dike repair.[94] The plan required landholders whose property abutted damaged dikes to finance and provide labor for the repairs. Landholders with property inland from the dikes (*hulpkarspelen*) would assist. *Stad Groningen*, which had escaped the flood relatively unscathed, supported the proposal because it absolved them of financial responsibility. Advocates pointed to historical precedent set in 1687 when *hulpkarspelen* supported reconstruction following the St. Martin's Flood.[95]

Ommelander regents rejected the proposal and its justification. Faced with what they deemed an unreasonable burden upon an already beleaguered population, they argued that their "miserable residents" could not be expected to "help other, equally as unfortunate people, also up to their necks in water."[96] In this situation, the example set in 1686 seemed an impractical precedent. The financial cost and amount of labor necessary to rebuild dikes was much higher than in the past and nearly impossible for landowners to shoulder themselves. They framed their arguments in defense of custom. Relying on *hulpkarspelen* ran contrary to their interpretation of traditional dike responsibilities. *Ommelander* requests went still further. They proposed an alternative that included remission from taxation and paid for repairs out of the provincial treasury. Considering the unprecedented scale of the disaster and economic turmoil, they wanted taxes broadly forgiven. In the view of the city, this attempt to expand the scale of financial responsibility was a break from precedent. Disputes prevented reconstruction efforts throughout the spring and summer of 1718.

To negotiate this impasse, the Estates of Groningen sought mediation from outside bodies on two occasions. In May, they invited delegates from the Estates of neighboring Friesland (which had wrestled with a similar issue in the sixteenth century) to advise. After this failed, they requested mediation again in August, this time from both Friesland and

---

[93] Resolution (31 December) 1717. *GrA*. Staten van Stad en Lande. 1.30. The most complete account of this dispute is Tonko Ufkes, "De Kerstvloed van 1717: Oorzaken en gevolgen van een natuurramp" (PhD diss., Rijksuniversiteit Groningen, 1984).

[94] Resolution. 10 January (1718). *GrA*. Staten van Stad en Lande, 1594-1798. 1.30.

[95] Milja van Tielhof terms the temporary use of *hulpkarspelen* "ad-hoc communalization." Van Tielhof, "Forced Solidarity," 326–9.

[96] Resolution. 10 January, 8 April (1718). *GrA*. Staten van Stad en Lande, 1594-1798. 1.30.

the Estates General in The Hague.[97] To the dismay of the *Ommelanden*, both conferences favored *Stad Groningen*'s position. The *Ommelanden* negotiated from a position of relative weakness. The risk of further flooding in the quickly approaching winter storm season disproportionately affected its territories and so they grudgingly accepted the plan. The *hulpkarspelen* would be required to assist in repairs with help from the militia. By the time these disputes were resolved, it was already late summer, leaving only a few months to repair dikes before the beginning of the storm season. All these conflicts presented an unwelcome delay for Van Seeratt.

3.4 RESISTANCE AND TECHNOCRATIC OPTIMISM

Van Seeratt began outlining his plans to rebuild and improve the dikes in February 1718 when the province ordered him to conduct a survey of the devastation.[98] He had already been convinced of the inadequacies of Groningen's sea defenses since the fall of 1716 when he faced his first flood emergency. Only three months into his appointment, a dike breached in eastern Groningen.[99] It was a minor storm and minor breach, one of many mundane events that reminded Groningers of their continued flood risk. Despite this, Van Seeratt included an ominous warning in his report to the province. "With much respect," he explained, "the eastern dikes as well as those of the entire province ... are, in general, in a very terrible and miserable state." The earthen bodies of the dikes were too steep, and the wooden posts and groynes that protected their seaward sides were insufficient to protect against the erosive force of the waves. Furthermore, landholders had removed so much earth on the landward sides of the dikes "that there was as much a sea inside as outside the dike."[100] The relatively minor 1716 flood was both a managerial and technological failure that reflected systemic issues facing Groningen's flood security.

Provincial administrators were initially unconvinced. Van Seeratt was advocating a comprehensive and expensive program of improvement and his brief tenure as *commies provinciaal* lent him limited credibility.[101]

---

[97] C. Tolle, "Tweedracht maakt zacht. Conflicten binnen een Fries waterschap (1533–1573)," *Leidschrift* 28, no. 2 (2010): 41–57.
[98] Van Seeratt, "Journaal," 32.   [99] Ibid., 4.   [100] Ibid., 7–8.
[101] Bureaucrats had historically occupied the position of *commies provincial*. State surveyors and engineers compensated for this gap in expertise. Gea van Essen, *Bouwheer en*

His first challenge was convincing them of his expertise. Van Seeratt cited his experience as a sea captain and his grounding in mathematics. "Most of my life has been spent at sea," he argued, and "whenever a person, through God's grace, has a reasonable understanding and skill in mathematics," his authority should be recognized.[102] Apparently convinced, they sent Van Seeratt to Friesland in early 1717 to assess their dikes with a view to adapting their own. He reported back impressed. Their largest dikes were twenty "feet" in height with a gently sloping seaward face. Large wood and stone barriers and groynes protected the most vulnerable sections. In marked contrast to Groningen, "the storms of last December ... caused no damage to the Frisian dikes but remained in a good state."[103] Van Seeratt argued that Frisian dike designs should be emulated, but he made little headway during the summer or fall of 1717. The Christmas Flood provided a new opportunity to enact these plans.

Van Seeratt began inspecting dikes as soon as relief work ended.[104] He conducted a complete assessment, section by section, noting the extent and location of damage. He also noted design failures of the broken dikes. He contrasted their steep slopes with Frisian dikes, already formulating a rationale for improvement. "If a dike's slope stands like a wall against the roaring sea it must always break." "If flat, the sea plays upon it with no violence."[105] Just as Van Seeratt completed this preparatory work, progress ground to a halt. The financial disputes between *Stad Groningen* and the *Ommelanden* were boiling over. In his journal, Van Seeratt lamented that "all of the provincial works were zealously proposed and prepared, but the actual dikes remained without hands to lay them."[106] Nearly half of the work season was over, and much of the landscape remained vulnerable to inundation.

When presented with provincial orders to begin reconstruction in June, he faced new resistance, this time from his labor force. *Ommelander* villagers from the *hulpkarspelen* refused to assist with reconstruction. Like their *Ommelander* representatives, they chafed at the growing power of the city in water management. They saw provincial orders to present themselves for work as an example of this breach with tradition. For Van Seeratt, this meant further delays. If villagers reported for duty at all, they

bouwmeester: Bouwkunst in Groningen, Stad en Lande (1594-1795) (Assen: Koninklijke Van Gorcum, 2010), 52.
[102] Van Seeratt, "Journaal," 8.
[103] Ibid., 11. Twenty feet was approximately six meters. Verhoeff, *Maten en gewichten*.
[104] Ibid., 27.   [105] Ibid., 39–40.   [106] Ibid., 40.

refused to work or destroyed tools.[107] In October, disaffected villagers from one *Ommelander* region, the *Westerkwartier*, even mounted a short-lived farmer insurrection to protest this coercion. Nobody could have denied the increasing threat of calamity as the storm season approached, but this was the second time that the province had foisted financial responsibility upon the *hulpkarspelen* for dike repair since 1686. In the meantime, farmers had contended with poor harvests in 1709–10 and a cattle plague. Their passive and violent resistance reflected their economic and political dislocation.[108]

Van Seeratt understood this resistance differently. From his perspective, refusal to work was merely evidence of ignorant fatalism. He identified a "dangerous discourse," widely held among his workers, that dikes only protected communities from smaller floods like the one that occurred in 1716. No dikes could withstand the largest floods. From this perspective, Van Seeratt's ambition to improve dikes was folly. According to Van Seeratt, resistance was simply a pessimistic rejection of his dike designs. Why improve dikes when "floods came once every thirty years and nobody could construct dikes against them?"[109]

Van Seeratt's criticism spoke to a cognitive dissonance in providential thinking about flooding. Dikes memorialized workers' piety and industry, but dikes breaches signified the opposite.[110] God ultimately governed both success and failure. According to Van Seeratt, his workers' response underscored their deep-seated pessimism about the control of water. Flood *memoria* promulgated this rhetoric as well. "No dike, however strong, remained standing, but was washed across the country by the flood," wrote Crous in his *Historis-Verhaal*.[111] Adriaan Spinniker echoed this pessimism in his poem on the flood. "No industry or effort may here avail the angry waves," he proclaimed. "The strongest dikes were penetrated, the spring flood empties without limit, and an innumerable number of villages are consumed by its fury."[112] Groningen's coastal culture of flooding accommodated the

[107] Ibid., 63–4.
[108] Cattle plague ultimately resulted in a 47 percent reduction in tax income. L. S. Meihuizen, "Sociaal-economische geschiedenis van Groningerland," in *Historie van Groningen Stad en Land*, eds. Wiebe Jannes Formsma and M. G. Buist (Groningen, 1981), 314. Tonko Ufkes, "De opstand in Aduard in 1718," *Groningse Volksalmanak* (1985): 91–101. Sundberg, "Gemeenschappelijke Verantwoordelijkheid," 32–49.
[109] Van Seeratt, "Journaal," 86–7.
[110] Allemeyer, "The World According to Harro," 66.   [111] Crous, *Opregt*, 62.
[112] Adriaan Spinniker, *Gods gerichten op aarde, Vertoond in den Schrikkelyken Storm en hoogen Watervloed, op den 25 en 25sten van Wintermaand in 't 1717 de Jaar voorgevallen* (Groningen: Jurjen Spandaw, 1718), 8, 12.

contradictory imperatives to control water and accept limits to that control, but Van Seeratt's claim reveals their potential for conflict.

Van Seeratt, of course, rejected this pessimism. Dikes were a source of pride and technocratic optimism. The design proposals he submitted to the province in summer 1718 depicted structures larger than any previously built in Groningen. Van Seeratt understood the skepticism his plans engendered, both from his labor force and from provincial administrators already divided over the question of financing. Van Seeratt doubled down on his optimistic rhetoric. His dikes would repel even the highest floods. His journal, sketches, and notes outlined and explained their advantages. As evidence, he pointed to the severity of the Christmas Flood. Dike breaches and social disruption revealed the inadequacies of past approaches and warranted improvement.

Van Seeratt explicitly contrasted his plans with the shortcomings of historic approaches in a series of dike schematics that depicted his improved designs superimposed over older dikes. His new dikes visually overwhelmed their historic equivalents. He emphasized their gently sloping seaward faces, which contrasted with earlier, steeper designs. He noted other planning failures as well. Many dikes breaches in 1717 resulted from poor siting. According to Van Seeratt, the "sandy ground" supporting the sea dike to the north of Leens, for instance, undermined its stability. The mixture of sand and clay in other sections appeared equally problematic.[113] Some of these areas had breached in 1686 and they remained vulnerable.[114] Dikes would need to be moved, not simply improved. Ideally, dikes would also assist reclamation efforts. In the Christmas Flood, Van Seeratt saw an opportunity to achieve each of these ambitions.

With the storm season fast approaching, Van Seeratt saved time and effort by promoting smaller dikes that survived the storm to sea dike status. These smaller dikes (*kadijken*) protected reclaimed territory from summer, tidal inundations (Figure 3.3). They had been built in the mid-seventeenth century and extended across much of Groningen's coast with the Wadden Sea. Simply upgrading these dikes solved several problems at once. It saved time and money, since part of the dike was already in place. It also encouraged further reclamation.[115] The *kadijk* had been part of the long-term project to convert this coastal area into productive land. Although they repulsed high water during the summer, the dikes also trapped silt, which eventually formed the foundation of new land. Walling this territory off from the waves encouraged resumption and then expansion of that project.

---

[113] Van Seeratt, "Journaal," 254. [114] Ibid., 83–4.
[115] Van de Ven, *Man-made Lowlands*, 145.

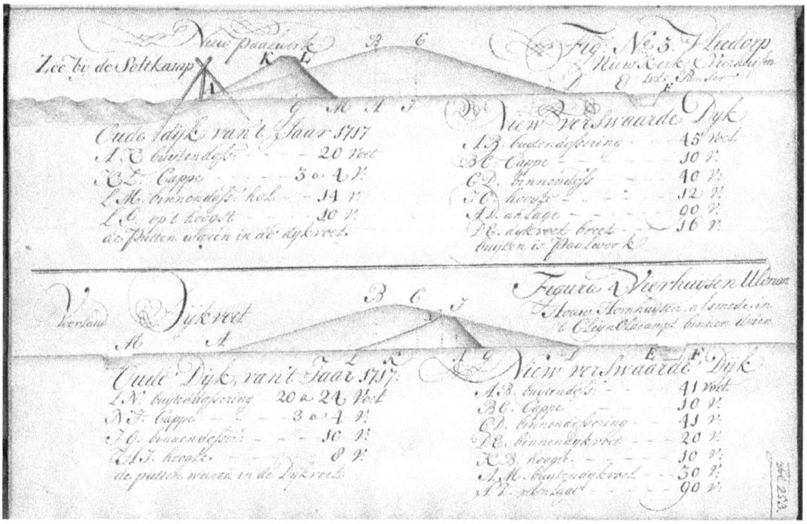

FIGURE 3.3. Post-1717 dike design improvements. *Source*: Thomas van Seeratt, "Journaal Van De Commies Provinciaal Thomas van Seeratt," GrA. 1.818. fol. 253.

Van Seeratt's design sketches highlight this as well. He depicts two examples of groynes (*kisthoofden*) extending from a dike. Predictably, he contrasted his designs with an historical counterpoint. The older design, which showed the groynes extending perpendicular from the dikes "traps the stream flow and washes the ground away"[116] (Figure 3.4). In Van Seeratt's new proposal, the *kisthoofd* thrust into the sea at an angle, causing no whirlpool action and accreted new sediment for reclamation.[117] Van Seeratt's designs played upon two components of Dutch coastal culture – they appealed to the economic sensibilities of the provincial authorities just as it reinforced the cultural imperative to win land from the sea.

3.5 CONCLUSION

On 1 January 1719, another strong winter storm hit Groningen. Waves overtopped dikes across northern Groningen. Those completed according to Van Seeratt's specifications held; the remainder suffered serious damage. Van Seeratt interpreted this as divine validation of his ideas. "God desired this storm," he proudly noted. It "opened the eyes of all those who are sincere" about the value of his improved designs, which had been completed a little more than one month earlier. In Van Seeratt's moral

[116] Van Seeratt, "Journaal," 262. [117] Ibid., 263.

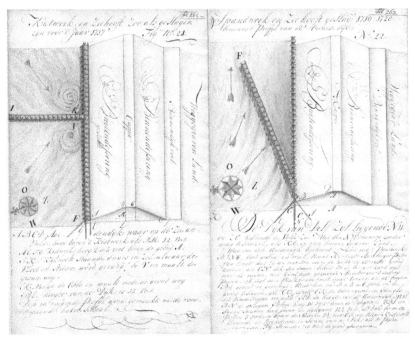

FIGURE 3.4. Diagrams of *kisthoofden*. *Source*: Thomas van Seeratt, "Journaal Van De Commies Provinciaal Thomas Van Seeratt," GrA. 1.818. fol. 262–3.

geography of the flood, God favored his ingenuity. Naturally, he conceded, "my own credit was not slightly increased."[118] To Van Seeratt, these dikes improved Groningen's flood protection and cemented his reputation.

This final statement of vindication was self-serving, though also a fitting reminder that the past loomed large in every reaction to the Christmas Flood. Van Seeratt's journal was an overt attempt to claim space in this cultural memory of flooding. His dikes were an even more visible legacy and his strongest statement for posterity.[119] Moralists used

---

[118] Ibid., 86–7.
[119] To a certain extent, he was successful. Van Seeratt is the central figure in a film adaptation of the Christmas Flood, *De Kerstvloed* (2017); his story was a central element in recent remembrances in Groningen (www.kerstvloed1717.nl/), and Rijkswaterstaat memorialized his contributions in 1919 by naming a water board after him. The Thomas van Seeratt Waterschap (1919–74) is now part of the waterschap Noorderzijlvest. K. A. M. Engbers and A. L. Hempenius, *Verzamelinventaris van de archieven van de "Kust"waterschappen inliggend in het waterschap Hunsingo en het waterschap Reitdiep (1805) 1856–1990 (1994)* (Groningen: Laurentius Archief & Geschiedenis, 2011), 159.

the history and memory of flooding as both coping strategy and rubric for evaluating flood severity. Flood chronicles comforted readers with the knowledge that *Neerlands Volk* were a resilient chosen people. At the same time, the exceptionality of the flood and its catastrophic timing encouraged moralists to seek meaning and assign responsibility for localized flooding in provincial or even broader terms. For many, the Christmas Flood reinforced a declensionist narrative that warned of a widening rift with the more prosperous Golden Age. Financial debates between city and countryside over dike reconstruction also hinged upon conflicting interpretations of scale and responsibility. In each case, memory, history, and precedent proved malleable, yet indispensable, tools in disaster response. Even as people turned their attention to the future, they drew legitimacy from the past.

Van Seeratt's statement was also a reminder that moral meaning remained embedded in the business of building and governing dikes – and in state response to disasters more generally. Van Seeratt's confidence reflected an optimistic vision of Dutch flood culture that drew pride from the control of water. Providence reinforced his technocratic ambitions. The belief that disasters could improve or redeem society was integral to Dutch flood culture. Redemption was not guaranteed, however. Each new natural disaster amplified public anxiety. On November 3, 1720, the deputies of the Estates of Groningen published a general call for thanksgiving and prayer. They had much to be thankful for. Van Seeratt's dikes, now completed, withstood the storm surges of 1719 and 1720 that had devastated communities across much of northern Germany.[120] The call also reminded readers that "God's striking hand" remained raised over much of the Republic. Declensionist rhetoric was still nascent and remained largely linked to moral concerns, but it continued to build. The Christmas Flood would pass, but other disasters including "terrible inundations, lengthy sickness and death among the cattle, barriers to commerce, and general lessening of prosperity" remained on the horizon.[121]

---

[120] Jakubowski-Tiessen, *Sturmflut 1717*, 42.
[121] Proclamation (3 November) *GrA*. Gedeputeerde Staten van Stad en Lande 1720. 1.477.

4

A Plague from the Sea

The Shipworm Epidemic, 1730–1735

In late December 1732, the engraver Hendrik de Leth purchased a newspaper advertisement for "a new accurate map" of West Friesland.[1] The map depicted a bucolic rural region of North Holland that bordered the Southern Sea called the *Vier Noorder Koggen* (Figure 4.1). More decorative than practical, it commemorated the accomplishments of the region's water board. The cartouches draped along its upper edge, which contained the coats of arms of local and regional water authorities spoke to this primary purpose. The map was impressive in artistry and detail. It included symbols for roads, footpaths, and canals. Hash marks along the northern coastline indicated wooden wiers and wave breakers, and tiny icons showed where windmills pumped polder water through sluices out to sea. The clean, rectilinear order of the landscape bespoke its intensive, skillful management. Every detail reflected the water board's core task: to maintain the balance between land and water.

Yet, at the time De Leth advertised the map, that balance appeared tenuous. A hitherto little-known marine mollusk had unexpectedly appeared along the coast of Zeeland two years earlier, burrowing into the wooden barriers that protected its dikes. Inspections in subsequent years revealed that the mysterious wormlike creature had spread across much of the coastal Netherlands, from Zeeland to Friesland. Slim creatures that could grow up to 60 cm long, *Teredo navalis* gnawed into submerged wooden infrastructure, creating a honeycomb of passages that undermined Dutch dikes. Without wooden groynes, palisades, and wave

---

[1] Advertisement. December 22 (1732). *'s Gravenhagse Courant*.

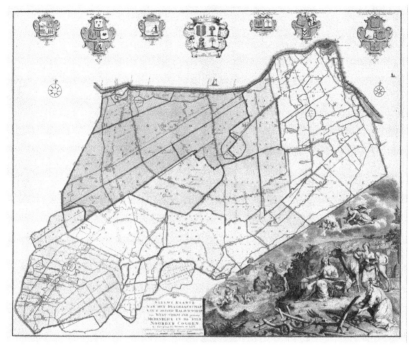

FIGURE 4.1. De Leth's map of the "Vier Noorder Coggen" water board. *Source*: Pieter Straat, *Nieuwe kaarte van het dijkgraafschap van 't Ooster Baljuwschap van West-Vriesland, genaamt Medenblick en de Vier Noorder Coggen*, Map, 1730, Zuiderzeemuseum, Enkhuizen, http://hdl.handle.net/21.12111/zzm-collect-12589.

breakers, the earthen bodies of dikes stood exposed to the erosive force of the sea.

By the time De Leth's advertisement appeared in winter 1732, the "shipworm epidemic" had grown into an existential disaster that affected multiple provinces, and many argued the Netherlands as a whole. Newspapers across the Republic printed exposés on the outbreak, noting its extent and what was then known (and not known) about the animal. Moralists published pamphlets that probed its providential meaning, and scholars produced the first scientific treatises on their natural history. Most people had no prior experience with shipworms, and no similar outbreak had ever occurred in Dutch waters. De Leth's map, like much of the literature that emerged during the shipworm panic, responded to the lack of information about the threat.

The map also emphasized a second, more troubling set of concerns about the infestation. Although localized dike breaches occurred with some frequency, shipworms presented a uniquely comprehensive threat to Dutch safety and prosperity. The Christmas Flood had revealed the catastrophic consequences of dike failure on a provincial scale, but what would happen if sea dikes failed across the Republic? Observers predicted "total ruin" of the maritime economy, inundation of its great cities, and the devastation of its coastal communities should adequate solutions fail to materialize.[2] An image inset into the bottom-right corner of De Leth's map underscores this point. It contrasted effective water management with the threat of flooding (Figure 4.2). In the foreground, personifications of productivity and wisdom sit contentedly, protected by an earthen dike covered with grass and shrubs. The dike also shields Dutch prosperity, symbolized by a cow, agricultural tools, and a horn of plenty. Beehives represent industriousness, and a rooster indicates vigilance – both critical virtues for water managers. Surrounding the scene, classical figures representing wind and water temper this otherwise confident message of Dutch mastery of nature. Waves, propelled by Aeolus and Neptune, burst through a broken dike protected by wood palisades.[3] Shipworms make no appearance on the map despite De Leth's description of the region as "most heavily ravaged by the plague of worms."[4] De Leth likely first printed the map around 1730 before the public recognized its extent and severity. He advertised the map again two years later because he understood how many viewers would interpret the threat. Shipworms embodied the flood hazard.

Framing shipworms as a flooding hazard proved a powerful response to the disaster because it tapped into the Dutch coastal culture of flooding. Like cattle plague and the Christmas Flood before it, observers understood that shipworms were divine retribution for sin. This interpretation appeared in public media, newspapers and sermons, poetry, enlightened "spectatorial" essays, and even water management reports. The dramatic manifestation of monstrous "worms" reinforced the conviction that the Dutch Republic suffered moral decay. Dutch sin seemed perfectly embodied in the worms themselves, which literally devoured the economic and moral foundations of the Republic. Unlike earlier disasters, however, moralists causally tied shipworms to a single sin – sodomy – and to a wave of trials and executions taking place across the Republic.

---

[2] J. Bos, F. R. C. Burghardt, J. K. H. van der Meer, and F. Timmerman, eds., *Handschrift Schoemaker: een achttiende-eeuwse kijk op de Drentse geschiedenis* (Assen: Van Gorcum, 2004), 95.

[3] Marc Hameleers, *West-Friesland in oude kaarten* (Wormerveer: Stichting Uitgeverij Noord-Holland, 1987), 93.

[4] Advertisement. December 22 (1732). *'s Gravenhagse Courant*.

FIGURE 4.2. Detail from Straat, *Nieuwe kaarte*. Decorative wall maps often included allegorical symbols that alluded to the power and responsibility of water boards, as well as potential consequences of failure. The dike in this image protects symbols of prosperity from untamed elements. *Source*: Zuiderzeemuseum, Enkhuizen, http://hdl.handle.net/21.12111/zzm-collect-12589.

Shipworms may have symbolized flooding, but they were not floods. The unique nature of this threat presented new interpretive possibilities and demanded new responses.

The novelty of the shipworm epidemic produced an unprecedented social and technological response in the Dutch Republic. Prior to the 1730s, shipworms remained unexamined and largely unknown animals. Their obscurity invited diverse interpretations of its origins, vulnerabilities, and ultimate meaning, all of which affected Dutch response. The connections to sodomy persecutions were the most unique displays of a much broader social reaction to this novelty. If previous disasters had amplified the quiet anxieties of Dutch communities suffering from economic stagnation and moral uncertainty, shipworms created a panic. According to critics, sodomy, like the shipworms themselves, revealed a society in decline.

At the same time, the existential threat of dike failure prompted a wave of scientific and technological responses. The novelty of the threat precluded any reliable solutions drawn from the cultural memory of previous disasters. The outbreak thus sparked a systematic examination of shipworm biology, their environmental tolerances and origins, and innovative strategies to combat them. Provincial authorities in Holland solicited advice from across Europe and tested a suite of novel solutions to the threat. Their costly decision to rebuild entire stretches of coastal infrastructure reflected years of trial and error, new knowledge sparked by the infestation, and the fear of total inundation and decline should their plans fail. Shipworm novelty motivated response and action during each stage of the epidemic. At no point was this more apparent than the shipworms' initial appearance following an otherwise unremarkable storm in 1730.

4.1 THE (RE)DISCOVERY OF SHIPWORMS

The North Sea storm that sparked the shipworm epidemic was in no way extraordinary. When it hit the island of Walcheren in the province of Zeeland in the late fall of 1730, it caused no major damage and little to no loss of life or property. The dike that protected the westernmost edge of the island, called the Westkapelle sea dike, had weathered far worse storm events since its construction in the early fifteenth century.[5] It was typical of many coastal dikes in Zeeland in the eighteenth century. Little if any

---

[5] Zeeland suffered two serious flood disasters in living memory (1682 and 1715), both fueled by heavy gales from the northwest and a high spring tide. Numerous additional

land stood between the sea and large sections of the dike, so Zeelanders employed large wooden piles as wave-breaking wiers (*hoofden*) or wood and stone buttresses (*staketwerken*) and used straw, seaweed, and brushwood cushions (*krammaten*) to shield the dikes' earthen body.[6] These constructions were expensive to build and maintain, but they protected the dike and ordinarily withstood the force of minor storms.

A dike inspection of the Westkapelle sea dike on November 23 reported no significant damage to its earthen body. Inspectors noted with surprise, however, that a large number of wooden piles lay strewn across the beach. Many had been recently installed and they were broken in the middle, not at the base, as was typical following a storm. Ice was not responsible. During cold weather, sea ice sometimes shattered piles or loosened *staketwerken*, but a warm spell had descended upon the Netherlands the previous winter and had remained through fall 1730.[7] In his report to the local governing council, dike official Eduardus Reynvaan described this unusual situation and offered an even stranger explanation. "The piles," he noted, "are full of worms."[8] Within two weeks, inspections across the island revealed that worms had spread to "most of the piles and brushwood mats around the island."[9] By December, the clerk of the Estates of Walcheren characterized the outbreak as a disaster. "It cannot be seen as anything" other than an event that is of "the utmost consequence, if not total ruin of the island."[10] News of this alarming development in Zeeland spread to other parts of the Republic, and, by fall 1731, dike inspectors in the provinces of Holland and Friesland too found evidence of worm infestation (Figure 4.3).

Although Dutch reports referred to the species as a "worm," it was in fact a mollusk today commonly known as the naval shipworm (*T. navalis*), a marine bivalve of the family Teredinidae. Its outward appearance

---

storm events dating back to the early fifteenth century were part of the collective memory. De Kraker, "Two Floods Compared," 186–7, 198.

[6] Leo Hollestelle, "De zorg voor de zeewering van Walcheren ten tijde van de Republiek,1574–1795," in *Duizend jaar Walcheren: Over gelanden, heren en geschot, over binnen- en buiten beheer*, ed. A. Beenhakker (Middelburg: Koninklijk Zeeuwsch Genoots, 1996), 106.

[7] Buisman, *Duizend jaar*, Vol. 5, 573–6.

[8] November 23 (1730). ZA. Notulen van de Staten en Gecommitteerden van de Breede Geërfden, 1511–1812. 3000.20.

[9] Ibid. "Bericht wegens de plaage der wormen in het paalwerk der dykagien van Holland en Zeeland," *Europische Mercurius* (1732), 297.

[10] December 28 (1730). ZA. Notulen van de Staten en Gecommitteerden van de Breede Geërfden, 1511–1812. 3000.20.

FIGURE 4.3. Shipworm infestations in the Netherlands (1730–2). Early infestations clustered in highly saline waters near major ports for the WIC, VOC, and Admiralty. Boundaries modified from Boonstra, NLGis shapefiles.

resembles a worm, save for a small anterior helmetlike shell that it uses to burrow into wood.[11] Adult shipworms have high temperature and salinity tolerances, can produce up to two million larvae per cycle, and, depending on environmental conditions, live for several years. Their fecundity and resilience mean, in ideal conditions, *T. navalis* multiplies at astonishingly high rates. In their larval stage, they can travel hundreds of kilometers in water currents, and far greater distances when they bore into the wooden hulls of ships. These characteristics also make *T. navalis*

---

[11] Linneaus classified *T. navalis* in the tenth edition of his *Systema naturae* in 1758. Carl Linné, *Systema naturae, per regna tria naturae: secundum classes, ordines, genera, species cum characteribus, differentiis, synonymis, locis* (Holmiae: Salvius, 1758).

a successful colonizing species, capable of expanding their range quickly.[12] They range from sizes of up to 25 cm in cooler waters to 60 cm in warmer waters and spend almost their entire lives in submerged wood.[13] Eighteenth-century observers were often surprised that one piece of wood housed multiple mollusks. Cross sections of wooden piles sketched by contemporaries emphasized the obvious damage multiple mollusks could inflict, yet even large shipworms left holes barely visible from the outside. Shipworms were often only discovered when the wood broke (Figure 4.4).

Most observers professed little awareness of shipworms in the 1730s, but Europeans (including the Dutch) in fact had a long, shared history with marine woodborers, including *T. navalis*. The earliest European accounts of shipworms appear in antiquity, but the number and frequency of reports dramatically increased during the early-modern period as maritime trade expanded to the tropics. Indeed, European colonialism and trade likely provided a windfall opportunity for their distribution.[14] Wooden ships transported shipworms across oceans, and their high ecological tolerances and fecundity increased the likelihood of successful introduction. The expansion of coastal infrastructure undoubtedly accelerated this biotic transfer and opened new beachheads for invasion. Wooden groynes, sluice gates, and dike revetments presented perfect ecosystems for shipworms to flourish in foreign environments.[15] Maritime trade resulted in entire marine communities transported across

---

[12] Luísa M. S. Borges, Lucas M. Merckelbach, Íris Sampaio, and Simon M. Cragg, "Diversity, Environmental Requirements, and Biogeography of Bivalve Wood-Borers (Teredinidae) in European Coastal Waters," *Frontiers in Zoology* 11, no. 1 (2014): 7–8.

[13] K. N. Hoppe, "Teredo Navalis – The Cryptogenic Shipworm," in *Invasive Aquatic Species of Europe; Distribution, Impacts, and Management*, eds. Erkki Leppäkoski, Stephan Gollasch, and Sergej Olenin (Dordrecht: Springer, 2002), 116–19. Shipworms can tolerate temperature and salinity gradients of between 0–30°C (with optimum temperatures between 15–25°C and 7–39 PSU, respectively. N. B. Nair and M. Saraswathy, "The Biology of Wood-Boring Teredinid Molluscs," *Advances in Marine Biology* 9 (1971): 447–56.

[14] John Scarborough, *Medical and Biological Terminologies: Classical Origins* (Norman: University of Oklahoma Press, 1992), 65. Courtney A. Rayes, James Beattie, and Ian C. Duggan, "Boring through History: An Environmental History of the Eextent, Impact and Management of Marine Woodborers in a Global and Local Context, 500 BCE to 1930s CE," *Environment and History* 21, no. 4 (2015): 483–6. James T. Carlton, "The Scale and Ecological Consequences of Biological Invasions in the World's Oceans," in *Invasive Species and Biodiversity Management*, eds. Odd Terje Sandlund, Peter Johan Schei, and Åslaug Viken (Dordrecht: Kluwer Academic Publishers, 1999), 195–212.

[15] Hoppe, "Teredo Navalis," 117. James T. Carlton, "Molluscan Invasions in Marine and Estuarine Communities," *Malacologia* 41, no. 2 (1999): 439–54.

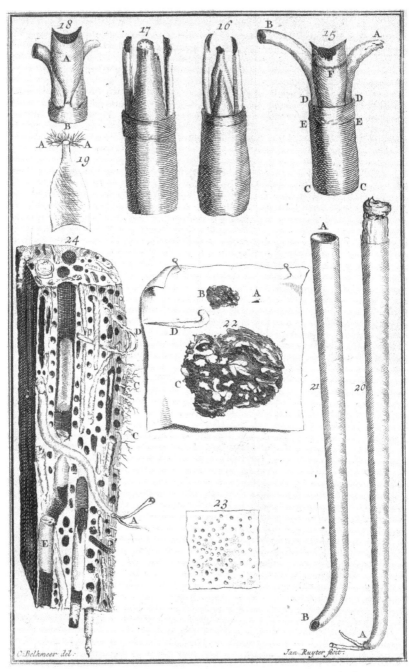

FIGURE 4.4. Image number 24 depicts a piece of wood riddled with shipworms and number 23 indicates how inconspicuous shipworm infestation appears from the outside. *Source*: Jan Ruyter, *Studie van wormen*, Print, 1726–44, Rijksmuseum, Amsterdam, http://hdl.handle.net/10934/RM0001.COLLECT.342890.

the globe and ensured their establishment upon reaching distant ports. Today, *T. navalis* is among the most widely dispersed marine species, and this cosmopolitanism was likely an historical consequence of oceangoing travel and trade.[16]

Mariners' journals, ship logs, and maintenance reports described these stowaways, particularly when they resulted in costly damage or delay. Christopher Columbus famously lost two vessels to shipworms on his fourth voyage to the West Indies, and European shipbuilders began protecting ships' hulls from shipworms in the early sixteenth century.[17] In 1590, the Dutch merchant Jan Huygen van Linschoten reported in his famous *Itinerario* that the Portuguese were forbidden from travelling to Brazil due to the danger of the "worm that damages the ships."[18] Seventeenth-century Dutch sailors returning from the West Indies attributed the loss of their vessels to worm attacks in the Caribbean.[19] The seventeenth-century naval officer Witte de With noted shipworm damage on the coast of Africa and later in Brazil. Numerous VOC reports indicated their presence in East Indies. Shipworms, in other words, were well distributed and consequential. Indeed, *T. navalis* was one of the principal hazards for oceanic travel throughout the early-modern period.[20]

---

[16] Cosmopolitan refers to species with broad geographic distribution, a term that has recently been challenged because it obscures anthropogenic influences on biogeography, particularly before the modern era. John A. Darling and James T. Carlton, "A Framework for Understanding Marine Cosmopolitanism in the Anthropocene," *Frontiers in Marine Science* 5 (2018): 293.

[17] James T. Carlton and J. Hodder, "Biogeography and Dispersal of Coastal Marine Organisms: Experimental Studies on a Replica of a 16th-Century Sailing Vessel," *Marine Biology* 121, no. 4 (1995): 722. Dutch East India Trading Company records reveal shipworm adaptations, including multiple methods of hull sheathing. Wendy van Duivenvoorde, *Dutch East India Company Shipbuilding: The Archaeological Study of Batavia and Other Seventeenth-Century VOC Ships* (College Station: Texas A&M University Press, 2015). Woodshole Oceanographic Institution, "Marine Fouling and Its Prevention," Prepared for the Bureau of Ships, Navy Department (Annapolis, MD: U.S. Naval Institute, 1952), 211–23.

[18] J. H. Linschoten, *Itinerario: voyage ofte schipvaert van Jan Huygen van Linschoten naer d'ost ofte Portugaels Indien* (Amsterdam: Cornelis Claesz, 1596), 154.

[19] Joannes Goudsbloem, March 17, 1644; December 14, 1645; January 1, 1645; December 31, 1644. SA, Archief van de Notarissen ter Standplaats Amsterdam. 5075.1925A. Notarial accounts of West Indies freighters also noted the effect of shipworms in "English Virginia." Notary Pieter Capoen (September 15, 1645) SA, "Minuten Van Bevrachtingscontracten," Archief van de Notarissen ter Standplaats Amsterdam. 5075.1587. Many thanks to Nicholas Cunigan for these documents.

[20] Derek Lee Nelson and Adam Sundberg, "Shipworm Ecology in the Age of Sail," in *Maritime Animals: Ships, Species, Stories* (forthcoming). Anne Doedens, *Witte de With 1599–1658: Wereldwijde strijd op zee in de Gouden Eeuw* (Hilversum: Uitgeverij

Shipworms appeared in Western Europe as well. They bored into the hulls of the Spanish Armada as it prepared in Spanish and Portuguese ports to invade England in 1588. Evidence of infestations in dikes in East Friesland (now Northwest Germany) appears in the seventeenth century. Dutch newspapers reported shipworm damage to ships returning from the East Indies as early as 1673. In the 1680s and '90s, shipworms damaged the French military harbor at Toulon. Dike inspections in North Holland indicated the presence of a woodboring species as early as 1680. Later reports indicated limited damage to dikes in Zeeland in 1706 and the Southern Sea in 1721 and 1725.[21] Shipworms presented costly challenges for early-modern commerce, warfare, and at times water management, yet despite their significance, they appeared only sporadically in the historical record. They were an unavoidable but manageable cost of doing business and living near the sea. Even among the relatively few with direct experience of shipworms, none had experienced a European outbreak before.

The shipworm epidemic of the 1730s forced early-modern observers to reckon with the geographic origins of *T. navalis*, and that debate continues today. Scientists consider it as a classic cryptogenic species.[22] Scholars are divided between those who favor Atlantic indigeneity and

Verloren, 2008), 23, 80. Robert Parthesius, Dutch Ships in Tropical Waters: The Development of the Dutch East India Company (VOC) Shipping Network in Asia 1595–1660 (Amsterdam: Amsterdam University Press, 2010), 102.

[21] Hoppe, "Teredo Navalis," 117. Ernst Siebert, "Entwicklung des Deichwesens von Mittelalter bis zur Gegenwart," in *Ostfriesland im Schutze des Deiches: Beiträge zur Kultur-und Wirtschaftsgeschichte des Ostfriesischen Küstenlandes*, Vol IV, ed. Jannes Ohling (Leer: Gerhard Rautenberg, 1969), 313. For newspaper accounts, see *Amsterdamse Courant*, February 11, 1673; *Oprechte Haerlemsche Courant*, May 7, 1675; *Amsterdamse Courant*, July 31, 1698. The infestation in France forced engineers to modify local hydrology to increase fresh water into the harbor. Raphael Morera, email to author, June 25, 2015. Theodorus Speeleveldt, *Brieven over het eiland Walcheren* (The Hague: Immerzeel en comp., 1808), 65. H. Schoorl, *Zeshonderd jaar water en land: bijdrage tot de historische geo- en hydrografie van de Kop van Noord-Holland in de periode ± 1150–1750* (Groningen: Wolters-Noordhoff, 1973). The Dutch historian P. C. Hooft recorded damage to dikes in 1580. The *Journal des Sçavans* reported damage in 1665, and J. F. Martinet recorded damage to herring buses in 1714 and 1720. J. F. Martinet, *Katechismus der natuur: met plaaten*, Vol. 2 (Amsterdam: Allart, 1778). José Mouthaan, "The Appearance of a Strange Kind of Sea Worm at the Dutch Coast, 1731," *Dutch Crossing* 27, no. 1 (2003): 4. Michiel Bartels, "Het bolwerk tegen de woede van de zee," in *Dwars door de dijk: Archeologie en geschiedenis van de Westfriese Omringdijk tussen Hoorn en Enkhuizen*, ed. Michiel Bartels (Hoorn: Stichting Archeologie West Friesland, 2016), 146–8.

[22] Borges et al., "Diversity, Environmental Requirements," 13. James T. Carlton, "Biological Invasions and Cryptogenic Species," *Ecology* 77 (1996): 1653–5.

those who consider it an invasive brought to European shores via voyages to the tropics.[23] The most recent genetic barcoding analyses suggest it was introduced to the North Sea via the Dutch Republic, but these findings are not conclusive.[24] Many favoring Atlantic origins point to earlier evidence of woodborers in European waters and question why a sudden outbreak would appear after centuries of prior exposure. *T. navalis* was not the only woodborer in European waters, however, and these earlier sightings may have referred to other species such as *Psiloteredo megotara* or *Teredo norvagica*.[25] Introductions also frequently experience lags between their initial arrival and later expansion and rapidly changing habitats are common triggers for exponential growth.[26] The explosive character of the outbreak suggests introduction.

The location of infestations also hints at exotic origins. Invasion ecologists have identified increasing trade as among the most significant predictors of species introductions.[27] Between the late seventeenth and first half of the eighteenth century, ship arrivals from the tropics increased[28]

---

[23] The most recent historical work claiming Atlantic indigeneity is Albert van Brakel, "De paalworm in Hollandse zeedijken," *Tijdschrift voor Waterstaatsgeschiedenis* 24, no. 2 (2015): 70–81. For work asserting exotic origins, see Stephan Gollasch, Deniz Haydar, Dan Minchin et al., "Introduced Aquatic Species of the North Sea Coasts and Adjacent Brackish Waters," in *Biological Invasions in Marine Ecosystems: Ecological, Management, and Geographic Perspectives*, eds. Gil Rilov and Jeffrey A. Crooks (Berlin, Heidelberg: Springer Berlin Heidelberg, 2009), 507–28.

[24] Ronny Weigelt, Heike Lippert, Ulf Karsten, and Ralf Bastrop, "Genetic Population Structure and Demographic History of the Widespread Common Shipworm *Teredo navalis* Linnaeus 1758 (Mollusca: Bivalvia: Teredinidae) in European Waters Inferred from Mitochondrial COI Sequence Data," *Frontiers in Marine Science* 4 (2017): 9–10.

[25] Wim J. Wolff, "Non-Indigenous Marine and Estuarine Species in the Netherlands," *Zoologische Mededelingen Leiden* 79, no. 1 (2005): 85–6.

[26] Jeffrey A. Crooks, "Lag Times and Exotic Species: The Ecology and Management of Biological Invasions in Slow-Motion," *Ecoscience* 12, no. 3 (2005): 316–29. Charles Howard Edmondson, "Teredinidae, Ocean Travelers," *Occasional Papers of Bernice P. Bishop Museum* 23, no. 3 (1962): 45. N. Balakrishnan Nair, "Ecology of Marine Fouling and Wood-Boring Organisms of Western Norway," *Sarsia* 8, no. 1 (1962): 1–88.

[27] B. S. Galil, Agnese Marchini, Anna Occhipinti-Ambrogi et al., "International Arrivals: Widespread Bioinvasions in European Seas," *Ethology Ecology & Evolution* 26, no. 2–3 (2014): 152–71. Petr Pyšek, Vojtěch Jarošík, Philip E. Hulme et al., "Disentangling the Role of Environmental and Human Pressures on Biological Invasions across Europe," *Proceedings of the National Academy of Sciences* 107, no. 27 (2010): 12157–62.

[28] Atlantic data is more fragmentary than VOC arrivals, but the total number of ships arrived from the Western hemisphere increased in the decades leading up to the 1730s. Victor Enthoven, "An Assessment of Dutch Transatlantic Commerce, 1585–1817," in *Riches from Atlantic Commerce: Dutch Transatlantic Trade and Shipping, 1585–1817*, eds. Johannes Postma and Victor Enthoven (Leiden: Brill, 2003), 406. At least one contemporary account noted this correlation with increasing shipping traffic. In a 1732 letter written in response to the request

(Figure 4.5). It was likely no coincidence that shipworm infestations clustered near major ports that welcomed ships returning from the tropics, including the island Texel in North Holland where East Indiamen frequently docked to stock up on freshwater (Figure 4.3). Rotterdam and Delfshaven (which serviced Delft) were the exceptions, but they sat far enough up the Rhine–Meuse delta that salinity was likely low enough to withstand substantial shipworm activity.[29] With every new ship arrival, the likelihood of introduction and outbreak of marine woodboring species grew.

Shipping traffic from the tropics increased the likelihood of introductions, but the conjuncture of long- and short-term environmental changes laid the foundation for disaster. *T. navalis* populations likely expanded during the 1730s due in part to rising temperatures. Average temperatures in the Dutch Republic in all seasons increased beginning in the first decades of the eighteenth century, signaling the end of the Maunder Minimum. This relatively warm intermezzo would last until the Dalton Minimum (1790–1830), the final great cold period of the Little Ice Age.[30] Temperatures in shallow Dutch coastal waters are highly influenced by the surrounding air, resulting in significant seasonal impacts on its ecosystems.[31] The years around

---

for shipworm remedies, H. van der Dussen noted the possibility that shipworms were "brought from the East Indies and the West Indies, which for the past several years have received many more ships, in particular from Suriname." "H. van der Dussen," September 11 (1732). NA. Gedeputeerden van Haarlem ter Dagvaart. 3.01.09. 1238.

[29] Adam Sundberg, "Molluscan Explosion: The Dutch Shipworm Epidemic of the 1730s," Environment & Society Portal, *Arcadia* (2015): 14. Rachel Carson Center for Environment and Society. https://doi.org/10.5282/rcc/7307.

[30] Stefan Brönnimann, Sam White, and Victoria Slonosky, "Climate from 1800 to 1970 in North America and Europe," in *The Palgrave Handbook of Climate History*, eds. Sam White, Christian Pfister, and Franz Mauelshagen (London: Palgrave Macmillan, 2018), 312. Buisman, *Duizend jaar*, Vol. 5, 929–30. KNMI, "Monthly, Seasonal, and Annual Means of the Air Temperature," http://climexp.knmi.nl/data/ilabrijn.dat. Luterbacher et al., "European Seasonal and Annual Temperature Variability," 1500–2. H. M. Van den Dool, H. J. Krijnen, and C. J. E. Schuurmans, "Average Winter Temperatures at De Bilt (The Netherlands): 1634–1977," *Climatic Change* 1, no. 4 (1978): 326–7.

[31] Hydrographic changes in the North Atlantic are also significant. C. G. N. de Voys, "Expected Biological Effects of Long-Term Changes in Temperatures on Benthic Ecosystems in Coastal Waters around the Netherlands," in *Expected Effects of Climatic Change on Marine Coastal Ecosystems*, eds. J. J. Beukema, W. J. Wolff, and J. J. W. M. Brouns (Dordrecht: Kluwer Academic, 1990), 78–80. Hans J. Lindeboom, "Changes in Coastal Zone Ecosytems," in *Climate Development and History of the North Atlantic Realm*, eds. Gerold Wefer, Wolfgang H. Berger, Karl-Ernst Behre, and Eystein Jansen (Berlin: Springer, 2002), 448–9. Shipworms' boring activity concentrates in warmer waters at or just below the tidal zone. Roberto Lopez-Anido, Antonis P. Michael, Barry Goodell, and Thomas C. Sandford, "Assessment of Wood Pile Deterioration due to Marine Organisms," *Journal of Waterway, Port, Coastal, and Ocean Engineering* 130, no. 2 (2004): 72–3.

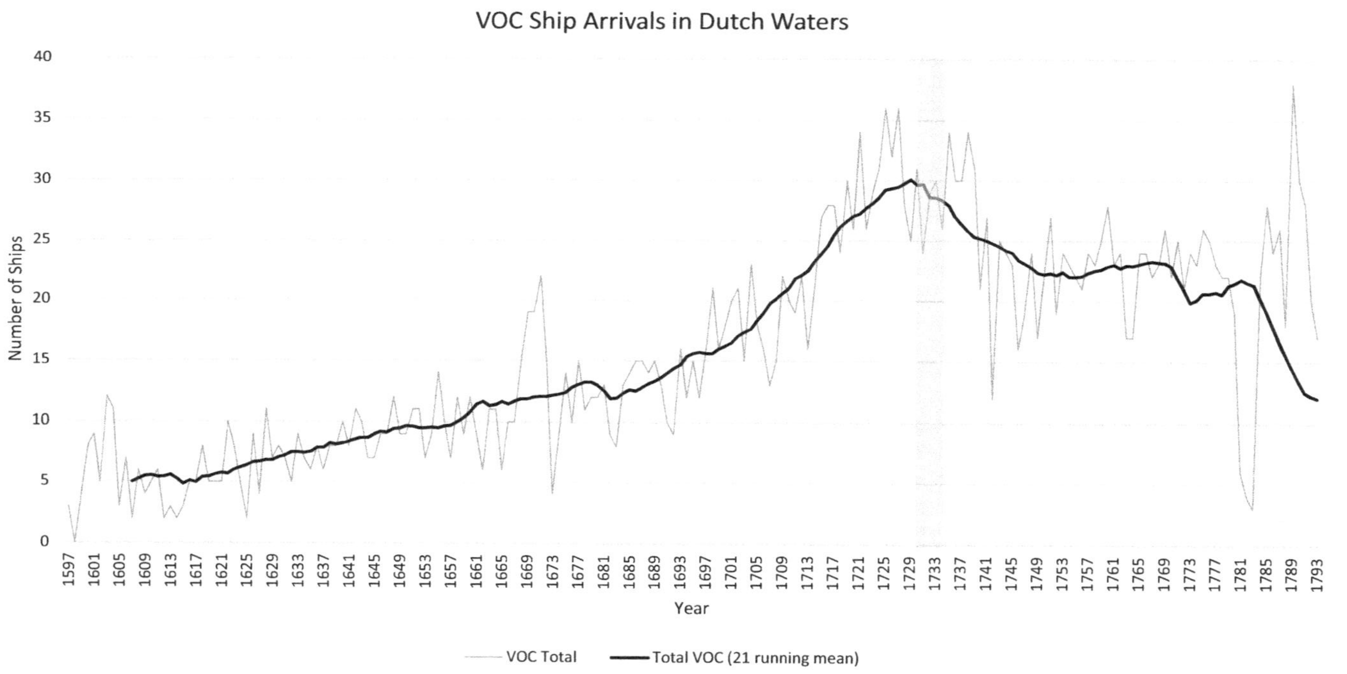

FIGURE 4.5. Total VOC ship arrivals in Dutch waters. Includes twenty-one-year running mean. Vertical grey bar indicates initial period of shipworm discovery. *Source:* J. R. Bruijn, F. S. Gaastra, and I. Schöffer, *The Dutch East India Company's shipping between the Netherlands and Asia 1595–1795*, Huygens Instituut voor Nederlandse Geschiedenis, http://resources.knaw.nl/das. Accessed July 2018.

1730 were unusually warm, including in the spring and summer, the seasons most significant for shipworm growth and recruitment[32] (Figure 4.6).

Salinity was just as important as temperature for shipworm growth and distribution. A combination of long-term changes to the coastline and river mouths and shorter-term changes in temperature and precipitation in the 1730s created ideal ecological conditions for an outbreak. Between the fifteenth and the eighteenth centuries, inlets between Wadden Sea islands eroded, pushing the highly saline waters of the North Sea into the Southern Sea. The silting up of the mouth of the IJssel, a distributary of the Rhine, amplified these effects in the Southern Sea by reducing discharge of freshwater. The higher evaporation rate and lower water levels in the IJssel during the months of greatest shipworm activity (March to September) contributed to these conditions as well.[33] These increases in salinity would have aided shipworm growth and recruitment.

Finally, a series of droughts between 1729 and 1737 likely affected water levels in the Rhine and Meuse, triggering the shipworm explosion. Dry summers reduced discharge out of the rain-fed Meuse. Rhine water levels, by contrast, depend on precipitation and alpine snowmelt.[34] Dry winters in the alps or dry summers in the lower basins of the Rhine, thus resulted in lower discharge. Winter precipitation across Europe generally decreased between 1720 and 1740, although extreme droughts were rare in the Alps. Spring and summer rainfall in the river basins, however, experienced high decadal variability across the first half of the eighteenth century, and reconstructed drought indices revealed repeated dry periods between 1727 and 1740, with a peak in 1731.[35] Dutch precipitation data from this era is scarce, but one short

---

[32] P. D. Jones and K. R. Briffa, "Unusual Climate in Northwest Europe during the Period 1730 to 1745 Based on Instrumental and Documentary Data," *Climatic Change* 79, no. 3/4 (2006): 364. Christin Appelqvist and Jonathan N. Havenhand, "A Phenological Shift in the Time of Recruitment of the Shipworm, *Teredo navalis* L., Mirrors Marine Climate Change," *Ecology and Evolution* 6, no. 12 (2016): 3862–70.

[33] C. Baars, "De paalwormfurie van 1731–1732 en de schade aan de West-Fries zeedijk," *Waterschapsbelangen* 73 (1988): 810. Van Brakel, "De paalworm," 78–9.

[34] Van de Ven, *Man-made Lowlands*, 24–5.

[35] Andreas Pauling, Jürg Luterbacher, Carlo Casty, and Heinz Wanner, "Five Hundred Years of Gridded High-resolution Precipitation Reconstructions over Europe and the Connection to Large-Scale Circulation," *Climate Dynamics* 26, no. 4 (2006): 393–4. Robert J. S. Wilson, Brian H. Luckman, and Jan Esper, "A 500 Year Dendroclimatic Reconstruction of Spring–Summer Precipitation from the Lower Bavarian Forest Region, Germany," *International Journal of Climatology* 25, no. 5 (2005): 611–30. Casty et al., "Temperature and Precipitation Variability," 1859–1863. Old World Drought Atlas. Accessed July 23, 2020. http://drought.memphis.edu/OWDA/. Emmanuel Garnier, "Historic Drought from Archives: Beyond the Instrumental Record," in Drought: Science and Policy, eds. Ana Iglesias, Dionysis Assimacopoulos, and Henny A. J. van Lanen (Hoboken, NJ: Wiley, 2019), 45–67. The first half of the eighteenth century also featured fewer floods. Günter Blöschl, Andrea Kiss, Alberto Viglione et al., "Current European Flood-Rich Period Exceptional Compared with Past 500 Years," *Nature* 583, no. 7817 (2020): 560–6.

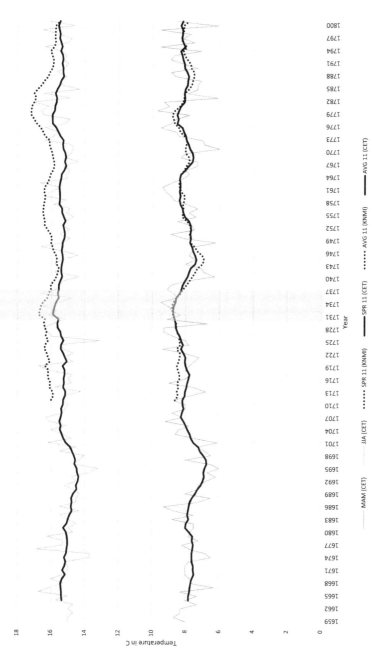

FIGURE 4.6. Spring (MAM) and summer (JJA) temperatures in central England (CET) between 1659 and 1799, along with eleven-year moving averages for spring (SPR 11) and summer (AVG 11). Similar averages based on documentary evidence extend back to 1706 in the Netherlands (KNMI). Grey bar indicates period of drought. *Sources:* MET Office, "Seasonal Central England Temperature, 1659–2020," www.metoffice.gov.uk/hadobs/hadcet/ssn_HadCET_mean.txt. KNMI, "Monthly, Seasonal, and Annual Means of the Air Temperature," http://climexp.knmi.nl/data/ilabrijn.dat. Accessed July 21, 2020.

series from Utrecht confirms this trend. Total precipitation between May and September 1731 amounted to only 176 mm, significantly less than the summer average of 300 mm. With the exception of a wet summer in 1734, conditions would remain dry until 1737.[36] This combination of long- and short-term changes in temperature, precipitation, and the resulting salinity of coastal waters laid the ecological groundwork to transform shipworms from a commercial nuisance into an existential threat.

4.2 FINDING THE LIMITS OF INFESTATION AND PRIOR EXPERIENCE

The discovery of shipworms in 1730 did not spark an immediate public panic. More than a year and a half elapsed before the outbreak appeared in newspapers or pamphlets. During these early months, discussion remained confined to Zeeland's water boards and provincial committees. They exchanged letters reporting on the comprehensive nature of the infestation and the animal's behavior. They noted with consternation that no type of wood seemed resistant and worried that all submerged wood, and any lands behind them, were vulnerable. "The fear is that the worms will cause inconceivable destruction because they are in every sort of wood, whether pine, oak, birch, willow, or alder," one report noted.[37] The shipworm threat grew month after month. By January 1731, inspectors discovered mollusks over a foot long and identified infestations in nearly every section of Walcheren's dikes. "Walcheren had never before experienced destruction like this," they concluded.[38]

The Estates of Zeeland responded in a deliberate manner. Shipworms presented a unique challenge, but not one that necessarily precluded time-tested responses. They ordered a systematic investigation of possible solutions drawing on strategies developed for ships.[39] On January 18, water boards installed new piles "in those areas most affected by the

---

[36] A. Labrijn, "Het klimaat van Nederland gedurende de laatste twee en een halve eeuw," *Mededeelingen en Verhandelingen* 102 (De Bilt: KNMI, 1945), 86, 101.
[37] Report. December 14 (1730). ZA. Notulen van de Staten en Gecommitteerden van de Breede Geërfden, 1511–1812. 3000.20.
[38] Ibid. Twelve *duim* are roughly 31 cm or 12.2 in. Verhoeff, *Maten en gewichten*.
[39] Adam Sundberg, "Dikes, Ships, and Worms: Testing the Limits of Envirotechnical Transfer during the Dutch Shipworm Epidemic of the 1730s," in *Disaster in the Early Modern World: Examinations, Representations, Interventions* (forthcoming).

worms to see after a period of a few months which measures work the best."[40] Test piles were lightly burned and covered in harpuis, a boiled pine resin that prevented the growth of algae and shellfish on ship hulls. They also tried substituting wood varieties. In subsequent months, they tested other maritime strategies, including coating wood in pitch and tar. The engineer Adriaan Bommenee suggested applying quicklime. A timber merchant from Aalst (in what is now Belgium) suggested the water board purchase supposedly worm-resistant Flemish oaks.[41] Nothing seemed to work. By the start of the storm season in fall 1731, the "calamitous state of the dikes" on Walcheren was impossible to ignore.[42] No solution appeared acceptable, and by winter the outbreak spread beyond Zeeland and assumed a much more dramatic character.

The second outbreak of shipworms appeared in 1731 following another small storm. Drechterland, a water board in the West Frisian region of northern Holland, began receiving reports that worms had infested a large section of their northern dike (called the *Drechterlandse noorderdijk*). An inspection in October revealed that shipworms had colonized almost 95 percent of dikes.[43] A second investigation from the neighboring water board *Vier Noorder Koggen* reported that their dikes had also been "gnawed" by a "certain unknown sort of Worm."[44] The dikes held fast during a December storm, but water boards saw the writing on the wall. "It is truly to be feared," they noted, "that in the case of a persistent storm, everything could be washed away if not prevented."[45]

---

[40] Report. January 18 (1731). ZA. Notulen van de Staten en Gecommitteerden van de Breede Geërfden, 1511–1812. 3000.20.
[41] Reports. January 18, February 1, March 15, May 17, June 7, June 28, July 5, October 25, November 29 (1731). ZA. Notulen van de Staten en Gecommitteerden van de Breede Geërfden, 1511–1812. 3000.20.
[42] Ibid., Report. October 25 (1731).
[43] Cornelis Alard Abbing, *Geschiedenis der stad Hoorn, hoofdstad van West-Vriesland gedurende het grootste gedeelte der 17e en 18e Eeuw, of vervolg op Velius Chronyk*, Vol. 2 (Hoorn: Gebr. Vermande, 1842), 243. "Rumors" had spread from Texel and Den Helder that their own piles had been broken and "gnawed through." This may have been the final piece of evidence needed to push for the special inspection.
[44] *Beschryvinge, van de schade en raseringe aan de zee-dyken van Noort-Holland en West-friesland, door de worm in de palen, en de daar op gevolgde storm, en vervolgens* (Hoorn: Jacob Duyn, 1732), 9.
[45] Ibid., 11.

Like in Zeeland, water boards in Holland treated the shipworm outbreak as a flooding hazard. Officials noted the location and severity of the outbreak and estimated the cost of repairs.[46] Shipworms were not storms, however, and dike officials quickly recognized that dike design and location determined shipworm impact. They only appeared in dikes with wooden infrastructure and could not survive in freshwater or areas that remained dry during low tide. Even so, significant coastal stretches lay exposed to infestation, especially in West Friesland.

Dike construction in North Holland differed significantly from the southwestern Delta region of Zeeland, but both regions employed wooden barriers. Holland's water boards employed wooden palisades (*paalschermen*) or installed wooden frames that supported compacted eelgrass cushions and stone (*wierriemen*) to shield their dikes from the erosive impact of waves[47] (Figures 4.7 and 4.8). These wier dikes (*wierdijken*) were costly to build and maintain. Their oak and pine timbers had to be imported at significant cost, and the eelgrass cushions in wier dikes required constant maintenance and complete replacement every thirty years.[48] Dikes with significant stretches of land on their seaward side (*voorland*) did not need these barriers and were unaffected by shipworms.

Unfortunately, centuries of environmental change in West Friesland had necessitated greater and greater reliance upon them. Drainage-induced subsidence behind the dikes lowered the level of land relative to the sea, increasing the severity of inundations if dikes breached. At the same time, the natural expansion of Wadden Sea tidal inlets elevated tides and storm surges that eroded *voorland*, reducing this critical buffer zone.[49] The combination of these long-term natural and cultural changes increased coastal vulnerability and incentivized communities to rely on wier dikes. The earliest evidence of wier dikes appears in the fourteenth century, but their use increased by the seventeenth century, as the rate and

---

[46] "Rapportage van de dijkbesturen van de Westfriese ambachten," *WFA*. Ambacht van West-Friesland genaamd Drechterland en hoofdingelanden van West-Friesland (Drechterland). 1562.1442.

[47] Bartels, "Het bolwerk," 134–43.

[48] C. P. Schrickx, "Een kommerlijkste toestand en groot gevaar: Archeologie en historie van de Westfriese Omringdijk tussen Hoorn en Schellinkhout," *West-Friese Archeologische Rapporten* 16 (2010): 14–15. C. M. Lesger, *Hoorn als stedelijk knooppunt: Stedensystemen tijdens de late middeleeuwen en vroegmoderne tijd* (Hilversum: Verloren, 1990), 54.

[49] Piet Boon, "Voorland en inlagen: de Westfriese strijd tegen het buitenwater," *West-Frieslands Oud en Nieuw* 58 (1991): 78–113.

FIGURE 4.7. Pile dikes (*paaldijken*) fronted water at high and low tide. Piles driven into the nearshore served as wave-breaking palisades. *Source*: Anon., *De slechte toestand van de zeedijk vanaf Diemen aan Jaap Hannes* (eerste deel), Print, 1705, Rijksmuseum, Amsterdam. http://hdl.handle.net/10934/RM0001.COLLECT.472532.

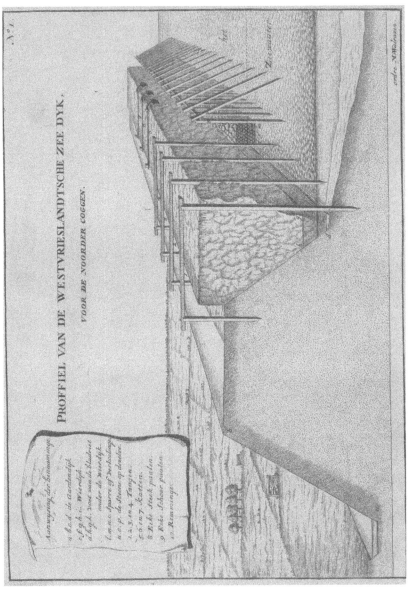

FIGURE 4.8. This "Profile of the West Frisian sea dike" depicts the use of wood and eelgrass cushions in "wier dike" (*wierdijk*) construction, which protected the earthen dike body. Shipworms infested the wooden framework as well as the seaward support piles. *Source*: M. Walraven, "Profiel van de Westvrieslandtsche zee dyk voor de Noorder Coggen," Print, 1732, *NHA*, Technische tekeningen, NL-HlmNHA_492_0055_M.

awareness of these environmental changes deepened.⁵⁰ The fear of inundation was not enough to diminish the economic and cultural imperative to exploit these coastal landscapes. Installing wooden infrastructure offered a technological fix that mitigated the risk of flooding. In combination with short- and long-term environmental changes that coalesced in the 1730s, the expansion of wooden infrastructure contributed the final piece of the puzzle for a shipworm outbreak.

The widespread use of wooden infrastructure across Holland presented a daunting problem. The flurry of correspondence between West Frisian water boards and the province in 1732 revealed a confused and tense situation. Inspectors produced detailed accounts of the damage and estimated the cost of replacement, but dikes could not be repaired absent consideration of the shipworms.⁵¹ Water boards initially sought advice from within their own ranks, including practical advice from dike bosses who had worked for decades to maintain this infrastructure.⁵² Nobody knew how to eradicate them, and, as a result, the true cost of repair proved elusive. Each strategy entailed a different, often substantial investment of labor and resources that West Frisian landholders could not bear alone. Dike reconstruction on this scale depended upon provincial assistance, including subsidies and remission from taxation. Early proposals, as a result, emphasized their cost-saving benefits. One promising strategy proposed removing the wooden frame that surrounded the wave-breaking cushion and, instead, stapling the cushion to the dike. This was cheaper than installing a wooden frame, and when tested, it

---

⁵⁰ C. P. Schrickx, "Tot behoud van stad en vaderland. Hout en houtgebruik in de Drechterlandse dijk," in *Dwars door de dijk: Archeologie en geschiedenis van de Westfriese Omringdijk tussen Hoorn en Enkhuizen*, ed. Michiel Bartels (Hoorn: Stichting Archeologie West Friesland, 2016), 87–92. Guus J. Borger, Philippus Breuker, and Hylkje de Jong, eds. *Van Groningen Tot Zeeland: geschiedenis van het cultuurhistorisch onderzoek naar het kustlandschap* (Hilversum: Verloren, 2010), 91. Erik Walsmit, Hans Kloosterboer, Nils Persson, and Rinus Ostermann, eds. *Spiegel van de Zuiderzee: Geschiedenis en cartobibliografie van de Zuiderzee en het Hollands Waddengebied* (Utrecht: HES & De Graaf, 2009), 19–23.

⁵¹ C. Baars, "Het dijkherstel onder leiding van de Staten van Holland," *Waterschapsbelangen* 74, no. 6 (1989): 196–204. "Voorslaagen en Calculatien, dienende tot een Plan, hoedanig en op welke manieren de door Wormen beschaadigde Zeedijken van Dregterland, vier Noorder Coggen, Geestmer Ambagt, Schaager en Niedorper Coggen, het best te repareeren en te onderhouden soude zijn tegen het geweld en doorbraake der Zee," Resolution May 29 (1732). SA. Archief van de Burgemeesters-Resolutions van de Staten van Holland. 5038.169.

⁵² Attestation 8 (June 1732). Notarissen in West-Friesland tot 1843, 1552–1843. WFA. 375.131.

prevented infestation. Unfortunately, the first strong waves washed the stapled cushions away.[53] Even if political will could be mobilized to test adaptations, tried-and-true strategies proved difficult to adapt to the unique challenges of the 1730s.

By 1732, the belief grew that this novel threat required a new type of dike. Seger Lakenman, secretary of Drechterland, conceded that water authorities may need to consider "entirely new forms of dike construction."[54] Water boards had gathered a significant amount of information on the shipworms, especially regarding their environmental tolerances, but the danger had not lessened. In desperation, Drechterland sent a letter to the Leiden professor and polymath, Willem Jacob 's Gravesande, in 1732, asking for any suggestions or insight. Sadly, 's Gravesande reported back that he could offer little useful advice, having "no experience relative to the matter, and having found nothing in the books that can help me."[55] The stakes could not have been higher. In a letter to the Estates of Holland, the water board of the Vier Noorder Koogen proclaimed their sea dikes a "barrier and fortress that protects the entirety of Holland against their shared, restless enemy." Any breach would lead to widespread inundation.[56] This conviction grew overwhelming during the next phase of the outbreak, when awareness and fear of shipworms spread beyond water boards and provincial commissions to the broader public.

4.3 SEEKING NEW SOCIAL AND SCIENTIFIC MEANING

Few in the public were aware of the shipworm threat in early 1732 even though it had been more than a year since they appeared in Zeeland. Their discovery in Holland, especially after inspectors found infestations near Amsterdam, created a panic. Pamphleteers published exposés revealing the monstrous "worms" to an astonished public. Newspapers

---

[53] C. Baars, "Nabeschouwing over de paalwormplaag van 1731–32 en de gevolgen daarvan," *Waterschapsbelangen* 75 (1990): 504.

[54] Seger Lakenman, *Het wonder Oordeel Godts, ofte een kort verhaal van de ongehoorde bezoeking dezer Provintie door zekere plage van Zee-wormen* (1732), 15.

[55] Wijnandt Nieuwstadt to William Jacob 's Gravesande – "over het voorkomen van de paalworm in Noord-Holland." KB. 128 D3.

[56] "Verbaal doen maken bij den Dijkg ... aan de Hoog Edele Heeren" (1732) WFA. Ambacht van Westfriesland genaamd de Vier Noorder Koggen. 1558.1820. "Fortresse" was an oft-repeated analogy in design proposals and requests for provincial assistance. L. T. and J. B., *Nederlant ten toneele, van Godts regtvaardige wormstraf gekozen* (Groningen: Laurens Groenewout, 1732). Lakenman made a similar claim in 1732. Lakenman, *Het wonder Oordeel Godts*.

and news digest accounts largely eschewed sensationalism, preferring to reprint eyewitness accounts or provincial decrees. In its 1732 "Report on the Plague of Worms in the Pileworks of Dikes in Zeeland and West Friesland," the Dutch news digest *Europische Mercurius* presented readers with a short history of the outbreak that drew directly upon dike inspections and offered several causal explanations for the disaster. The publishers enlivened the text with maps and woodcut imagery and even included images of shipworms on the digest frontispiece.[57] Reality was sensational enough. Even the most matter-of-fact descriptions acknowledged the mysterious origins of the worms and the unprecedented character of the disaster. The obvious failure of water authorities to solve the problem fueled public interest and fear.

These failures notwithstanding, newspapers and pamphlets used information gleaned from official reports to counter the troubling spread of media misinformation. The publishers of the *Mercurius* worried about foreign newspapers publishing egregious falsehoods about the disaster that had potentially damaging economic implications. One German story reported that "the great and beautiful city (Amsterdam) is near destruction, all of the houses which stand upon stakes thrust in the ground, are in very great danger of being destroyed by the worms that rapaciously eat through them."[58] Relatives and business partners living abroad wrote letters home, inquiring about the truth of these claims.[59] The *Mercurius* assured its readers this was fiction. Dike inspectors had already established that shipworms lived only in saltwater and certainly could not survive in the soils that surrounded the wooden foundations of Amsterdam's buildings. Similarly, a report from Bern greatly exaggerated the size of the worms and the extent of devastation. It claimed that two-foot long worms ate through sections of the earthen dike bodies leaving "uncommonly large holes."[60] Images published abroad reinforced these claims, depicting shipworms as literal sea monsters eating into the

---

[57] "Bericht," *Europische Mercurius* (1732). Joop Koopmans argues that the inclusion of a domestic disaster on the frontispiece, a rarity for this publication, indicates the perceived marketability of shipworm news to the Dutch public. Joop Koopmans, "The Early 1730s Shipworm Disaster in Dutch News Media," *Dutch Crossing* 40, no. 2 (2016): 139–50.
[58] *Europische Mercurius* (1732), 107. Dutch newspapers also reported on falsehoods printed in Nuremberg and Cologne. January 12 (1733) *Amsterdamse Courant*.
[59] J. R. de Missy, *Aanmerkingen over den oorsprong, gesteltheit, en aard der zeewormen: die de schepen en paal-werken doorboren door den hr. Rousset* (Lyden: Gysbert Langerak, 1733).
[60] *Europische Mercurius* (1732), 107.

FIGURE 4.9. Baeck's image emphasizes the extent, origin (West Indies), and monstrousness of the worms. *Source*: Elias Baeck, *Abildung deren höchst schädliche unbekandten See-Würmer*, Print, 1732, Frankfurt a. M.: Stadt- und Universitätsbibliothek, 2002, http://sammlungen.ub.uni-frankfurt.de/4360341.

foundations of the Dutch Republic (Figure 4.9). Rumors about shipworms, exaggerated or not, spread panic and undermined confidence in Amsterdam's commercial vitality.

The danger of misinformation assumed geopolitical significance by 1733. The Estates General of the Netherlands acknowledged the danger posed by this "uncommon plague of destructive sea worms" in a proclamation in February but countered that "through God's goodness [they have] not spread nearly as far or as much as the unsubstantiated rumors would have one believe."[61] The *Mercurius* attributed these rumors to a "snowball" effect that compounded untruths and exaggerated reports.

[61] Proclamation. February 17 (1733). *SA*. Archief van de Burgemeesters: Publicaties van de Staten-Generaal en van de Staten van Holland en West-Friesland. 5022.11.

Other commentators were less gracious. The Huguenot journalist and historian Jean Rousset de Missy, who lived in the Netherlands, contended that "false Reports on this Subject, which have been spread either by the Enemies or the Enviers of the Wealth, Prosperity and Happiness of the Republick" [sic] may have been to blame.[62] Misinformation, malicious or not, fed off the mystery surrounding the animal. The lack of information, combined with the real and continuing exposure of Dutch coasts to further infestation, sparked an outpouring of scientific literature about shipworms.

4.4 DISCOVERING THE NATURAL HISTORY OF SHIPWORMS

The first systematic investigations of shipworms appeared in 1733. This early work catalogued their traits, described their life cycle, and hypothesized about their origins. The Prussian natural philosopher Gottfried Sellius produced the most complete and influential treatise, entitled *Natural history of the shipworm or wood eater* (*Natuurkundige histori van den zeehoutworm ofte houtvreeter*). He based his findings on empirical analysis of infested wood sent to him from the Republic as well as in situ investigations performed during later visits. He drew from a network of international scholars interested in the subject and prior research conducted in the Netherlands.[63] Sellius discovered that the "worms" were in fact mollusks. His work, like many that appeared after 1733, reflected an Enlightenment-era emphasis on categorization, order, and assigning meaning to the natural world.[64]

The early science of shipworms largely linked outbreaks to environmental change. Sellius connected the explosive outbreak in the early part of the decade to coastal increases in salinity and temperature.[65] Jean Rousset de Missy agreed and demonstrated that environmental factors also affected their growth, size, and the timing of outbreaks. Severe winters limited their growth and spread. Colder temperatures "destroy'd

---

[62] Jean Rousset de Missy, *Observations on the Sea- or Pile-Worms Which Have Been Lately Discover'd to Have Made Great Ravages in the Pile-or Wood-Works on the Coast of Holland* (London: J. Roberts, 1733), 1.

[63] Gottfried Sellius, *Natuurkundige histori van den zeehoutworm, ofte houtvreeter, zynde koker- en meerschelpigh, inzonderheit van den Nederlantschen ... : Eerste Deel* (Utrecht: H. Besseling, 1733).

[64] Pierre Massuet, *Wetenswaardig onderzoek over den oorsprongk, de voortteling, de ontzwachteling, het maaksel, de gedaante, de gesteltheit, den arbeidt, en de verbazende menigte der verscheide soorten van kokerwurmen die de dykpalen en schepen van enige der vereenigde Nederlandsche provintsien doorboren* (Amsterdam: Wor, 1733).

[65] Baars, "paalwormfurie," 809.

both them and the Seeds of them, before they became of any considerable Big-ness."⁶⁶ Unfortunately, the early 1730s featured unusually warm winters. Sellius and De Missy presented their findings as objective products of expertise and largely reinforced causal stories that linked the shipworm epidemic to natural change.

Scholars invariably framed their biological studies in the social context of the shipworm problem. Shipworm novelty precluded effective control, but the science of shipworms might reveal new strategies to combat them. Several presented "remedies" or "measures against the worms" based on their insights. Naturalist Zacharias L'Epie, for instance, performed a series of experiments to test the environmental limits of the shipworms. His solution – grandiose even by later standards – involved reclaiming large portions of the Southern Sea, depriving shipworms of their habitat and opening new lands for agricultural exploitation.⁶⁷ Cornelis Belkmeer's remedies drew upon his study of shipworm reproduction. "The reproduction and increase of the animals," he stated, should be combatted by using "iron scrapers and stiff brushes" to clean off the slime that coated the interior of shipworm passages.⁶⁸ L'Epie, Belkmeer, and others self-consciously promoted knowledge as a weapon to combat the outbreak.

These experts had no monopoly on experimentation, nor were they the only people interested in developing practical solutions to the shipworm problem. In July 1732, Drechterland purchased a newspaper advertisement soliciting advice and remedies from "anyone who had experimented or developed inventions against the boring of worms into dike piles." This rare invitation for outsiders to contribute insights reflected the limits of institutional knowledge. When the *Europische Mercurius* picked up and disseminated this unusual request, a wave of pamphlets, patent requests, and remedies from "upright patriots" with "hearts and hands lifted to the almighty" flooded into Holland from across the Republic and Europe.⁶⁹

---

⁶⁶ De Missy, *Observations*, 7.
⁶⁷ Zacharias L'Epie, *Onderzoek Over de oude en tegenwoordige Natuurlyke Gesteldheyd van Holland en West-Vriesland, Desselfs rivieren en landen; Aanwas, Ophooping, Zakking, Laagte en Dykagie; Mitsgaders eene Verhandeling over de ... Zee- of Kokerwormen ... Als mede de Middelen tot Verbetering en Versterking der Zeeweeringen ...* (Amsterdam: Jacobus Hayman, 1753), 190–6.
⁶⁸ Cornelis Belkmeer, *Naturkundige verhandeling of waarneminge, betreffende den houtuytraspende en doorboorende zee-worm* (Amsterdam: J. Ratelband en Compagnie, 1733), 44.
⁶⁹ Advertisement. July 5 (1732). *Amsterdamse Courant*. "Letter from Frederick Duim," December 12 (1732). NA. Gedeputeerden van Haarlem ter Dagvaart. 3.01.09 1238.

The Shipworm Epidemic, 1730–1735

Most proposals followed a standard rubric published in the *Mercurius*. Proposals should: "1. repair [the dikes] in a way that prevents further damage from the worms 2. Protect against the violence of the sea so that people can live peacefully 3. Do so using the least costly methods."[70] Many believed that retaining the basic design of the dikes, including the use of wood, would appeal to water boards. They submitted proposals to protect the wood by mechanically removing shipworms or kill them using toxins, such as mercury. One Silesian submission suggested applying sulfur oil to combat putrefaction.[71] Others proposed design changes. Henricus Engelhardt submitted a letter and later published a pamphlet outlining his design for a "sea wall" built of iron and stone.[72] Engelhardt argued his design would reduce costs by employing no masonry and incorporating recycled iron from anchors or bridges.

A commission from the Estates of Holland vetted the proposals. They rejected some outright, either because they were too costly or infeasible. They rejected others that displayed an obvious ignorance of shipworm biology. They threw out the proposal of a "chevalier engineer of France" because he could not identify the right species of shipworm. The commission rejected another by J. C. Brunner because "the author showed not the least bit of knowledge about dikes or the worms."[73] After 1732, in other words, authorities no longer worked from a position of frustrated ignorance. They knew enough about the shipworms to evaluate the feasibility of proposals.

This period of experimentation integrated the interpretation of shipworms as a flooding hazard with a much broader, public emphasis on the animal itself, drawing on new scientific descriptions. The appeal for

---

[70] *Europische Mercurius* (1732), 304–5.
[71] "Charles Frederick de Baudiss, from Numptsch in the Dukedom Brieg in Silesia," January 23 (1733). NA. Gedeputeerden van Haarlem ter Dagvaart. 3.01.09. 1238.
[72] Henricus Engelhardt, *Goede suffisante, Gods verleende, uitgevondene middelen, omme yzere, en steene zeemuuren, met paalen zonder metzelwerk, tegens het schadelyke zeegewormte te maken, dewelke in plaats van zeedyken, dammen, en in de zehavens zoude kunnen dienen* ('s Gravenhage: Laurens Berkoske, 1733), appendix 2.
[73] "J. C. Brunner," SA. Gedeputeerden van Haarlem ter Dagvaart. 3.01.09. 1238. Reports of tests were also printed by the Estates of Holland. Resolution May 22 (1733). SA. Archief van de Burgemeesters-Resolutions van de Staten van Holland. 5038.170. Holland also collected trials from other provinces. "Slaan van proefpalen en beschouwingen van E. Reynvaan te Middelburg, J. H. Eekholt te Groningen en J. Bushall over het optreden van de paalworm en manieren omhout tegen deze wormen te beschermen, 1732–1734," WFA. Ambacht van West-Friesland genaamd Drechterland en hoofdingelanden van West-Friesland (Drechterland). 1562.1443.

popular proposals after 1732 demonstrated how serious the shipworm problem had become. Despite this democratized response, the shipworm threat also reinforced the authority of water professionals.[74] The Estates of Holland turned to the hydraulics experts Cornelis Velsen and Nicolaas Cruquius, both of whom worked on water management problems in southern Holland, to advise shipworm commissions.[75] The shipworm threat, in other words, opened new space for untrained practitioners and formally trained technical experts. Their willingness to examine even the most unorthodox solutions acknowledged the limitations of existing water management strategies without undermining their faith in technological solutions. Neither, however, did it displace the important role of divine providence.

4.5 PROVIDENCE AND DECLINE

Providential rhetoric permeated nearly every response to the shipworm threat, from natural history treatises to state resolutions. Shipworms, like earlier disasters, revealed God's displeasure and portended more devastating calamities. Shipworms appeared the perfect symbols of providential judgment. They arrived in dramatic numbers with little explanation and no ready response. "Who has seen a stranger plague," one anonymous author asked, this being such "a rare pest, from God's vengeful hand?"[76] Most documents referred to shipworms as "strange" or "unexpected" or the epidemic nature of the threat as "uncommon" or "previously unknown." This characterization was an important element of moral interpretation because it supported the argument that the worms were a supernatural expression of special providence.[77]

Some moralists framed shipworms as embodiments of the flooding hazard. Water boards had already spent two years establishing this connection and moralists adapted it to their purposes. They understood how flooding resonated in the public consciousness. The cultural significance of flooding ran far deeper and therefore had greater instructive potential than any other natural disaster. Shipworms signaled the most comprehensive realization of the flooding hazard to date. When pamphlets like *The Netherlands Exhorted to Penitence* (*Nederlandt*

---

[74] Davids, *Rise and Decline*, 507–8.   [75] See Chapter 5.
[76] *Nederlant aengemaemt tot boetvaerdigheit* ('s Gravenhage: G. Block, 1732), 3.
[77] Adam Sundberg, "An Uncommon Threat: Shipworms as a Novel Disaster," *Dutch Crossing* 40, no. 2 (2016): 122–38.

aengemaendt to boetvaardigheit) explained that these "uncommon pests" threatened to unleash the "violence of an unshackled sea," they tapped into this deep and familiar cultural anxiety.[78]

Shipworms acquired significant providential meaning independent of flooding as well. Eighteenth-century Dutch literature was replete with allusions to the real and symbolic significance of "worms." Natural histories treated them as wondrous examples of divine order. Religious literature used worm analogies to emphasize human insignificance.[79] In the broad spectrum of divine creation, humanity's position relative to God was no greater than a worm. Shipworms reinforced all these interpretations. Dutch pride and confidence, particularly in the realm of water management, seemed woefully misplaced if such a lowly animal could undermine the very foundations of the Republic.

The most powerful interpretations saw shipworms as evidence of decline. The outbreak's timing, arriving on the heels of previous disasters, proved critical to this narrative. Floods, plagues, and other misfortunes were unique in the own right, but they also portended catastrophes yet to come. Shipworms confirmed their darkest predictions. In his 1733 shipworm sermon, entitled *God's Almighty and Righteous Striking Hand (De Almagtige en Regtveerdige Slaande Hand Gods)*, Johannes Mobachius reminded readers of God's previous punishments, including the floods of 1686 and 1717 and the cattle plague that "so graciously and happily has ended."[80] Dutch sinners had not reformed and "now God has again with his angry rod struck the entirety of the Netherlands, with worms in these evil days in the pileworks that stand in front of the dikes."[81] Shipworms became the latest link in an increasingly long causal chain that linked Dutch sin, disaster, and declining morality.

Mobachius published repeatedly on natural disasters and was part of a group of civic-minded ministers and other moralists who advocated reform and linked moral recovery to the future of the "fatherland" (*vaderland*).[82] Their broad focus accurately captured the scope of

---

[78] Ibid.
[79] Inger Leemans and Gert-Jan Johannes, "Gnawing Worms and Rolling Thunder: The Unstable Harmony of Dutch Eighteenth-Century Literature," in *Discord and Consensus in the Low Countries, 1700–2000*, eds. Jane Fenoulhet, Gerdi Quist, and Ulrich Tiedau (London: UCL Press, 2016), 27–8.
[80] Johannes A. Mobachius, *De almagtige en regtveerdige slaande hand Gods, Ter Besoeking van het Land en deese Provincie met de verderfelyke Plage der Zeewormen ...* (Groningen: Jurjen Spandaw, 1733), 43.
[81] Ibid., preface.   [82] Edwards, "Varieties," 40.

the panic. "Now the danger threatens more than one location!" one 1732 pamphlet noted. "Along Zeeland's coast, north Holland's fat pastures, on Friesland's ground it is already heard groaning."[83] So much of the Republic's wealth was tied to maritime trade and commerce that fear of shipworms extended to inland provinces. Moralists across the Republic reiterated the anxieties voiced in Holland. "Already this century, you were admonished with war," one pamphlet in Groningen declared. The lessons of the Christmas Flood and cattle plagues had fallen on deaf ears. Now, a disaster "threatens the corruption of the entire Church and State."[84] Shipworms confirmed that the decline of Dutch moral character was a problem broadly shared across the Republic, and the stakes were enormous.

For the most part, state and religious authorities responded in a predictable manner, drawing from a playbook practiced over previous decades. Provinces released proclamations that mandated fasting, thanks, and prayer. Their texts linked war, natural disaster, and economic adversity together. The broad scale of the shipworm threat prompted a similar proclamation from the Estates General in March 1733. "God's striking hand" again threatened the "lessening of navigation and commerce, through storm floods, the uncommon sicknesses among men and beasts, and lastly through an unusual plague of destructive worms in the piles and woodwork."[85] These documents identified the providential nature of the affliction, connected disasters together as evidence of decline, and called for the moral renewal of Dutch society. In this respect, the providential response to shipworms was similar in type, if greater in scale, than the earlier disasters of the eighteenth century. In one way, however, shipworms bucked continuity. By 1733, moralists began linking the shipworm epidemic to sodomy.

4.6 SHIPWORMS AND SODOMY

In the wake of natural disasters in the early eighteenth century, moralists had little trouble identifying sins likely culpable for catastrophic events. The suite of reprehensible behaviors ranged from pride and excess to avarice and lying, to drunkenness and fornication. Reformed Pietist

---

[83] R. B., *Nederlands klachte over Gods naekende oordeelen; Duidelyk te bespeuren in het knaegen der wormen aen de paelen der Nederlandze zeedyken* (Harlingen: Folkert Jansz van der Plaetz, 1732), 6.

[84] L. T. and J. B., *Nederlant ten toneele.*   [85] Kist, *Neêrland's bededagen*, 327.

ministers like Jacob Harkenroht warned against the decline of faith and parishioners' frequent absences from church.[86] During previous disasters like the cattle plague and Christmas Flood, the list of sins had been long, diverse, and for the most part disconnected from the character of specific calamities. The shipworm epidemic produced a different response. It nearly coincided with an unusually high number of sodomy trials.[87] For some moralists, the form and timing of the disaster seemed an unmistakable sign that shipworms, sodomy, and moral decline were part of the same causal story.

Sodomy persecutions began in March 1730, several months before the discovery of shipworms in Zeeland. Although the trials began as a limited affair, they rapidly expanded after authorities in Utrecht arrested and tried Zacharias Wilsma, a twenty-two-year-old Catholic soldier who testified about his sexual contacts in several Dutch cities. For the next two years, a wave of persecutions targeted sodomitical networks in courts across the Netherlands.[88] Between 1730 and 1732, Dutch authorities prosecuted 350 men, some of whom were upper class, and sentenced 80 to public execution. Although largely an urban phenomenon, one of the more brutal episodes occurred in the rural village Faan in Groningen in 1731, where a local magistrate named Rudolphe de Mepsche spearheaded the investigation, trial, and execution of twenty-four men and boys. The sodomy trials of the 1730s occurred across the Netherlands and were one of the deadliest in early-modern European history.[89]

---

[86] Ton van der Schans, "Als een straf van God, weekdiertje dedreigde onze dijken in de achttiende eeuw," *Kleio* 55, no. 4 (2014): 5–8.

[87] Sodomy has an interpretive lineage stretching back to the biblical stories of Sodom and Gomorrah and included any sexual practice not intended for procreation. By the eighteenth century in the Dutch Republic, it referred almost exclusively to anal intercourse between males, ending in ejaculation. Theo van der Meer, "The Persecutions of Sodomites in Eighteenth-Century Amsterdam," *Journal of Homosexuality* 16, no. 1/2 (1988): 265.

[88] L. J. Boon, *Dien godlosen hoop van menschen: Vervolging van homoseksuelen in De Republiek in de jaren dertig van de achttiende eeuw* (Amsterdam: De Bataafsche Leeuw, 1997). L. J. Boon, "Those Damned Sodomites: Public Images of Sodomy in the Eighteenth Century Netherlands," *Journal of Homosexuality* 16, no. 1/2 (1988): 239.

[89] W. T. Vleer, *"Sterf sodomieten!": Rudolf de Mepsche, de homofielenvervolging, het Faanse zedenproces en de massamoord te Zuidhorn* (Norg: Veja, 1972). Trials took place in several provinces. A. H. Huussen Jr., "Prosecution of Sodomy in Eighteenth Century Frisia, Netherlands," *Journal of Homosexuality* 16, no. 1/2 (1988): 249–62. A. H. Huussen Jr. "Sodomy in the Dutch Republic during the Eighteenth Century," in *Hidden from History: Reclaiming the Gay and Lesbian Past*, eds. Martin Duberman, Martha Chauncey, and George Vicinus (Ann Arbor: University of Michigan Press, 1990), 141–9.

The origins of the sodomy trials were manifold, but the appearance of shipworms in the 1730s was not one of them. The trials began prior to their discovery, and moralists only connected them once shipworm anxiety reached fever pitch in 1732. Trials and executions for sodomy rarely took place in the seventeenth-century Republic, and sodomy scarcely appeared in theological writings or state records. Following the social and political tumult of the 1670s, however, prosecutions to promote "moral discipline" gradually increased. This was partly due to the influence of the Further Reformation, which sought to translate the grievances of Dutch Pietism into a comprehensive program of renewal in all aspects of life based on individual and public morality.[90] Criminalization of sexual crimes rose consequently.[91] A specific catalyst for the trials of the 1730s, however, has been more difficult to pin down. Some historians argue they emerged out of a milieu of cultural and moral uncertainty, particularly the danger stemming from new, early Enlightenment ideas.[92] Others argue they arose as a reaction against perceived social excess, "lesser sins" like card playing and debauchery, and resulting decline of the Republic from its chosen status as a "new Israel."[93] Contemporaries emphasized this last point. Sodomy was an individual crime that nevertheless implicated the community and spoke to broader concerns about the future of the Republic. If Dutch virtue laid the foundation of the Golden Age, its descent into sodomy affirmed its decline.

Moralists interpreted sodomy as both the cause and consequence of social, economic, and moral decline. In the words of one historian, they blamed sodomites for "commercial decline, rising unemployment, the demise of strict church practice, the rising influence of papism and, concomitant with papism, the overwhelming influence of French and

---

[90] Willem Jan op 't Hof, "Lusthof des Gemoets in Comparison and Competition with De Practycke ofte Oeffeninghe der Godtzaligheydt: Vredestad and Reformed Piety in Seventeenth-Century Dutch Culture," in *Religious Minorities and Cultural Diversity in the Dutch Republic*, eds. August den Hollander, Mirjam van Veen, Anna Voolstra, and Alex Noord (Leiden: Brill, 2014), 138–9.

[91] H. Schilling, *Civic Calvinism in Northwestern Germany and the Netherlands: Sixteenth to Nineteenth Centuries* (Ann Arbor: University of Michigan Press, 1991). D. J. Noordam, "Homosocial Relations in Leiden (1533–1811)," in *Among Men, Among Women: Sociological and Historical Recognition of Homosocial Arrangements*, eds. Mattias Duyves, Gert Hekma, and Paula Koelemij (Amsterdam: University of Amsterdam, 1983), 218–23.

[92] Leemans and Johannes, *Worm en Donder*, 625.

[93] Theo van der Meer, "Sodom's Seed in the Netherlands," *Journal of Homosexuality* 16, no. 1/2 (1988): 7–8.

Italian culture and *mores*."⁹⁴ Pamphlets, sermons, and poems framed sodomy as *crimen contra natura* (an offense against nature), representative of and responsible for Dutch decline. Authors contrasted Golden Age prosperity with this new fallen state. In his 1730 book *Sodom's Sins and Punishments* (*Sodoms zonde en strafe*), the minister Leonard Beels described how God blessed the Republic with "grassy pastures, fertile fields, bedecked with every sort of tree, cut by rivers, and washed by sea, which allowed trade to blossom and made them the market for the world." "But oh!" he went on, they now "despise all the riches of His goodness."⁹⁵ In the closing of his 1731 book on sodomy, *The Heinous Sodom Punished* (*Het gruwelijk sodom gestraft*), the theologian Koenraad Mel noted, "Fire, plague, contagious diseases, war and famine, are the arrows and angry rods with which [God punishes] the people."⁹⁶ Perhaps no moralist was as influential as Groningen minister Henricus Carolinus van Byler, whose 1731 book *Hellish Wickedness of the Horrible Sin of Sodomy* (*Helsche Boosheit of Grouwelyke Sonde van Sodomie*) inflamed persecutorial zeal in the Faan. Van Byler had earlier penned a pamphlet on cattle plague, and he interpreted sodomy as yet another expression of decline. "Have there not been judgments on the Land for our sins?" he asked. "What accounts for the wars, cattle plagues, floods and loss of commerce? Are they not because of sins? Has the State not weakened? Its splendor extinguished?"⁹⁷ Sodomy and disasters were symptomatic of this same fall from grace, and persecutions ritualistically cleansed and redeemed society.

When awareness of shipworms grew to panic, the discovery of sodomitical networks provided moralists a dramatic and obvious causal story. The similarities between the two phenomena seemed too apparent to ignore. They saw both as insidious threats that undermined the foundations of Dutch society. Both appeared suddenly, inexplicably, and unnaturally. Like the shipworms, sodomitical networks appeared to have

---

⁹⁴ Boon, "Those Damned Sodomites," 241.
⁹⁵ Leonard Beels, *Sodoms zonde en straffe of streng wraakrecht over vervloekte boosheidt, en loths vrouw, verandert in een zoutpilaar* (Amsterdam: Adriaan Wor, en de erve G. Onder de Linden, 1730), 125–6.
⁹⁶ Koenraad Mel, *Het gruwelijk sodom gestraft* (Middelburg: Stichting de Gihonbron, 2004), 92.
⁹⁷ Vleer, "*Sterf sodomieten*," 62–73. Van Byler, Historis-Verhaal. Henricus Carolinus van Byler, *Helsche boosheit of grouwelyke zonde van sodomie, in haar afschouwelykheit, en welverdiende straffe uit Goddelyke, en menschelyke schriften tot een spiegel voor het tegenwoordige, en toekomende geslagte openlyk ten toon gestelt* (Groningen: Jacobus Sipkes, 1731), 43–4.

manifested where none existed before. Contemporaries considered sodomy a Catholic vice, by some accounts imported into the Republic via Catholic ambassadors during the peace negotiations of 1713. Most agreed it was a product of a recent turn to decadence. Golden Age piety and virtue would have prevented its manifestation. This was easy for eighteenth-century readers to believe because sodomy trials, executions, and burials during the seventeenth century took place largely outside public view.[98] The near concurrent and very public exposure of sodomitical networks alongside the shipworm outbreak could not be coincidence.

The most striking illustration of the connection between shipworms and sodomy emerged, not out of Pietist sermons or state proclamations but rather "spectatorial" essays. The *Hollandsche Spectator* (1731–5) was an influential periodical and the most important Dutch source of Enlightenment thought in the early eighteenth century. It was written by Justus van Effen, a writer and member of the British Royal Society, who modeled his publication on earlier English examples of the genre.[99] Spectatorial essayists wrote in the vernacular, tackled intellectual debates, and generally targeted learned, burgher audiences. Van Effen's work emphasized reason and the perfectibility of mankind. He also emphasized specifically Dutch themes – in particular the anxiety about Dutch decline.[100]

The *Hollandsche Spectator* was Van Effen's platform to explain the Republic's loss of stature, and he did so by contrasting it with Golden Age splendor. He was a staunch supporter of what he considered traditional Dutch society and identity: industriousness, frugality, and moral virtue. "It is my conviction that our nation really merits esteem, even more than

---

[98] Theo van der Meer, "Sodomy and Its Discontents: Discourse, Desire, and the Rise of a Same-Sex Proto-Something in the Early Modern Dutch Republic," *Historical Reflections/Réflexions Historiques* 33, no. 1 (2007): 46–8.

[99] Joseph Addison and Richard Steele created the genre in 1709 with the publication of *The Tattler*. This was followed in England by *The Spectator* (1711) and *The Guardian* (1713). The *Hollandsche Spectator* was only the first in a long list of spectatorial publication in the Netherlands in the eighteenth century. Maartje Janse, *De geest van Jan Salie: Nederland in verval?* (Hilversum: Verloren, 2002), 26.

[100] Wijnand Mijnhardt, "The Dutch Enlightenment: Humanism, Nationalism, and Decline," in *The Dutch Republic in the Eighteenth Century: Decline, Enlightenment, and Revolution*, eds. Margaret Jacob and Wijnand W. Mijnhardt (Ithaca, NY: Cornell University Press, 1992), 197–223.

any other nation even," he declared. "Given the choice, I would be compelled by reason to choose being a Dutchman."[101] He was proud of the Republic's Golden Age accomplishments. That legacy appeared under threat in the eighteenth century. Historian Wyger Velema argues that both the glorification of the Golden Age as a period of virtuous progress and the caricature of the eighteenth century as stagnant and morally corrupt stem from Van Effen's spectatorial essays.[102] Notwithstanding the earlier moralizing arguments that contrasted the moral shortcomings of their present with glory of the past, this interpretation rightfully illustrates how powerful the decline discourse had become.

The shipworm panic coincided with the earliest years of the *Hollandsche Spectator*. To Van Effen, shipworms revealed God's displeasure with a society that tolerated sodomitical networks, which he believed "stretched across all of our provinces in a chain of Godlessness."[103] Van Effen went further than critiquing the moral source of the disaster. He sharply criticized other moral and material responses. He condemned the use of Catholic rituals to protect the dikes and bemoaned the overweening emphasis on "invention" and reliance upon "ingenious men."[104] He believed that pious faith and conversion would ultimately produce a solution. Van Effen was not a fatalist. His confidence in human progress melded neatly with his Reformed faith. The conviction that the Dutch Republic required spiritual rehabilitation was characteristic of the moderate Dutch Enlightenment.[105] To Van Effen, Dutch decline was real, and shipworms certainly revealed God's displeasure, but further decay was neither inevitable nor irreversible. This sentiment grew in influence, even if Van Effen's critique faltered, during the final phase of the shipworm episode when two ingenious men proposed a plan to reshape the Dutch coastline.

4.7 STONE SLOPES AND THE LANGUAGE OF INNOVATION

By 1733, Dutch water authorities proceeded with dike improvements with far greater confidence than the first two years of the outbreak.

---

[101] Ibid., 208.
[102] Wyger R. E. Velema, *Republicans: Essays on Eighteenth-Century Dutch Political Thought* (Leiden: Brill, 2007), 77–91.
[103] Justus van Effen, *De Hollandsche Spectator*, Vol. 106–50, 76.
[104] Justus van Effen, *De Hollandsche Spectator*, Vol. 196–240, 214.
[105] J. de Jong, L. Kooijmans, and H. F. de Wit, "Schuld en boete in de Nederlandse Verlichting," *Kleio* 19 (1978): 241.

Subsequent experience had revealed the limits of shipworm environmental tolerances, and scientific descriptions of the mollusk had done much to dispel the cloud of mystery surrounding them. Based in part on these insights, water authorities in Zeeland, Holland, and Friesland experimented with and enacted several dike adaptations. In Zeeland, they eventually embedded iron "worm nails" (*wormspijkers*) into exposed woodwork. Worm nails had large, flat heads that rusted in saltwater, sheathing the piles from infestation. VOC ships had employed a similar strategy with some success, and authorities in West Friesland would later install *wormspijkers* to protect some of their harborworks as well.[106] Water boards in the south and west of Friesland built inland "sleeper dikes" (*inlaag-* or *slaperdijken*) to protect the interior should their first line of defense fail.[107] Although none were employed on a large scale, the flood of proposals sent to Holland between 1732 and 1733 had expanded the realm of testable solutions. Holland eventually settled upon a design proposed by two water authorities from West Friesland, Pieter Straat and Pieter van der Deure. Their design, although grounded in preexisting dike practice, signaled the beginning of a dramatic restructuring of the Dutch coastline.

Straat and Van der Deure published a short, ambitious pamphlet in 1733 that outlined their strategy to combat the "unknown sea worm ... and at the same time strengthen and bring [the dikes] to a formidable state again, enabling them to resist the violence of the sea."[108] They proposed laying boulders and large stones on the seaward side of the wier dike. This would protect the vulnerable dikes from erosion and prevent the shipworms from reaching the wooden components behind them. This design had been inspired by a close inspection of the West Frisian dikes. Those dike sections already protected by boulders and stones seemed

---

[106] Adriaan Bommenee, "Testament van Bommenee," ZA. Verzameling Handschriften Gemeentearchief Veere. 2854.59, fol. 212–23. Bartels, *Dwars door de dijk*, 154. J. Gawroński, *De "Equipagie" van de Hollandia en de Amsterdam: VOC-bedrijvigheid in 18de-eeuws Amsterdam* (Amsterdam: De Bataafsche Leeuw, 1996), 288–29. Michiel H. Bartels, Peter Swart, and H. de Weerd, "Wormspijkers in het Medemblikker havenhoofd: Archeologisch en historisch onderzoek naar de maatregelen tegen de paalworm in het noordelijk havenhoofd van Medemblik, West-Friesland," *West-Friese Archaeologische Rapporten* 80 (2015).

[107] Baars, "Nabeschouwing," 505.

[108] Pieter Straat and Pieter van der Deure, *Ontwerp tot een minst kostbaare zeekerste en schielykste herstelling van de zorgelyke toestand der Westfriesche zeedyken ... Met een nader ontwerp Hoe men de Dyken, daar de grootste dieptens zyn op de zekerste, minst kostbaarste en schielykste wyze kan herstellen ...* (Amsterdam: J. Oosterwyk, 1733), 6.

surprisingly well preserved despite the presence of shipworms. They reasoned that if the province built a "stone slope" along the entire West Frisian coast, coastal dikes would be more durable, water boards would save the cost of replacing wooden components, and, most importantly, this new design would resist infestation from shipworms. They knew this approach would be incredibly costly and estimated a total investment of 1,787,500 guilders.[109] This accounted for the enormous amount of stone the project required. Stone was expensive and would have to be imported, so only the most at-risk areas without *voorland* warranted immediate adaptation.[110] Straat and Van der Deure prepared for the inevitable resistance by performing a trial of their design near the West Frisian town of Enkhuizen and reported that it performed admirably.[111]

Enthusiasm for Straat and Van der Deure's plan grew despite challenges from prominent figures in West Friesland. Secretary Seger Lakenman criticized its high cost preferring his own plan to build a series of inland "sleeper" and "transverse" dikes as insurance against inundation. The great disadvantage of his design was that it forced landholders to give up significant acreage to water protection rather than productive use – a concession few would tolerate. Zacharias L'Epie believed Straat and Van der Deure's plan too logistically difficult and questioned their accounting of costs. He estimated it would cost over two and a half million guilders to retrofit one section of the sea dike alone.[112]

Despite these reservations, support grew. Straat and Van der Deure's success can be partly attributed to its inherent advantages. By laying stone on the outside of existing wier dikes, and later atop more gently sloping clay and rubble, they lost no productive territory to the sea, reinforced the stability of the dike, and deprived shipworms of their habitat. They were also effective marketers. Many of the earliest proposals sent to Holland arrived as handwritten letters. Authors eventually professionalized their

---

[109] Ibid., 15.
[110] An entirely new transportation infrastructure (and host of regulations) emerged in the port cities of Enkhuizen and Hoorn to source this stone, first from the eastern region of Drenthe and later from Norway. Stone merchants initially pillaged the megalithic prehistoric settlements of Drenthe for their stones. This prompted a law in 1734 against the taking of stones from these hunnebeds – the third law in the world to protect antiquities. J. A. Bakker, *Megalithic Research in the Netherlands, 1547–1911: From "Giant's Beds" and "Pillars of Hercules" to Accurate Investigations* (Leiden: Sidestone Press, 2010), 62. Over the course of seventy years, 1.2 million tons of stone were imported into West Friesland. Johannes Jouke Schilstra, *In de ban van de Dijk: De Westfriese omringdijk* (Hoorn: West-Friesland, 1982), 85.
[111] Straat and Van der Deure, *Ontwerp*, 6–7.   [112] L'Epie, *Onderzoek*, 248.

FIGURE 4.10. Straat and Van der Deure's "stone slope" retained the use of wier, which would later be phased out. *Source*: Pieter Straat and Pieter van der Deure, *Ontwerp tot een minst kostbaare zeekerste en schielykste herstelling van de zorgeliyke toestand der Westfriesche zeedyken*, Print, 1733, Zuiderzeemuseum, Enkhuizen, http://hdl.handle.net/21.12111/zzm-collect-4718.

proposals and published them as formal pamphlets. They described firsthand observations of the dikes, demonstrated their knowledge of shipworms, and included diagrams of their proposed changes. Straat and Van der Deure's pamphlet was among the finest representations of this trend. Their pamphlet not only conveyed the benefits of their design but also accounted for its total cost down to specific subsections of the dikes and included an illustration that depicted the proposed "new slope of stone" dividing a placid sea from bucolic agricultural fields (Figure 4.10).

Their 1733 pamphlet also employed rhetoric popularized during the previous two years of the shipworm epidemic. They promoted their design as the "least expensive, quickest, and most certain" plan for the repair of the West Frisian dikes – drawing from the rubric popularized during the public appeal for shipworm remedies.[113] Straat and Van der

---

[113] Straat and Van der Deure, *Ontwerp*, 6.

Deure argued that theirs was "a new manner of dike building that has never been practiced before" – tapping into the institutional emphasis on innovation that had been growing since 1731.[114] In reality, dike builders had advocated and implemented similar solutions since the sixteenth century, and Cornelis Velsen had even tested a similar design in West Friesland in 1732–3, although it used smaller stones and retained wooden piles as a barrier between the wier cushion and stone slope.[115] Much like Van Seeratt following the Christmas Flood, Straat and Van der Deure found that the rhetoric of "improvement" remained powerful even if precedent existed. Their design diverged from Velsen's in one important way, however. Larger stones added stability and removed the need for wooden piles, the source of the shipworm threat. Even if their claim to innovation stretched the truth, their adaptations yielded significant advantages.

Perhaps most importantly, Straat and Van der Deure's pamphlet tapped the deepest existential fears of the shipworm plague. Water authorities had warned of the danger that dike breaches in North Holland could present if the "fortress" of West Friesland's dikes fell. In this doomsday scenario, water would spill over dikes and into North Holland's polders, flooding entire regions and perhaps even spilling into the economic heartland of South Holland. Straat and Van der Deure's pamphlet played into this fear. They imagined a powerful storm and dike breaches turning the area between the cities of Haarlem, Leiden, and Amsterdam into "its own sea."[116] Further infestation had continued during the period between 1730 and 1733, not only along its coastal dikes but also along the sluice gates that drained its productive interior.[117] "In these worrisome times," their plan would respond effectively to the "urgent need" of the Republic. This was the original interpretation of

---

[114] Ibid.
[115] Andries Vierlingh had already designed sloping, stone-clad dikes in the sixteenth century, and Gottfried Sellius suggested a similar approach in his shipworm treatise. Baars, "Het dijkherstel," 197.
[116] Ibid., 29.
[117] Sluice gates often presented more complicated problems for technological and political reasons; water boards in Holland and Zeeland employed a variety of shipworm proofing strategies, including *wormspijkers*, copper plating, and reconstruction with stone. P. Harting, W. Vrolik, D. J. Storm Buysing et al., *Verslag over den paalworm: Uitgegeven door de Natuurkundige afdeeling der Koninklijke Akademie van Wetenschappen* (Amsterdam: C.G. van der Post, 1860), 50–1. Johannes Jouke Schilstra, *Wie water deert: Het Hoogheemraadschap van de Uitwaterende Sluizen in Kennemerland en West-Friesland, 1544–1969* (Wormerveer: Meijer, 1969), 136.

shipworms as the flooding hazard merged with the existential threat of total inundation and decline. They responded with providential and patriotic zeal calling on "the help of heaven to prevent dike breaches" for which "the Beloved Fatherland's prosperity and salvation depend."[118] Straat and Van der Deure had mastered each dialogue of shipworm response.

4.8 CONCLUSION

Between 1733 and 1746, the Dutch Republic underwent the most comprehensive reconstruction of its coastline in the early-modern period. Beginning in West Friesland, water boards replaced wooden revetments with stone across Holland and eventually across the Republic, employing variations of Straat and Van der Deure's design. Friesland began laying stone at the foot of their dikes by 1734. Water boards near Amsterdam rebuilt the Diemer dike with a stone slope between 1735 and 1737. The neighboring Bunschoter dike underwent a similar process beginning in 1737. Dike designs were neither universally applied nor uniform in appearance, but the risk of inaction prompted water boards to enact dramatic changes across the Netherlands.[119] Historians have long recognized the power of coastal flooding to prompt innovation in dike management and design. No flood, however, had come close to catalyzing so dramatic a response. The widespread reconstruction of coastal dikes after the 1730s was a testament to the unique nature of the shipworm threat.

Shipworm novelty affected Dutch response in numerous ways. Water boards typically relied on their extensive experience with flooding to prescribe solutions, but the cultural memory of disaster provided no template for shipworm response. Although not completely unknown, shipworms had never been systematically studied, and no similar outbreak had ever occurred in Dutch waters. Dike inspectors immediately gathered information about the threat and disseminated it within a small cadre of water authorities. The inadequacy of their collective experience prompted provinces to bring in formally trained experts to advise their commissions, and, by 1733, natural historians contributed the first

---

[118] Straat and Van der Deure, *Ontwerp*, 6.
[119] Pieter Straat would be consulted about the state of the Diemerdijk in 1735. Fransen, *Dijk onder spanning*, 219, 230. Heleen Kole, *Polderen of niet? Participatie in het bestuur van de waterschappen Bunschoten en Mastenbroek vóór 1800* (Hilversum: Verloren, 2017), 136–40. Davids, *The Rise and Decline*, 69. In Zeeland, the province ultimately opted to protect their wooden piles with *wormspijkers*. Hollestelle, "De zorg," 103–21.

systematic descriptions of the mollusk and its outbreak. This new synergy between state authorities and outside experts modeled a pattern that would be repeated during subsequent disasters.

The novelty of the animal and the unprecedented nature of the outbreak also encouraged diverse interpretations of the disaster's meaning and implications. Natural histories and dike reports described the mollusks and the scope of the disaster in great detail, yet no single causal story emerged. Water boards coped with this uncertainty by transforming shipworms into the more familiar threat of flooding. In contrast to most floods, however, the shipworm threat lasted for years and threatened entire coastlines. No flood had presented so broad a danger to the Dutch Republic. Any wooden structure fronting the sea seemed vulnerable and any city or landscape behind it at risk. Shipworms posed a doomsday scenario where storms breached weakened dikes and floodwaters stretched into the heart of Holland. The combination of widespread infestation with an interpretation that transformed them into flooding hazards incentivized a technological response for a flooding threat that was ultimately far more consequential than any actual storm.

At the same time, the "previously unknown" shipworms encouraged other equally powerful interpretations. Most observers considered shipworms as a manifestation of divine providence and rebuke of the Republic's moral decline. Pietist ministers and early-Enlightenment moralists joined in common cause to condemn this trend. Unlike previous disasters, however, moralists tied shipworms to a single sin. The shipworm panic arose in the wake of a wave of sodomy trials in the Netherlands, and moralists quickly developed causal connections between the two phenomena. The novelty and timing of the outbreak reinforced this interpretation. Only God could have directed so bizarre an affliction arriving on the heels of the sodomy trials, war, cattle plagues, and floods.

The malleability of the shipworm as a symbol ultimately proved to be its most significant characteristic. Shipworms represented the apocalyptic threat of flooding, the unfathomable mystery of God's creation, and the certainty of moral decline. Each interpretation provoked dramatic and unprecedented responses, whether the first systematic descriptions of the mollusk, violence toward persecuted minorities, or promoting capital-intensive reorganization of coastal infrastructure. This last technological response yielded lasting economic implications that echoed through subsequent decades. Dike investment locked coastal communities into a protracted adaptive strategy, whose cost proved enormous and often exceeded the value of the lands they protected.[120] The ongoing economic recession in North Holland, and increasingly vocal criticisms about the Dutch Republic's financial standing, fueled greater anxiety about its future. As the 1730s neared its end, coastal dike reinforcement had not yet been completed when another series of flood disasters struck the United Provinces – this time reaching into the rural heart of the Republic.

---

[120] Bieleman, *Geschiedenis*, 106–07. Two recent studies reveal how stone slopes were built during an era of economic depression, both along the south-eastern coast of the Southern Sea. Kole, *Polderen of niet?*. Wim Hagoort, *Het hoofd boven water. De geschiedenis van de Gelderse zeepolder Arkemheen, gemeenten Nijkerk en Putten 1356 (806–1916)* (Nijkerk: Nabij, 2018).

5

"Increasingly Numerous and Higher Floods"

The River Floods of 1740–1741

Shortly after floodwaters receded around 's Hertogenbosch (Den Bosch) in the early spring of 1757, Jacob Pierlinck surveyed the extent of the damage. Pierlinck lived in the city and had witnessed the disaster firsthand. In a pamphlet printed several months later, he described "the sad state that I saw around the region of Den Bosch along the rivers Waal and Meuse and the ... great numbers of unlucky people in whose midst I lived that flooded from all sides into the city." He recounted miraculous rescues, identified dike breaches, and described Den Bosch as a refugee center. Pierlinck's reportage was detailed, but it served a greater purpose than mere description. He wanted to use the tragedy to understand river flooding and "in all zealousness to alert and investigate the origins of these disasters to the best of my limited ability."[1] This was a tall order in the 1750s, not only because the environmental context and provincial nature of Dutch river management produced widely variable conditions but also because the nature of the flooding itself seemed to be changing. To Pierlinck and many of his contemporaries, the flood of 1757 revealed a dangerous trend. River floods had become more frequent and more devastating.

Pierlinck was a respected surveyor, an *ordinaire ingenieur* of the Republic, and an astute observer of recent history. The flood of 1757 was only the latest in a series of truly disastrous inundations to affect the Dutch riverlands. Exceptional flooding produced disasters in 1709, 1726, 1740–1, 1744, 1747, 1751, and 1753. Pierlinck listed no fewer than twelve inundations or

---

[1] Jacob Pierlinck, De verschrikkelyke watersnood, langs de ... Waal en de Maas, voorgevallen in ... February des Jaars 1757 ... (Amsterdam: Isaak Tirion, 1757), 2. Many thanks to Toon Bosch for sharing this source.

floods resulting from ice dams or high water between 1653 and 1757. The scale of these disasters was astonishing – a point he emphasized on a hand-colored map of the 1757 inundations (Figure 5.1). Light, feathered lines burst outward from river dikes to indicate breaches; dark, thick lines clog riverways, indicating ice dams; and light shading over the land between the Meuse and Waal rivers signify the spatial extent of floodwaters. More importantly, the frequency and severity of disasters like the 1757 flood seemed to be increasing. "Who doesn't fear the sadder consequences," he asked, of "increasingly numerous and higher floods, especially since the year 1740?"[2] To Pierlinck, river flooding seemed to be getting worse.

Pierlinck was not alone in his belief that river flooding was worsening and that it warranted greater attention. State and popular interest in river flooding and management rose in tandem amid the repeated river disasters of the first half of the eighteenth century. This was a change from the seventeenth century when coastal flooding had dominated public discourse. River floods, although not infrequent during the seventeenth century, rarely resulted in the extreme loss of life that at once horrified and captivated readers of disaster media.[3] By the second quarter of the eighteenth century, the tide of public concern shifted and greater attention was focused on rivers. Repeated inundation, and also protracted environmental changes such as sedimentation and subsidence, convinced many observers that the river system itself was deteriorating. After a dike breach released floodwaters deep into the heart of Holland in 1726, fear intensified that an even greater flood might affect the urban core of the Republic. The Rhine and Meuse river floods during the winter of 1740–1 became the closest realization of this apocalyptic scenario to date. Contemporaries calculated that floodwaters inundated 165,550 *morgens* of land, affecting over 100,000 people.[4] Water stretched across Holland, Utrecht, the Generality Lands, and Gelderland (Figure 5.2). The 1740–1 disaster dwarfed the floods of 1726. The height, extent, and damage caused by the 1740–41 river floods served as a new measure against which river floods, past and future, would be compared.

---

[2] Ibid., 34.
[3] High flooding mortality was exceedingly rare, though the deadliest were coastal floods. Van Bavel et al., "Economic Inequality."
[4] *Nederlansch Gedenkboek: Of, Europische Mercurius* (Amsterdam: J. Ratelband, 1741), 122. In the Southern Netherlands, morgens could be as small as 0.66 hectares near the town of Breda in Noord Brabant or as large as 1.38 hectares in the Achterhoek of Gelderland. Verhoeff, *Maten en gewichten databank*.

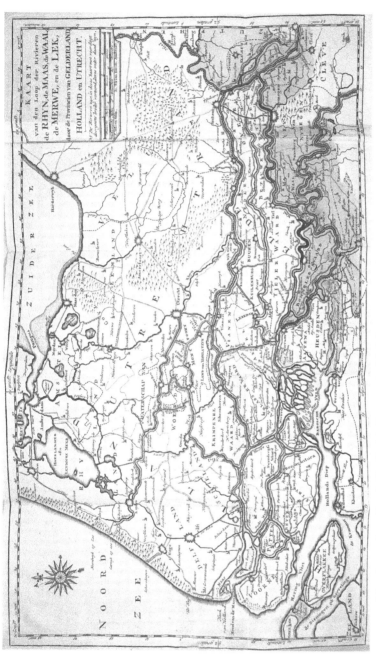

FIGURE 5.1. Map insert from Jacob Pierlinck's *De verschrikkelyke watersnood*. The shaded areas indicate the extent of the 1757 floods. Pierlinck borrowed the base map from Cornelis Velsen, *Kaart van den loop der rivieren de Rhyn, de Maas, de Waal, de Merwe, en de Lek, door de provincien van Gelderland, Holland en Utrecht*, Map, 1749, NA, Kaartcollectie Zuid-Holland Ernsting, http://proxy.handle.net/10648/af934a74-d0b4-102d-bcf8-003048976d84.

FIGURE 5.2. Map of territories affected by 1740–1 inundations, including major water boards, municipalities, cities, and towns. Hashed area indicates flood extent, approximated from Jan Goeree, *Kaart van Zuid-Holland waarop zijn aangegeven de overstromingen in de jaren 1726 en 1740–1741*, Map, 1741. Rijksmuseum Amsterdam, http://hdl.handle.net/10934/RM0001 .COLLECT.479253 and Buisman, *Duizend jaar weer*, vol. 5, 702–18. Boundaries modified from Boonstra, NLGis shapefiles.

The River Floods of 1740–1741

This chapter explores the changing nature of river flooding during the first half of the eighteenth century. It evaluates the environmental and social influences on flood frequency and severity, as well as the perceptions of technical experts, state authorities, and the broader public who believed those conditions were changing. The 1740–1 disaster catalyzed new considerations of both. State efforts treated the disaster as a symptom of a much broader disorder – the declining condition of the Dutch riverscape. Rivers discharged either too much or too little water, sedimentation elevated riverbeds and clogged river mouths, and dikes suffered from mismanagement and disrepair. Facing what it perceived as a set of complex and systemic challenges, the province of Holland spearheaded technocratic interventions into the river system. These provincial efforts largely stemmed from the perception that the province stood exposed to increasing flood risk, which formally trained hydraulic experts identified and described in catastrophic terms and visualized using diverse cartographic tools. Pierlinck himself was a representative of this first generation of water management specialists to whom state authorities consistently turned for guidance. These experts used the floods of 1740–1 and the fear of total inundation to promote innovation and interprovincial management of their unruly rivers.

At the same time, the 1740–1 river floods promoted new social and religious responses. The scale and severity of the disaster overwhelmed the local and regional institutions traditionally tasked with relief and, in the process, transformed Dutch flood culture. Flooding was only partly responsible for these changes, however. Many inundated communities had suffered acute financial losses in the preceding years, including cattle plagues, river floods, harvest failures, and an historically cold winter in 1739–40. These concurrent or sequential calamities produced a disaster cascade that rippled through rural society. Disaster victims, in their appeals for assistance, argued that the floods were merely the latest episode in a decades-long disaster that affected not only them but also the Republic at large. The floods reinforced the providential arguments of clergy that linked disaster to sin and decline, just as they prompted the first coordinated disaster relief efforts from outside affected regions in Dutch history. These new public relief efforts as well as technocratic state responses relied upon discourses of increasing risk.[5] Recent history seemed to confirm that disasters were getting worse, due in part to their

---

[5] "Risk" and "vulnerability" are fundamental concepts in disaster studies. Disaster risk is often interpreted as a function of the severity, frequency, and spatial extent of a natural hazard combined with social susceptibility to harm (vulnerability). Vulnerability is often

own interventions in the river system. This conviction, although grounded in an historical awareness of past flooding events, remained firmly focused on the future. The 1740–1 floods expanded debates about environmental and moral decline, but it also promoted new responses to both.

5.1 THE RHINE–MEUSE RIVER SYSTEM AND FLOOD FREQUENCY

Any understanding of the changing nature of extreme flooding along the Rhine and Meuse river system must begin with the rivers themselves. Few parts of the western and southern areas of the Dutch Republic were unaffected by the fluvial history of its rivers, and no two rivers were as important as the Rhine and the Meuse.[6] Indeed, much of the most highly urbanized, intensively cultivated areas of the Republic lay in their western delta. These deltaic landscapes, characterized by peat bogs interspersed with river sand and clay soils, were formed through the interaction of the rivers and the North Sea. Further east, the rivers assumed a more central role in landscape morphology. Much of the soil of the "riverlands" region of Utrecht, the Generality Lands in what is now North Brabant, and Gelderland is composed of river clay, sand, and sometimes peat. These conditions influenced later settlement, water management, and agricultural productivity.[7] Along with the sea, the Rhine and the Meuse are two of the most significant geophysical forces responsible for the formation of the Netherlands.

Human influences on the river landscapes began to rival the power of rivers and seas already by the Middle Ages. Reclamation, dam building, and embankment increased dramatically throughout this period.[8] These technological interventions were also self-reinforcing. Dam building, for instance, converted former branches of the Rhine into drainage channels

---

stratified via social inequalities in class, gender, and ethnicity. Ben Wisner, Jean-Christophe Gaillard, and Ilan Kelman, "Framing Disaster: Theories and Stories Seeking to Understand Hazards, Vulnerability and Risk," in *Handbook of Hazards and Disaster Risk Reduction*, eds. Ben Wisner, Jean-Christophe Gaillard, and Ilan Kelman (New York: Routledge, 2012), 22–4. Cannon, "Vulnerability Analysis," 19–20. Eighteenth-century observers did not interpret flood risk and vulnerability in these terms. Risk entailed recognition of differential exposure to hazards, but social vulnerability focused largely on moral concerns.

[6] Wildred Ten Brinke, *The Dutch Rhine: A Restrained River* (Diemen: Veen Magazines, 2005), vii, 1.
[7] Van de Ven, *Man-made Lowlands*, 45; Van Bavel, *Manors and Markets*, 21–2.
[8] Berendsen, "Birds-Eye View," 746–8.

The River Floods of 1740–1741 171

for reclamation. River dikes provided a degree of flood protection for these new polder landscapes as well as the towns and cities that sprang up near the dams.[9] The rivers connected these regions of rural and urban development together. Inundation remained a perennial challenge despite efforts to tame the rivers, and not only during extreme flooding events. In the river regions of Gelderland, for instance, drainage proved insufficient during the winter when discharge was greatest. The region typically remained inundated until spring.[10] Like all rivers, the Rhine and the Meuse were dynamic actors that resisted efforts to constrain or harness their power. By the seventeenth century, however, drainage and embankment technologies had improved, and communities were more dependent upon the Rhine and the Meuse than ever for reclamation, navigation, and commerce. Centuries of river modification had largely stabilized their paths to the sea; yet, these efforts also produced unexpected consequences that came to a head by the early eighteenth century.

The broad contours of the modern Rhine and Meuse system were already in place by the early eighteenth century (Figure 5.3). It consisted of multiple, meandering branches stretching from east to west across four provinces and the Generality Lands. Of the two, the Rhine was far longer, tapped a larger basin, and discharged much more water.[11] Almost as soon as the Rhine entered the Republic, it forked in two. The main branch of the Rhine continued north, but most water poured into the Waal River, which flowed westward toward Holland. The Rhine branched again near the city of Arnhem in Gelderland. Water diverted north flowed along the IJssel and eventually found its way to the Southern Sea. Water moving westward became the Nederrijn and eventually the Lek after it passed the town of Wijk bij Duurstede. Both the Waal and Lek branches eventually released their waters into the North Sea. The discharge of these branches was far from equitable, however. By 1700, 90 percent of Rhine water flowing through the Dutch Republic moved along the Waal, with the remaining 10 percent divided between the Lek and the IJssel. This created serious issues with sedimentation in the Lek and the IJssel that hindered navigation, but it also increased the risk of flooding along the swollen

---

[9] Van de Ven, *Man-made Lowlands*, 72–5.
[10] Anneke Driessen, *Watersnood tussen Maas en Waal: Overstromingsrampen in het rivierengebied tussen 1780 en 1810* (Zutphen: Walburg Pers, 1994), 44.
[11] Ton Burgers, *Nederlands grote rivieren: Drie eeuwen strijd tegen overstromingen* (Utrecht: Uitgeverij Matrijs, 2014), 20–1.

FIGURE 5.3. The Rhine–Meuse River System. Labeled rivers indicate water bodies that flooded in 1740–1 or were otherwise subjects of consistent anxiety for hydraulic experts. Boundaries modified from Boonstra, NLGis shapefiles.

Waal.[12] For these reasons, one of the principal challenges of river water management and flood prevention throughout the eighteenth century was the stabilization of the Rhine River distributaries.

The proximity of the Rhine and Meuse rivers created additional challenges. The Meuse and the Rhine began to converge almost as soon as the Rhine entered the Republic. They ran roughly parallel, before finally merging into the Merwede River north of the town of Woudrichem in Holland. A portion of the Merwede continued westward, but most of its water emptied into the Biesbosch, a freshwater tidal wetland that had been created during a series of coastal and river floods in the 1420s.[13] The complexity of this river system with its multiple mouths, sandbars, and man-made structures such as

---

[12] Ibid., 14–20. Karel Davids, "River Control and the Evolution of Knowledge: A Comparison between Regions in China and Europe, c. 1400–1850," *Journal of Global History* 1, no. 1 (2006): 63–4.

[13] Maarten G. Kleinhans, Henk J. T. Weerts, and Kim M. Cohen, "Avulsion in Action: Reconstruction and Modelling Sedimentation Pace and Upstream Flood Water Levels Following a Medieval Tidal-River Diversion Catastrophe (Biesbosch, the Netherlands, 1421–1750 AD)," *Geomorphology* 118, no. 1–2 (2010): 65–79. Jan Buisman, *Duizend jaar weer, wind, en water in de Lage Landen, 1300–1450*, Vol. 2 (Franeker: Van Wijnen, 1996), 450–6.

groynes and dams created both flooding and navigation hazards, depending on location. High water in one portion of the river system might also affect another. The region north of Den Bosch, for instance, had no dikes between the Meuse and the Waal. If the Waal overtopped its embankments, water could flow into the Meuse.[14] Similarly, breaches in the relatively elevated eastern part of the riverlands would often spill water westward, encouraging additional breaches. The structural nature of these conditions meant that breaches often occurred in similar places for similar reasons, especially during periods of elevated river discharge.

Extreme weather compounded these structural weaknesses in the river system. Peak discharge generally occurred during periods of heavy winter precipitation in the Meuse and Lower Rhine basins, and virtually all river flooding took place during this time of the year.[15] The Rhine was also fed by Alpine snowmelt that typically peaked in the early summer. Cool springs and summers delayed meltwater discharge and, when combined with winter precipitation, elevated water levels still higher.[16] Dike breaches rarely occurred, however, unless high-water conditions persisted for long periods of time. Prolonged and pronounced high water was dangerous for several reasons. Protracted high-water episodes resulted in dike slippage when water filtered into and under the earthen dike bodies.[17] Dike breaches might also form when high water overtopped the crowns of dikes and scoured the base of their interior slopes. These were precisely the conditions that occurred in 1740-1, as dikes near the Dutch–German border succumbed to persistent high water. Over the next three weeks, subsequent dike breaches resulting from high water and the influx of floodwater pouring through breaches to the east expanded the disaster all the way to southern Holland.[18]

Floods caused by heavy winter precipitation, however, were somewhat outside the norm when compared to inundations that occurred over the previous century. Nearly 70 percent of breaches in the seventeenth

---

[14] Driessen, *Watersnood*, 31-2.
[15] G. P. van de Ven, A. M. A. J. Driessen, W. Wolters, H. J. Wasser, and T. Stol, *Niets is bestendig ...: De geschiedenis van de rivieroverstromingen in Nederland* (Utrecht: Matrijs, 1995), 18-22.
[16] Van de Ven, *Man-made Lowlands*, 24-5; Sonu Khanal, Arthur F. Lutz, Walter W. Immerzeel et al., "The Impact of Meteorological and Hydrological Memory on Compound Peak Flows in the Rhine River Basin," *Atmosphere* 10, no. 4 (2019): 2.
[17] Paul Hudson, Hans Middelkoop, and Esther Stouthamer, "Flood Management along the Lower Mississippi and Rhine Rivers (the Netherlands) and the Continuum of Geomorphic Adjustment," *Geomorphology* 101, no. 1 (2008): 228.
[18] Buisman, *Duizend jaar*, Vol. 5, 702-18.

century had resulted from ice dams.[19] Ice dams formed when rain in upstream basins, often combined with meltwater from quickly thawing snowpack, pushed water into still-frozen stretches of the delta. Lingering ice coalesced into massive dams that slowed river discharge and added considerable pressure to the dikes. Ice dams raised water levels and, in some cases, forced water over their crowns.[20] Severe winters encouraged these river-freezing events, and frozen soils prevented infiltration of runoff.[21]

It is perhaps this connection between ice dams, river flooding, and severe winter weather that has promoted their association with the cooler conditions of the Little Ice Age.[22] Many of the most catastrophic floods were not connected to these freezing and thawing events, however, including the floods of 1740–1. Indeed, contemporaries considered the lack of ice during the 1740–1 floods one of its most unusual traits. One report from Culemborg in winter 1740–1 described the surprise that the river maintained its high-water levels despite being an "open river" free of ice.[23] The *Europische Mercurius* reported that the lack of ice "appears incomprehensible and has not happened in peoples' memory."[24] The absence of ice dams in the presence of extreme high water was perhaps the clearest evidence that this flood was unique.

The warmer winters that prevailed by the 1730s may point to other potential climatic influences on changing flood regimes. Proxy

---

[19] R. Glaser and H. Stangl, "Historical Floods in the Dutch Rhine Delta," *Natural Hazards and Earth System Sciences* 3, no. 6 (2003): 611.

[20] Van de Ven, *Made-Made Lowlands*, 26–8.

[21] Willem H. J. Toonen, Hans Middelkoop, Tiuri Y. M. Konijndijk et al., "The Influence of Hydroclimatic Variability on Flood Frequency in the Lower Rhine," *Earth Surface Processes and Landforms* 41 (2016): 1267.

[22] Rudolf Brázdil, Zbigniew W. Kundzewicz, and Gerardo Benito, "Historical Hydrology for Studying Flood Risk in Europe," *Hydrological Sciences Journal* 51, no. 5 (2006): 758. Adriaan M. J. de Kraker, "Ice and Water. The Removal of Ice on Waterways in the Low Countries, 1330–1800," *Water History* 9, no. 2 (2017): 109–28. Christian Rohr, "Ice Jams and Their Impact on Urban Communities from a Long-Term Perspective (Middle Ages to 19th Century)," in *The Power of Urban Water: Studies in Premodern Urbanism*, eds. Nicola Chiarenza, Annette Haug, and Ulrich Müller (Berlin: Walter de Gruyter, 2020), 197–212.

[23] "Kort Verhaal van de Hooge Watervloed, waar mede de Graafschap Culemborg, nevens andere nabuurige Landschappen, in December 1740, en January 1741, is bezogt geworden." In *Verzameling Van eenige geloofwaardige Berigten en Brieven Betreffende de Elende van de Opgezetenen der overstroomde landen in Nederland* (Amsterdam: Isaak Tirion, 1741), 23.

[24] *Nederlansch Gedenkboek* (1741), 118.

reconstructions based on dike damage reports reveal that average flood intensity increased during the warm interregnum between the Maunder and Dalton minima.[25] Flood frequency along the Lower Rhine also grew, although recent analysis indicates that the most significant increase took place between 1760 and 1800, the latter a cooler period and, on a continental scale, the most flood-rich era of the past 500 years.[26] These changes in flood intensity and frequency may have responded to large-scale climatic phenomena that govern the Atlantic climate system. Hydroclimatic research, for instance, has shown a high correlation between periods of "normal" flooding along the Lower Rhine and positive phases of the North Atlantic Oscillation (NAO), which typically brought mild and wet winter conditions to Western Europe.[27] On the other hand, the 1730s were also uncharacteristically dry, and winters would become even drier in the early 1740s.[28] These conditions would have decreased the likelihood of flooding. The more intense period of flood intensity and frequency after 1760 likely responded to a still different atmospheric circulation regime.[29] The relationship between climate and flooding was complex and often broke down on the scale of extreme events. Many of the largest floods, including those in 1740–1, occurred during climatic phases less conducive to their development.[30] The floods of 1740–1 remained exceptional in the first half of the eighteenth century, but just as Pierlinck feared, they would become increasingly characteristic after midcentury.

---

[25] Willem H. J. Toonen, "Flood Frequency Analysis and Discussion of Non-Stationarity of the Lower Rhine Flooding Regime (AD 1350–2011): Using Discharge Data, Water Level Measurements, and Historical Records," *Journal of Hydrology* 528 (2015): 496.
[26] Blöschl et al., "Current European," 561–2.
[27] Studies have also indicated high correlation between flooding periodicity and the Atlantic Multi-decadal Oscillation (AMO). Toonen et al., "The Influence of Hydroclimatic Variability," 1271–2. Heinz Wanner, Christoph Beck, Rudolf Brázdil et al., "Dynamic and Socioeconomic Aspects of Historical Floods in Central Europe," *Erdkunde* 58, no. 1 (2004): 3. Tree ring reconstructions of the AMO in the period between 1709 and 1763, however, showed no strong signal. Stephen T. Gray, Lisa J. Graumlich, Julio L. Betancourt et al., "A Tree-Ring Based Reconstruction of the Atlantic Multidecadal Oscillation since 1567 AD," *Geophysical Research Letters* 31, no. 12 (2004): 2–3.
[28] Jones and Briffa, "Unusual Climate in Northwest Europe," 367–9.
[29] Likely a positive AMO and negative NAO. Blöschl et al., "Current European," 564.
[30] Ensemble reconstructions indicate predominantly negative NAO values during the first half of the eighteenth century. Pablo Ortega, Flavio Lehner, Didier Swingedouw et al., "A Model-Tested North Atlantic Oscillation Reconstruction for the Past Millennium," *Nature* 523, no. 7558 (2015): 72.

Most flood regime reconstructions rely at least in part on dike damage reports, which further complicates this picture because dike breaches resulted from high waters and also maintenance issues. A variety of social variables, including rising or declining political and economic inequalities, affected the capacity of communities to invest in flood protection and thus influenced the likelihood that dike breaches took place. In Gelderland, for instance, river flooding grew more intense and common after the fifteenth century. This was due in part to a growing neglect of dike maintenance. At the same time, wealthy farmers and elites who valued profitable drainage schemes over flood protection increasingly dominated water management decisions. Resulting floods decreased the value of adjoining lands, encouraging accumulation of landownership within these same classes. The feedback loop this created increased the severity and frequency of flooding.[31] The agricultural recession in the first half of the eighteenth century compounded these challenges further, particularly in regions dominated by large absentee landowners, and the flooding only grew worse after 1740.[32] This toxic interplay between social inequalities and inflexible water management institutions increased the likelihood that high water would result in disaster. Individual flood disasters, and also long-term trends in their frequency and severity, were environmental *and* social phenomena.

These changes in climate and society did not factor into contemporary considerations of flood frequency and severity, but that did not preclude their awareness of historical, environmental, and social change, nor their relationship to flooding. For Pierlinck and others, the timing and extent of extreme flood events provided the clearest indications that flood regimes were changing. Writing in 1757, he could remember the consequences of six major flooding events since 1726, including the catastrophic floods of 1740–1. As with coastal floods, river flooding *memoria* contained an archive of historical information against which more recent changes could be assessed. These included flood markers indicating high-water levels, chronicles and pamphlets documenting past inundations, and, by the mid-eighteenth century, increasingly detailed maps that documented past inundation and indicated areas at greatest future risk.[33] This

---

[31] Van Bavel et al., "Economic Inequality."
[32] Paul Brusse, "Property, Power and Participation in Local Administration in the Dutch Delta in the Early Modern Period," *Continuity and Change* 33, no. 1 (2018): 70. Paul Brusse, *Overleven door ondernemen: De agrarische geschiedenis van de Over-Betuwe 1650–1850*(Wageningen: Landbouwuniversiteit Wageningen, 1999), 201–6.
[33] Buisman, *Duizend jaar*, Vol. 5, 894–6.

The River Floods of 1740–1741

documentation contextualized extreme events and encouraged comparisons over time and space. Based on this information, Pierlinck and his contemporaries concluded they were experiencing the high point of Dutch river flooding within living memory. This awareness gave significant weight to a discourse of increasing flood risk and corresponding need for action. The floods of 1740–1 amplified and transformed these responses.

5.2 THE RIVER FLOODS OF 1740–1741

Well-informed Dutch observers were already apprehensive about flooding in fall 1740. The previous winter had been bitterly cold across western and central Europe with heavy precipitation that fell as snow in the Alps.[34] The chilly spring and summer preserved this snowpack, which observers understood increased the likelihood of dangerous high waters along the Rhine. Newspapers and news digests regularly reported on extraordinary or ominous weather conditions at home and abroad, especially those that pertained to Rhine and Meuse flooding. Reports of the continental extent of the previous harsh winter and the "heavy rains that fell almost the entire summer" of 1740 left many with a sense of foreboding, and a wet, cold October added to their unease.[35] By November, "many feared that the dikes along the Lek, Meuse, and Waal" would overflow.[36] As news filtered into the Netherlands of high waters and dike breaches in Cologne and Cleves, cities and villages in the eastern Netherlands braced for the worst.

The first dike breach in the Republic occurred on December 15, 1740 along the northern bank of the Meuse River near the town of Nederasselt. Rising floodwaters threatened a large area in Gelderland and spread eastward into the richly productive agricultural region between the Meuse and Waal rivers. On December 18, a Rhine breach occurred at Keeken (in what is now Germany) and a dike near the Pannerdens Canal and the eastern border of the Republic collapsed two days later. Villagers downriver in the village of Heusden near Den Bosch

---

[34] Ibid., 698–9.
[35] Jan Wagenaar, *Kort en Opregt Verhaal van de Elende der Opgezetenen van de overstroomde Landstreeken in Nederland en byzonderlyk van den Alblasserwaard. Opgesteld uit verscheiden' geloofwaardige Brieven en Berigten* (1741), 3.
[36] *Verhaal van de droevige waternood en inbreuken van een groot gedeelte van Gelderland en Holland, in 't laatst van den Jare 1740. en in den beginne van 't Jaar 1741* (Amsterdam: Jan ten Houten, 1741), A2.

FIGURE 5.4. The dike breach at Elden. *Source*: Isaac Vincentsz. van der Vinne, *Het doorbreeken van den band-dyk voor Elden Anno 1740*, Print, 1741, Rijksmuseum, Amsterdam, http://hdl.handle.net/10934/RM0001.COLLECT.479380.

nervously watched the water level of the Meuse climb steadily. It was already extremely high by December 22 and continued to rise by "more than a thumb hour after hour." By Christmas Eve, it was "three quarters of a foot higher than anyone could remember."[37] Similar anxious reports in Gelderland, Utrecht, and Holland noted that water levels shattered records set in the seventeenth century.

Dikes breaches resulted from both prolonged and extreme high water. Two days after the dike broke at Elden in Gelderland, for instance, the magistrate of Wageningen reported that high water had overtopped the dike, creating the breach.[38] Contemporary images of the disaster emphasize this point as well. A print by Isaac Vincentsz. van der Vinne shows floodwater flowing through the breach at Elden (Figure 5.4). With water cresting at record levels along the entire river system and remaining high for weeks, further breaches seemed inevitable.

---

[37] *Nederlansch Gedenkboek: Of, Europische Mercurius* (Amsterdam: J. Ratelband, 1740), 316. Three quarters of a foot was approximately 23 cm. Verhoeff, *Maten en gewichten*.

[38] J. Van Zellem, "'Nooyt gehoorde hooge waeteren': bestuurlijke, technische en sociale aspecten, in het bijzonder de hulpverlening, van de overstromingsramp in de Over-Betuwe in 1740–1741," *Tijdschrift Voor Waterstaatsgeschiedenis* 12 (2003): 11–20.

Poor dike maintenance exacerbated these conditions. Dozens of local and regional water boards maintained the river dikes and managed the riverlands surrounding them. A dike report in eastern Gelderland in November 1740 noted the troubling absence of osier buffers (woven willow or elm mats) on dikes near the Pannerdens Canal. Less than a month later, it collapsed.[39] Similar reports pointed to poor maintenance in Holland along the Lek River. The surveyor Melchior Bolstra noted the poor condition of the Dief dike during the floods in 1740 and bemoaned its haphazard maintenance.[40] This transverse dike was all that prevented water flowing from inundated Gelderland into Holland. Considering its uneven maintenance record, it seemed a miracle to have survived intact. Any sense of relief was short lived. Breaches downriver soon circumvented the dike and floodwaters rushed into Holland.

Over the next two weeks, dikes breached and water spread along the great rivers, inundating thousands of hectares of land across three Dutch provinces and the Generality Lands.[41] The sudden inundation of the Betuwe and Tielerwaard regions of Gelderland on Christmas Eve night led some to term the disaster the "Second Christmas Flood." Sisters in the Huissen cloister lamented that "instead of being able to rest and prepare for [Christmas] services early the next morning, one was busy until midnight to saving provisions from the cellar by moving them upstairs."[42] That same evening, the Rhine and Meuse flooded Den Bosch and the polder lands at the border of Utrecht and Holland.

By the time the final breaches occurred in early January 1741 along the Linge River, contemporaries already understood it to be one of the most severe in the eighteenth century, and perhaps much longer. The *Mercurius* reported that water in the Alblasserwaard region of southern Holland stood thirty-three thumbs higher than in 1726.[43] Another report noted the water in Den Bosch reached a height "seven thumbs higher than men had seen here in 89 years."[44] A report from Culemborg on January 8 stated that "this flood is ten thumbs higher than in the year 1726 and seven thumbs higher than in the year 1658."[45] Flood comparisons tapped into the cultural memory of river flooding, and these shattered records indicated they were experiencing something unprecedented. From the perspective of those worried that flood severity was increasing, the past and present offered compelling evidence.

---

[39] Ibid., 14.   [40] Buisman, *Duizend jaar*, Vol. 5, 710.   [41] Ibid., 704–6.
[42] Ibid., 706.
[43] *Nederlansch Gedenkboek* (1741), 115. Roughly 83 cm. Verhoeff, *Maten en gewichten*.
[44] Ibid., 108. Roughly 18 cm.   [45] Ibid., 118. Roughly 27 cm.

Witnesses understood the immediate influence of weather conditions and high water on the incidence of river flooding. They knew that ice dams and poor maintenance practices exacerbated these hazards. Except for ice damming, each of these conditions played a role, but none fully explained the flood's severity. Water authorities therefore scrambled for alternative explanations. The province of Holland, guided by a formally trained cohort of hydraulic experts, spearheaded this effort. They argued that the Rhine–Meuse River System suffered from multiple systemic problems that ultimately resulted in the 1740–1 floods. Hydraulic experts used the disaster to fashion a new causal story for river floods and exert greater authority in river management.

5.3 RIVER MANAGEMENT AND EXPERTISE IN THE EIGHTEENTH CENTURY

Before the eighteenth century, provincial governments played little role in river management. Rivers lay outside their jurisdiction. River dikes and polder management worked through a patchwork of local institutions and powerful regional water boards. Riverbeds, by contrast, remained largely unregulated and subject to the interests of local landholders. Provincial administrators began to expand their influence in water management during the seventeenth century, especially when it concerned coastal flood infrastructure. Natural disasters such as the Christmas Flood and the shipworm epidemic accelerated these changes in Groningen and Holland. By the 1720s, provinces grew increasingly convinced that floodplains were far too valuable and rivers too volatile to entrust to local or even regional administration. The ungoverned riverbeds seemed an ideal pressure point for provinces to exert control.

Focusing on riverbeds deflected potential legal challenges from territorial water boards. It also allowed them to address three interrelated weaknesses in the river system.[46] The first and by many accounts the most important flaw in the river system was the unequal distribution of Rhine water. Because the vast majority of Rhine water flowed into the Waal, the Nederrijn-Lek and IJssel began silting up. This hindered navigation and

---

[46] Provinces claimed jurisdiction despite protest by water boards because riverbeds and channels had been under the authority of medieval sovereign lords. Van de Ven, *Man-made Lowlands*, 150. Centralization proceeded differently depending on location. Holland established provincial instruments after the floods of 1726, whereas Gelderland codified their authority in 1715. Driessen, *Watersnood*, 61–2. Bosch, *Om de macht over het water*, 29–31.

weakened the Republic's primary defense against land invasions. If water levels dipped too low (as they had during the *Rampjaar*), the Dutch water line would be ineffective. Authorities in Gelderland, Utrecht, and Overijssel constructed a new mouth to the Rhine in 1707 as an attempt to rebalance the water levels. Called the Pannerdens Canal, it followed the path of a *retranchement* built on the order of William III to protect the state from France. Unfortunately, the canal further destabilized the system and the increased flood risk along the Nederrijn-Lek, intensifying anxieties in Holland and Utrecht.[47]

Throughout the 1730s, hydraulic experts in Holland focused on a second issue – the deteriorating state of river mouths. Increasing problems with sedimentation resulted in unbalanced water levels, particularly where the Meuse and Waal intersected to form the Merwede River. Roughly 70 percent of this water, however, filtered into the narrow creeks of the Biesbosch, which slowed river flow downstream, increasing sedimentation and the likelihood of ice damming.[48] Making matters worse, cities along the rivers constructed dams across river mouths. This improved navigability but increased the likelihood of flooding. River cities were oftentimes at odds whether to prioritize commerce or flood protection.

The third object of scrutiny was the deterioration of the riverbeds and floodplains. Rivers were public spaces subject to a variety of interests beyond navigation, including drainage, sanitation, fishing, and commercial exploitation. By the eighteenth century, the combination of reclamation and exploitation of the embanked floodplains (*uiterwaarden*) for pasturage, industry, and arable agriculture narrowed the rivers, reducing their storage capacity. Slowing rivers with dams and weirs encouraged sediment accretion. These obstructions also raised water levels and promoted ice dams. Flood mitigation often directly competed with river use, and authorities struggled to balance these interests throughout the eighteenth century.[49]

During the 1720s and 1730s, provinces grew increasingly invested in river management. This was partly prompted by catastrophic flooding.

---

[47] G. P. van de Ven, *Aan de wieg van Rijkswaterstaat: Wordingsgeschiedenis van het Pannerdens Kanaal* (Zutphen: Walburg Pers, 1976), 24–7.

[48] Paul van den Brink, "River Landscapes: The Origin and Development of the Printed River Map in the Netherlands, 1725–1795," *Imago Mundi* 52 (2000): 70–1.

[49] Van de Ven, *Aan de wieg*, 227–9. Annika Hesselink, "History Makes a River: Morphological Changes and Human Interference in the River Rhine, the Netherlands" (PhD diss., Utrecht University, 2002), 23.

Holland formed its first "Hydraulic Department" to guide river management, for instance, after the river floods of 1726. This was also in keeping with the slow trajectory of centralization exhibited during earlier disasters in the eighteenth century. The shipworm epidemic had been a case in point. Holland financed dike repairs, absorbed the costs of tax remission, and commissioned expert consultants to vet proposals for new dike designs. Two of those experts, the surveyors Nicolaas Cruquius and Cornelis Velsen, would become central players in Holland's river management.

Provincial reliance upon hydraulic expertise also increased during this period. Cruquius produced one of the first comprehensive river management plans in the 1720s and later published a groundbreaking map of the Merwede River, drawing on his extensive hydrographical and meteorological observations. Cornelis Velsen was Cruquius's student and the intellectual heir to his systematic, empirical approach to water management. Both were products of Leiden University, which became a European center of applied science in the eighteenth century. Students at Leiden could train in the *Duytse mathematique*, which offered vernacular courses in engineering and surveying. They might also learn from luminaries in continental Newtonian science, such as the physician Herman Boerhaave and mathematician Willem Jacob 's Gravesande. Cruquius studied medicine at Leiden and was heavily influenced by both scholars. Leiden's balance of professional training and theory proved an effective combination. In fact, all the first generation of hydraulic experts traced their intellectual lineage back to Leiden. When Holland installed Velsen as principal officer of its Hydraulic Department in 1731, this formalized a long-developing trend in the professionalization of water expertise.[50]

Cruquius and Velsen's formal training deeply informed their vision of river management. Boerhaave's approach to medical diagnoses influenced Cruquius's interpretation of water problems and, by extension, his followers. Rivers, like the human body, were complex systems. An imbalance or illness in one part of that body might affect another, or perhaps even the whole. Rather than addressing localized problems in isolation, experts began to view rivers as integrated systems. In the process, river management transformed from a practical task informed by localized experience to a generalizable responsibility that relied upon theoretical

---

[50] Paul van den Brink, "*In een opslag van het oog*": *De Hollandse rivierkartografie en waterstaatszorg in opkomst, 1725–1754* (Alphen aan den Rijn: Canaletto/Repro-Holland, 1998), 16, 26–41.

understanding of river hydraulics.[51] This shift also affected interpretations of flooding. Causal stories that once focused on water levels and poor dike maintenance on the local level began to emphasize the deterioration of the river system itself. The record-breaking inundations of 1740–1 accelerated this transition.

5.4 TECHNOCRATIC RESPONSE TO 1740–1741 FLOODS

Despite the growing influence of the provinces, local and regional water boards still presented the first line of defense during a flood. Rijnland was one of the oldest and most powerful water boards in Holland and it wasted no time in 1740 evaluating the risk of dike failure in their domain.[52] Only one day after the first dike breach in Nederasselt and a week before the first breaches in Holland, they sent their Leiden-trained surveyor, Melchior Bolstra, to inspect the dikes bordering their territory. Bolstra focused his attention on the Lek River. The Lek was a continuation of the Nederrijn and one of the main distributary channels of the Rhine system. Between December and February 1741, he inspected its entire northern dike, which protected valuable terrain in Holland.[53] Although Rijnland did not border the Lek directly, the water board feared that floodwater might spill over adjoining polders into their territory.

Bolstra's dike assessment shocked Rijnland. He reported that the Lek dikes "are in a terrible state," and water had reached historic levels. This was doubly surprising because the Rhine was "an open river without any ice dams." The water was thirteen thumbs higher than in 1598 and higher even than 1726.[54] More ominously, Holland's exposure to flooding had increased. Centuries of peat extraction and drainage had lowered the surface level of the land relative to surrounding rivers. Windmills pumped water out of Rijnland's productive polders into surrounding bodies of water that served as reservoirs, including rivers. Sedimentation had raised the level of those reservoirs, lowering their storage capacity. In other words, as rivers rose and surrounding landscapes sank, Rijnland's ability to control its water levels grew increasingly constrained.

---

[51] Ibid., 24–5. Davids, "River Control and the Evolution of Knowledge," 71–3.
[52] Van Tielhof and van Dam, *Waterstaat in Stedenland*, 2006.
[53] Resolution. December 16 (1741) SA. Archief van de Burgemeesters-Resolutions van de Staten van Holland 5038.179.
[54] Resolution. February 11 (1741) SA. Archief van de Burgemeesters-Resolutions van de Staten van Holland 5038.179. Roughly 34 cm. Verhoeff, *Maten en gewichten*.

The implications terrified Bolstra. If the northern Lek dike collapsed, water would spill into surrounding water boards like Woerden, Amstelland, Delfland, Schieland, and Rijnland. "The entirety of Rijnland to the north of the Rhine shall be submerged and the damage that results will be irreparable," he declared. Water would flood the "greatest and best parts of Holland."[55] Some of the drained peatlands in Rijnland sat several meters lower than the Lek, and with reduced capacity in the reservoirs, water would have nowhere to go. Echoing a similar fear made during the shipworm episode, he worried that dike breach presented an existential threat for 200,000 hectares in the economic and cultural heart of Holland, from Rotterdam to Amsterdam. "There is more to fear," he went on, "about the Lek River than ever before."[56] This was not simply a local, or even regional problem. It concerned the entire province.

Holland responded by sending a commission including Bolstra and 's Gravesande to investigate the disaster. Ostensibly, their task was to "investigate whether, and what sorts of help and relief can be offered to the city Gorinchem and the Alblasserwaard," two of the most heavily impacted places in Holland.[57] In reality, their task was much broader. They assessed the declining state of the system as a whole, paying particular attention to the Lek. Like Cruquius before them, Bolstra and 's Gravesande justified their approach in medical terms. Only by "diagnosing the origins of a disease, one can best arrive at a remedy."[58] Their task was time sensitive. Persistent high water in January and February increased the risk of further breaches. In February, the Estates of Holland received a letter from the Dike Reeve of Rijnland on behalf of every major water board in southern Holland warning about the "imminent danger of dike breaches."[59] The southern Lek dike had already collapsed, and the northern dike seemed primed to fail.

Ultimately, the northern Lek dike escaped the 1740–1 flood season intact, but anxieties among Holland's water managers only intensified.

---

[55] Ibid.
[56] Resolution. February 4 (1741) *OAR*. Register van resoluties van dijkgraaf en hoogheemraden, en van dijkgraaf, hoogheemraden en hoofdingelanden, 1626–1803/1805. 1.1.1.28.
[57] Resolution. January 24 (1741) *SA*. Archief van de Burgemeesters-Resolutions van de Staten van Holland 5038.179.
[58] Ibid.
[59] Resolution. February 11 (1741) *SA*. Archief van de Burgemeesters-Resolutions van de Staten van Holland 5038.179.

What had caused the floods? What would happen if waters eventually broke through? Rijnland and Holland sponsored a series of visually stunning flood risk maps that explored potential causal stories. Bolstra was a leader of this cartographic project. He had already produced maps that diagnosed flood risk in and around Rijnland in the 1730s.[60] After 1741, his attention broadened. His most visually impressive map – *Figuratieve kaart van de situatie van Gelderland, Holland, Uytrecht en Overyzel* – illustrated the spatial risk of Holland and Utrecht to an inundation from the Lek (Figure 5.5). He divided it in two parts: a upper visualization that spanned the width of the Republic and a lower map that focused on the Pannerdens Canal. The upper map emphasized regional water challenges in Holland and Utrecht. Subsidence had increased Holland and Utrecht's flood risk. A panel along the left edge of the map noted the relative depth of polder areas north of the Lek River. Bolstra shaded these water boards in deep green. He also depicted risk in temporal terms. The panel measured seasonal differences in the water level of the Lek relative to sea level (called the *Amsterdamse Peil*). The cities most at risk, including Amsterdam, Rotterdam, and Leiden, appeared in bold red. Contrasted against an otherwise monochrome base map, the vibrant colors of exposed polders and cities left a clear message.[61] Flood risk along the Lek threatened the urbanized core of Holland and much of the rich, agricultural area surrounding it.

The extensive scale of the maps framed Holland's flood risk as an outcome of longstanding structural issues with the river system. The lower map depicts the Pannerdens Canal in Gelderland and the tangle of dikes and meandering distributaries surrounding it. According to Bolstra and 's Gravesande, the opening of the canal in 1707 had exacerbated flooding along the Lek. Although intended to balance distribution between the Rhine distributaries, it guided an excessive amount of water into the Nederrijn-Lek. Bolstra and 's Gravesande concluded that "as

---

[60] Paul van den Brink, "Rijnland en de rivieren: Inrichting en vormgeving van de Hollandse rivierzorg in de achttiende eeuw," *Tijdschrift voor Waterstaatsgeschiedenis* 12 (2003): 69.

[61] In the early eighteenth century, sea level was measured in local or regional contexts. Bolstra was an early proponent of a uniform standard, the *Amsterdamse Peil*. Now termed the NAP (*Normaal Amsterdamse Peil*) or the Amsterdam Ordnance Datum, it is the European standard for sea level. Bolstra used the *Amsterdamse Peil* for national mapping projects and is further evidence of his interest in transforming water management into a systematic, interprovincial project. Petra J. E. M. van Dam, *Van Amsterdams Peil naar Europees referentievlak: De geschiedenis van het NAP tot 2018* (Hilversum: Verloren, 2018).

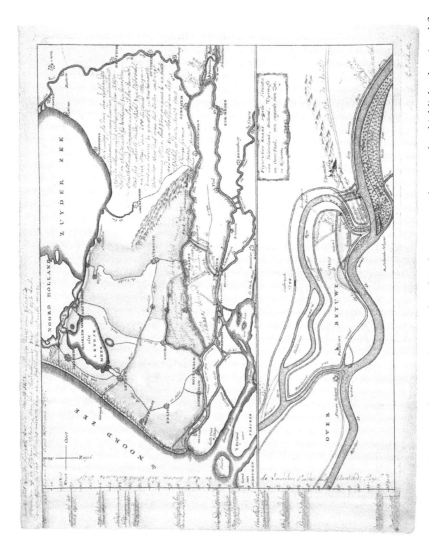

FIGURE 5.5. A later version of a series of maps Bolstra produced in the 1740s and 1750s that highlighted regional flood risk and connected it to interprovincial management challenges. *Source:* Melchior Bolstra, *Figurative Kaart vande Situatie van Gelderland, Holland, Uytrecht en OverYzel, ten regarde van Zee, en Rivieren,* Map, 1751, Rijksmuseum, Amsterdam, http://hdl.handle.net/10934/RM0001.COLLECT.482815.

long as this problem remained, and cannot be removed, it is a clear matter that no remedies can take place."[62] The maps demonstrated that any alteration would require interprovincial cooperation.

The floods of 1740–1 reinforced the conviction that the river system was deteriorating. Less than a year after the floods, the dike reeves of Rijnland, Schieland, and Delfland advised the Estates of Holland that if the Lek dike fell, water would spill from polder to polder, covering the roads that connected villages and cities, and "sweeping away the borders between the great water boards, leaving utter confusion."[63] If these problems were not resolved, "the danger will increase." "The dikes will rise higher and higher, they will grow more burdensome and costly, so much that the lands will not be able to bear their heavy weight."[64] Just as during the shipworm episode, their deepest fears concerned a disaster that had not yet come to pass. The record-shattering inundations of 1740–1 nevertheless intensified their search for systemic problems and incentivized involvement from provincial authorities. Provincial intervention proceeded slowly, however. In the meantime, the scale of the floods forced provinces to contend with a second problem – the unevenness and inadequacy of decentralized relief.

5.5 STATE AND PUBLIC RELIEF DURING THE 1740–1741 FLOODS

Victims faced different and more pressing challenges than water authorities worried about some future catastrophe. Their problems were immediate and concerned access to rescue, relief, and reconstruction. The severity of the flooding and the desperate condition of the affected regions (even prior to the floods) tested the limits of traditional relief and recovery operations. Some victims of the 1740–1 floods were still rebuilding in the aftermath of the floods of 1726, and many bore the financial strains of other concurrent disasters. Exceptional need forced some provinces to shoulder a portion of the financial burden and catalyzed new popular forms of relief. The 1740–1 floods were the first emergency where widespread, coordinated assistance arrived from outside affected regions,

---

[62] Resolution. January 24 (1741) *SA*. Archief van de Burgemeesters-Resolutions van de Staten van Holland 5038.179.
[63] Resolution. December 16 (1741) *SA*. Archief van de Burgemeesters-Resolutions van de Staten van Holland 5038.179.
[64] Ibid.

which, historian Toon Bosch has argued, was the first step in the long development of national disaster memory.⁶⁵ Disaster solidarity emerged as popular media framed the floods in the context of other disasters to appeal for support.

Like all disasters, people experienced this flood unequally. Floodwaters devastated some communities and spared others altogether. Some victims could flee to cities, while others remained trapped in their attics, upon rooftops, or in nearby trees. An image by the Amsterdam printmaker Jan l'Admiral showcased this differential experience of disaster. Water flows out of the Linge, Lek, Meuse, Merwede, and Waal, covering almost the entire landscape (Figure 5.6). The Bommelerwaard, one of the few regions that escaped inundation, appears as a sliver of dry land extending from the east. Elsewhere, desperate figures populate the illustration. A man falls upside down from the broken branch into the water below, representing the inversion of land and water. Villagers clamber to safety atop buildings barely above water. A man rests on a chimney, clasping his hands in supplication while a baby floats along calmly in a cradle.⁶⁶ Relief workers distribute bread and pull victims from the water. By many accounts, these tasks were overwhelming. Some victims would spend hours and even days waiting for relief. One contemporary pamphlet gave voice to their collective need, "[H]elp us! Oh, come to our aid," it proclaimed. "Can nobody hear us? Is nobody coming? If not, we must all drown and perish."⁶⁷

Prior to the 1740s, disaster relief assumed two forms: direct and indirect aid. Direct aid arrived primarily through local and city governments, water boards, and ad-hoc "help committees" funded by private donations. They acted as first responders, shuttling victims out of hazardous areas, providing food and other necessities. Water boards formulated standardized relief procedures that included the organization of dike labor, the acquisition and storage of emergency materials, internal communications, and external alarm. During severe disasters, they pooled

---

⁶⁵ Toon Bosch, "Waar is Johanna van Beek? Over het herinneren en vergeten van calamiteiten in het rivierengebied," *Streven: Cultureel maatschappelijk maandblad* 75, no. 2 (2008): 111–22.

⁶⁶ The baby in the cradle was a common trope in flood culture signifying the providential message of miraculous rescue via analogy to the Deluge or Moses. Hanneke van Asperen, "Charity after the Flood: The Rijksmuseum's St Elizabeth and St Elizabeth's Flood Altar Wings," *The Rijksmuseum Bulletin* 67, no. 1 (2019): 36–7, 40.

⁶⁷ François Kuypers, *De vreeslyke overstrooming van het Landt van Altena, voorgevallen tusschen den 24 en 25 December van 't jaar 1740, en op den 29 April 1741* (Gorinchem: Nicolaas Goetzee, 1741), 23.

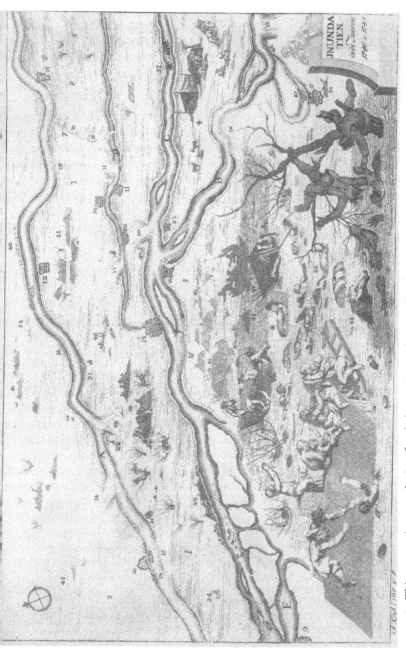

FIGURE 5.6. This composite map depicts flooded areas in southern Holland. The image highlights the immediate aftermath of inundation as people and animals struggle to escape the flood waters. Scenes of prayer, relief, and miraculous rescues indicate its providential message. *Source:* Jan l'Admiral, *Bedroefde Watervloed, Voornamentlyk van Het Land van Heusden, van Althona, de Alblasserwaard, Crimperwaard, en Tielerwaard, waar door meer dan 33500 Morgen Lands onder water staan, na 't leven afgetekend*, Print, 1741, Rijksmuseum, Amsterdam, http://hdl.handle.net/10934/RM0001.COLLECT.134298.

resources.[68] By the eighteenth century, cities and towns assumed important roles as well. They tended to occupy elevated lands so they could coordinate resources more effectively.[69] Authorities in Gelderland, for instance, depended on help committees in cities such as Arnhem and Wageningen.[70] In Den Bosch, guilds and local civic organizations organized rescue and relief.[71] Holland institutionalized the relief responsibilities of cities after the floods of 1726.[72] These types of decentralized responses were typical during Dutch river disasters in the early eighteenth century.

Residents of cities and towns weathered the disaster far more easily than rural regions. Cities like Dordrecht, Arnhem, and Gorinchem escaped the worst of the devastation in 1740–1. Other cities suffered only partial inundation. In Den Bosch, floodwaters inundated the lower lying portions of the city, including its central church. City magistrates struggled to evacuate their most vulnerable residents: the sick and disabled.[73] Nevertheless, Den Bosch operated as a refugee center. Flood victims poured in from the countryside. Bakers worked overtime to produce enough bread. People and livestock gathered at the central market in a hastily erected tent city. City officials sent boats into the countryside to distribute food and rescue rural residents. A help committee even took up a collection and raised 9,000 guilders.[74]

Most descriptions of the flood lauded these urban efforts to provide relief. Cities and towns were more likely and better able to provide relief than their hinterlands. One letter from Culemborg noted that rural people "were not forgotten by those bringing help; several brave citizens and skippers even offered to send the city's barges through the low fields, carrying men and livestock; so that here, few animals and no people

---

[68] Resolution. December 24 (1740) *SLHA*. Waterschap Hoge Maasdijk van Stad en Lande van Heusden. 0041.009. fol. 101. Water authorities from the water boards Oude Land van Altena and Hoge Maasdijk met to coordinate dike reinforcement.

[69] Toon Bosch, "Natuur en cultuur: modernising van hulpverlening na catastrofale overstromingen in de Nederlandse Delta, 1740–1861," *Tijdschrift voor Waterstaatsgeschiedenis* 21 (2012): 1–2, 39–47.

[70] Van Zellem, "Nooyt gehoorde," 16–17.

[71] Francien van den Heuvel, "'s-Hertogenbosch, een onneembare stad midden in een meer: Hulpverlening en preventie tijdens watersnoden, 1740–1795," *Tijdschrift voor Waterstaatsgeschiedenis* 29, no.1 (2020): 4–6.

[72] Resolution. February 9 (1726) *SA*. Archief van de Burgemeesters-Resolutions van de Staten van Holland. 5038.161.

[73] *Europische Mercurius* (1741), 109.   [74] Ibid., 110.

The River Floods of 1740–1741 191

died."⁷⁵ These cases were exceptional, however. The scale of the disaster severely limited relief efforts. Authorities in Den Bosch complained that the floods prevented communication with the countryside, and poor weather exacerbated these difficulties. Relief boats failed to reach the town of Leerbroek in Holland due to strong winds. Floodwaters and bitter cold left victims huddled in the village church for eight days near starvation.⁷⁶ One letter from Emmerik noted that "the destruction and misery resulting from the water that surrounds us is indescribable, the damage to the dikes and dams is unbelievably great and we have already waited for three weeks in the water."⁷⁷ The floods had largely overwhelmed the traditional system of relief.

In Holland, the severity of the floods forced the province to assume greater responsibility. Provincial committees received direct appeals for aid from cities and water boards swamped by rural refugees and lacking resources. The town of Woudrichem, for instance, suffered the flood as well as an influx of people from the countryside escaping even worse conditions. They appealed to the Estates of Holland for provisions and assistance so that "they be distributed to those most in need ... without which the people and animals will surely starve."⁷⁸ Hunger and exposure quickly became more hazardous than drowning, and the challenge of distributing goods proved just as problematic as supply.

A series of four composite maps by the printmaker Jan Smit conveys the extent, severity, and differential experience of the disaster. Each features an illustration of relief-related themes below a map of inundated territory. The illustrations take the viewers through each stage of the disaster experience, from its earliest moments through recovery. The first plate depicts a stretch of the Rhine, Linge, and Waal between Arnhem and Culemborg (Figure 5.7). Like l'Admiral's print, it depicts the initial moments of the flood. People and animals climb from the water into trees and atop buildings. It also references provincial assistance by the Estates of Holland. In response to direct appeals from victims in the Alblasserwaard (one of the most heavily impacted regions in Holland), the province sent relief boats loaded with provisions. In the print, their

---

⁷⁵ *Verzameling van eenige geloofwaardige berigten en brieven betreffende de elende van de opgezetenen der overstroomde landen in Nederland* (Amsterdam: Isaak Tirion, 1741), 26.
⁷⁶ *Europische Mercurius* (1741), 116.   ⁷⁷ Ibid., 105.
⁷⁸ Resolution. December 29 (1740) SA. Archief van de Burgemeesters-Resolutions van de Staten van Holland. 5038.178.

FIGURE 5.7. First composite map in series depicts inundations along the Rhine, Linge, and Waal rivers between Arnhem and Culemborg. The image highlights the immediate impacts of the flood. The text references relief efforts by the Estates of Holland. *Source:* Jan Smit, *Ware afbeelding na het Leven, van het overstroomen, der Revieren, en het Doorbreeken van den Dyck, by Elden ... tot aen de Stat Cuylenburg ... Anno 1740*, Print, 1741, Rijksmuseum, Amsterdam, http://hdl.handle.net/10934/RM0001.COLLECT.479254.

schuit (a small, flat-bottomed boat) remains in the distant background, but their "good and laudable assistance" appears in bold in the text.

The second plate emphasizes provincial assistance. The map depicts a stretch of Holland and Utrecht between the town of Vianen in the east and the city of Dordrecht to the west (Figure 5.8). Instead of floodwater teeming with bodies, large ships loaded with provisions navigate the Biesbosch in the foreground, and smaller ships populate the rivers. Despite "the hard wind and fast currents," a relief committee lands bearing provisions.[79] Victims crowd the right foreground, but the print emphasizes the committee. One member carries a bag of money and a basket of bread lies at his feet. Another hands coins to a supplicant while a minister looks on approvingly. Smit's depiction referenced Holland's decision in January 1741 to provide financial assistance for victims.[80] In the left foreground, a figure identified as "an engineer" scans the landscape with his spyglass and holds a map of the riverlands in his right hand. He undoubtedly represents experts like Bolstra. The message of this image was twofold: it extoled paternal, provincial leadership and promised recovery guided by hydraulic expertise.

The remaining two plates present alternative perspectives on flood relief. Plate three depicts flooded regions in Gelderland (Figure 5.9). Except for a hilly, elevated region stretching south from Nijmegen, the entire landscape is inundated. It depicts the repair and rebuilding of dikes. Rather than provincial assistance, authorities from the water boards direct laborers in their tasks. The scene is crowded and the work haphazard (tools spill out of a wagon), but it nevertheless promotes an alternate, local strategy to cope with disasters. The fourth plate is a study in contrasts. Smit returns the viewer to Holland, this time highlighting conditions in flooded (Land of Heusden and Altena) and non-flooded regions (the Bommelerwaard) (Figure 5.10). The illustration focuses on a fortified city (likely Heusden) and its efforts to prevent flooding and provide relief. Canons represent urban responsibility to warn the hinterlands of approaching floods. Townspeople crowd the edges of the scene flanked by soldiers maintaining order, while well-dressed burghers measure the flood level and direct the water-pumping operations with a horse-drawn chain mill.

---

[79] Jan Smit, *Tweede Plaat der Overstroomingen vande Provincien Gelderland en Holland in den Jaare 1740 en 1741* (Amsterdam: A. van Huyssteen en S. van Esveldt, 1741).
[80] The Alblasserwaard received 20,000 guilders. Resolution. January 26 (1741) SA. Archief van de Burgemeesters-Resoluties van de Staten van Holland. 5038.178.

FIGURE 5.8. The second composite map focuses on the territory between Holland, Utrecht, and Gelderland. The image emphasizes assistance by a relief committee. Figures in the foreground represent the victims and the paternalist response of church and state. The figure with a map and spyglass promises improvement in flood security. *Source:* Jan Smit, *Tweede plaat der overstroomingen vande provincien Gelderland en Holland in den jaare 1740 en 1741*, Print, 1741, Rijksmuseum, Amsterdam, http:/hdl.handle.net/10934/RM0001.COLLECT.479255.

194

FIGURE 5.9. The third composite map depicts inundations in Gelderland. The elevated hills south of Nijmegen remain dry. The water board directs dike repair efforts. *Source:* Jan Smit, *Derde Plaat der overstroominge inhoudende het Ryk Nimweegen . . . met alle desselfs doorbraken in 't Jaar 1740 en 1741*, Print, 1741, Rijksmuseum, Amsterdam, http://hdl.handle.net/10934/RM001.COLLECT .479260.

FIGURE 5.10. The fourth composite map depicts the Land of Heusden and Altena in Holland. The Bommelerwaard remains dry. Image of a town shows scenes of local responsibility and civic order. *Source:* Jan Smit, *De Vierde Plaat van der overstroomingen der Landen ... waar by gevoegd is een nette afbeelding van de Kettingmoolen die gebrukit is; om het water uit de Stad te maalen in den Jaaren 1740 en 1741*, Print, 1741, Rijksmuseum, Amsterdam, http://hdl.handle.net/10934/RM0001.COLLECT.479261.

Like the third plate, provincial authorities do not appear in the fourth map, even though Heusden received financial assistance in 1741. Instead, the fourth map focused on local responsibility. Although the 1740–1 floods forced Holland to provide relief, its efforts remained ad hoc and they channeled their assistance through local water boards and officials. In Gelderland, the province played no part in immediate relief. Smit's series of composite maps conveys the vast spatial scale of the flood, emphasizes differential victimhood, and justifies the paternalist relief efforts of state authorities. They also capture the human drama of the floods and valorize local solidarity. They do not depict the second major form of relief: structural assistance.

Structural responses like tax reductions, exemptions, and loans alleviated the economic consequences of river flooding. River flooding rarely resulted in extreme loss of life, but the loss of property and productivity was often severe. Rural residents were especially vulnerable. During floods like 1740–1, they lost their houses, crops, livestock, and potentially their annual income.[81] Temporary exemption from rent or taxation alleviated a substantial burden. The decentralized nature of disaster relief meant that each province or administrative region took different actions.[82] The Quarter of Nijmegen in Gelderland refused requests for the postponement or cancellation of outstanding debt. The Ambt of the Over-Betuwe in Gelderland also offered little in the way of tax relief, although they assisted in dike repair.[83] The Estates of Holland and the estates of the two regions in Gelderland (Arnhem and Zutphen) agreed to lower the import taxes on cattle.[84] Holland also lifted general taxes for six months in 1741 and extended exemptions for horses and cattle.[85] The province also provided subsidies and loans to communities such as Hedikhuizen, Huekelom, Heusden, Spyk, and Asperen to reconstruct their damaged river dikes.[86] Some of these remissions lasted for months. The Alblasserwaard continued requesting remissions into the summer of

---

[81] Anneke Driessen, "Hulpverlening na overstromingsrampen in het Nederlands rivierengebied," *Groniek* 33 (2000): 185–98.
[82] Van de Ven and Driessen, *Niets is bestendig*, 66.
[83] Van Zellem, "Nooyt gehoorde," 17.   [84] Ibid., 17.
[85] Resolution. March 17 (1741) SA. Archief van de Burgemeesters-Resolutions van de Staten van Holland 5038.178.
[86] Resolutions. January 14, January 18, January 24, January 26, February 18, March 17, March 28, June 20, July 20, August 24, September 6, October 21, October 25, December 2, December 23 (1741) SA. Archief van de Burgemeesters-Resolutions van de Staten van Holland. 5038.178.

1742.[87] These structural responses provided a necessary, albeit uneven, reprieve for affected rural regions. For victims, however, the floods had been only the latest in a long series of concurrent and sequential disasters that compounded their struggles.

5.6 THE MID-EIGHTEENTH-CENTURY PERIOD OF DISASTER

Prior to the first breaches in 1740, communities in the riverlands spent much of the first half of the eighteenth century recovering from disaster. Cattle plague had decimated herds between 1714 and 1720, and its economic effects were long lasting. In the Over-Betuwe, tenants were still receiving abatements of their debts in 1727 due to cattle losses.[88] Recurrent inundation undoubtedly amplified their troubles. Flooding due to seepage was a near-annual occurrence in the Over-Betuwe, especially in the winter when bloated rivers forced water through porous soils beneath dikes. River sedimentation elevated water levels above the surrounding landscape. According to historian Paul Brusse, their "battle against water was an almost constant struggle for existence."[89] This type of slow disaster did not receive the same attention as dramatic dike breaches, yet uncharismatic, perennial flooding was symptomatic of structural problems with the river system.

Extreme flooding along the Lek and Linge rivers in 1726, by contrast, had transfixed observers across the Republic. Ice dams produced a flood that inundated large regions of the western riverlands. Water stretched from Utrecht into the core of Holland. According to the *Europische Mercurius*, water flooded three thousand *morgens* of land in the Krimpenerwaard and Lopikerwaard. "Gouda, as the nearest city, was packed with fleeing people and livestock, so much so, that many could find no shelter."[90] From his vantage in the inundated town of Montfoort, the priest Godefridus Ram noted that "nothing was seen but water almost to the tops of trees and the rooves of houses." Water rose 17 feet in Culemborg, the highest level in the eighteenth century.[91] As water

---

[87] Resolution. July 7 (1742) *SA*. Archief van de Burgemeesters-Resolutions van de Staten van Holland. 5038.180.
[88] Brusse, *Overleven*, 169–79.   [89] Brusse, "Property," 60.
[90] *Europische Mercurius* (Amsterdam: Andries van Damme, 1726), 93–4. Roughly 2,500 hectares. Verhoeff, *Maten en gewichten*.
[91] "Doopboek of Godefridus Ram." (c. 1782) *RRL*. DTB Registers Montfoort. M084.564, fol. 4. *Vervolg der Beschryvinge van den zwaaren Ysgang, Dyk-breuken en Water-*

approached the core of Holland, catastrophe seemed imminent. Floods spared these valuable territories only through the combined efforts of three of Holland's water boards under the direction of Cornelis Velsen.[92] This near miss birthed the doomsday scenario that the shipworm epidemic and the 1740–1 floods popularized. For victims in the 170 villages and ten cities that did not escape, the 1726 floods resulted in enduring hardship.

The floods of 1726 affected many of the same regions in Holland and Utrecht that flooded in 1740–1. The Alblasserwaard was a case in point. Prior to the 1726 flood, it had also been inundated in 1709. The 1740–1 floods intensified their problems. A letter from Gorinchem described how the 1726 flood had "so impoverished the inhabitants that, notwithstanding all their industriousness, made it impossible to recover from the damage, [and] have now been ruined by this last lamentable flood [in 1740–41]."[93] Floods had grown more severe, and the economic consequences compounded. Because dike maintenance relied on local or sometimes regional financial support, flood disasters created a positive feedback that increased the likelihood that disasters would return, often in the same location. It was no accident that dike breaches in 1726 and 1740–1 occurred in some of the same places. A breach of the Linge at the town Kedichem flooded the Alblasserwaard in 1726 and 1741. In regions like this, the floods of 1726 lay the foundation for disaster almost two decades later.

The bitterly cold winter of 1739–40 amplified these hardships in the riverlands. Average winter temperatures dipped to among their lowest levels of the eighteenth century. Unlike the last great frost in 1708–9, however, this cold winter manifested during years of otherwise exceptional warmth. This extreme change likely exacerbated the experience. Frigid weather killed livestock, froze rivers and lakes, and disrupted daily life across the Republic. In Overijssel, Aleida Leurink noted in her diary that the cold prevented people from attending church, "people could not go outside or sit down and coffee cups froze fast to their tables."[94] The chronicler Zacheus de Beer reported that "several people froze to death"

---

vloeden voorgevallen in het begin van het Jaar 1726. In veele Landen en Steden van Europa (Haarlem: Aäron van Hulkenroy, 1726), 9–10.

[92] Report. March 5 (1726) OAR. Register van resoluties van dijkgraaf en hoogheemraden, en van dijkgraaf, hoogheemraden en hoofdingelanden, 1626–1803/1805. 1.1.1.28.

[93] *Europische Mercurius* (1741), 115.   [94] Van Deinse, "Uit Het Dagboek," 545.

in Haarlem.⁹⁵ Instrumental evidence supports one anonymous writer's assertion that "nobody has ever lived through such a remarkable year as this."⁹⁶ Beyond bitter temperatures, the duration of cooler-than-average conditions compounded these effects (Figure 5.11). Cool weather lasted from October 1739 through the following spring. The Amsterdam diarist Jacob Bicker-Raye noted that it was still wet and cold in May.⁹⁷ In Utrecht, the city administrator Thadeus François Quint observed that the worst of the weather lasted up to May 23, and the oldest people could not believe that "after the severity of the winter, the spring remained incredibly cold."⁹⁸

The cold winter and subsequent chilly, wet spring and summer affected the entirety of the Netherlands. Indeed, it likely contributed to harvest failures and a pronounced mortality spike across Western Europe. The Low Countries remained relatively insulated from its worst effects, but even here harvest shortfalls and illness resulted in a 22 percent mortality increase between 1740 and 1742.⁹⁹ The Zaanstad miller Gerrit Jacobszoon Nen reported price increases for food in the cities driven by scarcity.¹⁰⁰ The winter saved its harshest impacts for the countryside. Cold, wet spring weather reduced grain production and ruined the hay harvest. Livestock starved because grass grew poorly the following spring, and farmers were forced to slaughter their cattle.¹⁰¹ In his handwritten journal, the Delft farmer Paulus van der Spek interpreted these disastrous

---

⁹⁵ "Korte Kronijk Door Zacheus J. G. De Beer" (1740) NHA. Oudemannenhuis te Haarlem, 1607–1866. 3295.45.

⁹⁶ This was the first period of strong winter weather widely recorded with instrumental measurements. Buisman, *Duizend jaar*, Vol. 5, 683. *Een historische beschrijving van duure tijden, en hongersnoden* (Amsterdam: Arent van Huyssteen and Steeve van Esveldt, 1741), 184.

⁹⁷ Jacob Bicker-Raye, *Het dagboek van Jacob Bicker Raye 1732–1772*, eds. F. Beijerinck and M. G. de Boer (Amsterdam: H. J. Paris, 1935), 79.

⁹⁸ Thadeus François Quint, "Nieuwe Tijdingen, Aantekeningen Betreffende Gebeurtenissen Uit de Jaren 1733–1744" (1744) UA. C. Berger, als burgemeester van Utrecht. 753.229. fol. 67.

⁹⁹ John D. Post, "Climatic Variability and the European Mortality Wave of the Early 1740s," *Journal of Interdisciplinary History* 15, no. 1 (1984): 1–30. More recent studies have lent greater clarity to Post's findings. Mortality was regionally variable. Amsterdam experienced a 21 percent increase, Rotterdam a 10 percent increase, and cities in North Holland decreased. Curtis et al., "Low Countries," 130.

¹⁰⁰ Gerrit Jacobsz. Nen, "Aantekeningen Betreffende Het Weer En de Gevolgen Daarvan, Overlijden En Andere Familieberichten over Kennissen En Familie, Gebeden, Gelegenheidsrijmen, Spreuken Enz. Enkele Bladzijden Zijn Uitgescheurd" (1742) GaZ. Persoonlijk archief Honig. 10.147.

¹⁰¹ Van Bath, *Samenleving*, 514. Bieleman, *Boeren op het Drentse*, 320.

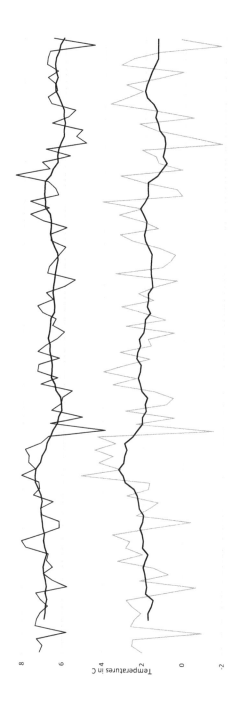

FIGURE 5.11. Winter and "long winter" temperatures between 1706 and 1800. Seasonal averages in winter (DJF) temperatures (bottom) and a nine-month average of fall, winter, and spring (LW) temperatures (top) indicate that the "long winter" of 1739/40 was the coldest of the eighteenth century. Bold lines (DJF 11 and LW 11) indicate eleven-year running means. *Source*: Surface air temperature at De Bilt, Labrijn Series, KNMI Climate Explorer and KNMI, http://climexp.knmi.nl/data/ilabrijn.dat.

conditions though a providential lens. "People speak as if God prevented the eastern wind and the summer dryness from entering ... and the same believe the Lord God punishes the land and the people this year."[102] Cold weather and untimely rains affected all of the Dutch society, but especially those directly reliant upon the harvests.

To make matters worse, the cold winter and chilly spring and summer preserved snowpack in the Alps, increasing flood risk along the Rhine.[103] Contemporaries understood this environmental relationship. The eighteenth-century historian Jan Wagenaar concluded that the winter of 1740 "brought forth this disaster among other miseries."[104] Others saw economic connections. "So many disasters, like bloody wars, pestilence among men and animals ... produced a despondent state," one anonymous pamphleteer declared. No disaster produced as much hardship as the "miserable, hard winter of the year 1740 ... the city dwellers and farmers, through loss of trade and decline in agriculture, have become weakened; while the winter crops died, the land in the spring remained under water from continual rains, one could not plant crops in the earth, and as a result had a late and short summer, so that one could not harvest; this is the cause of the ongoing scarcity and poverty."[105] The winter of 1739–40 had catalyzed a disaster cascade, amplified by the ongoing agricultural recession and extreme weather, that rippled through the Dutch society. Pamphleteers described a period of unrelenting disaster driven by environmental change and intensifying economic hardship.

Providence wove these interdependent environmental and economic disasters into a coherent narrative of increasing exposure and risk. It permeated reportage and appeared in state proclamations. "If we consider the terrible calamities to afflict our beloved Fatherland and threaten

---

[102] Paulus Abrahamse van der Spek, "Het Boek Der Geschiedenisse van Alderhande Oordelen En Segeninge Die de Heere Godt Ons Heeft Toegesonde Beginnende Met Den Jare 1738 Aangetekent Door Paulus Abramse van der Spek." GaD. B17.664 91 C1, fol. 8. Van der Spek was Pietist and interpreted the outbreak as a direct test of his faith. Fred A. van Lieburg, *Living for God: Eighteenth-Century Dutch Pietist Autobiography*, trans. Annemie Godbehere (Lanham, MD: Scarecrow Press, 2006).
[103] Carlo Casty, Heinz Wanner, Jürg Luterbacher et al., "Temperature and Precipitation Variability in the European Alps since 1500," *International Journal of Climatology* 25, no. 14 (2005): 1855–80.
[104] Wagenaar, *Kort en Opregt Verhaal*, 1741.
[105] *Beknopte beschryving, van de noodlydende en oversrtoomde landstreeken in Zuidholland, en een gedeelte van Gelderland* (Gorinchem: Jacob van Wijk 1741), 20.

to continue," a proclamation from the Estates of Holland declared, "such as the severe and lengthy winter, the extraordinary cold and barrenness of the previous year, and the resulting dearth and lack of food, and on top of that the terrible storm winds, and flooding of lands and houses, with loss of people and livestock, everything appears from Gods striking hand and wrath." They called for a general Thanks, Fasting and Prayer day to "forgive the county's sins and inequities" and beg for "further blessing over this our Fatherland."[106] Providential interpretations explained the extraordinary nature of the 1740-1 floods and their arrival amid other disasters and portended future events.

François Kuypers, a minister from Woudrichem, agreed with this assessment and expanded upon it. Moral failures explained ongoing events and exacerbated future risk. Disasters, after all, seemed to be increasing in severity. The rivers near his hometown had risen in past decades, a fact he noted his congregation had failed to recognize. "People did not understand or think," he explained, "that God's punishments are now becoming greater and more terrible through worsening conditions."[107] Record flooding shattered this false sense of security. Kuypers linked the declining state of Dutch rivers to Dutch decline. In a series of sermons, he condemned the Republic's continued moral decay. Disasters provided ample evidence of divine displeasure. The *Rampjaar*, cattle plagues, the shipworm epidemic, and the floods of 1740-1 all pointed to the same ominous conclusion. "O who knows what judgments God has prepared for the Netherlands?" he asked. "The downfall of our Land has long been threatened, which our Governments has long seen and acknowledged, and faithful moral instructors have many times decried. But now it seems to have drawn near, now there are already some signs and beginnings of this long-threatened decline."[108] Moralists like Kuypers published their sermons to reach broad audiences who might be sympathetic to the plight of the disaster victims and were equally anxious about the moral state of the Republic. Given the recent history of disaster, the limited and uneven relief efforts during the flood, and moral imperative to repent, extraordinary assistance seemed warranted.

---

[106] Ibid., 17–18.
[107] Kuypers, *De vreeslyke overstrooming van Het Landt van Altena*, 17.
[108] François Kuypers, *Het Lant van Altena gewaarschouwt en ten laatste zwaarlyk gestraft* (Gorinchem: Nicolaas Goetzee, 1741), 94, 148, 151.

5.7 PRIVATE RELIEF DURING THE 1740–1741 FLOODS

News of the disaster filtered slowly out of the riverlands. The first newspaper reports appeared a week after the earliest dike breaches.[109] Jan Wagenaar recalled this initial lack of information. "Rumors of the dike breaches and floods in the different parts of our Fatherland had not spread; people had no knowledge of the bitter disaster in which many inhabitants of the lands had found themselves."[110] By January, however, newspapers regularly covered the events and letters describing the disaster filtered out of the affected regions. A group of wealthy concerned citizens from Amsterdam and other cities in Holland travelled to the disaster zones in January to take stock. They brought back harrowing reports of the devastation. Notwithstanding relief efforts, the flood victims remained desperate. They needed "food, clothing, stockings, hats, and other coverings" to ward off the cold.[111] More than that, they required structural assistance to rebuild their homes, purchase new livestock, and replace lost possessions.

In view of these reports, wealthy citizens in Amsterdam and Rotterdam organized collections in support of the flood victims. Monetary donations arrived from across Holland and Zeeland, from "the rich and the less well-off, even widows."[112] The group raised 55,000 guilders, which they used to purchase goods and immediately began distributing them. Most relief remained in Holland, but as letters from victims arrived from outside the province, the scope of relief expanded, and assistance continued for months. Early donations focused on the immediate needs of victims, while later donations focused on structural assistance, including timber to rebuild houses and funds to support the purchase of livestock and dike reconstruction. By their own accounting, they helped rebuild 400 houses and donated almost 150 heads of cattle. In total, they claimed to have assisted at least 10,000 victims of the flood in 100 villages.[113]

---

[109] December 22 (1740). *Amsterdamse Courant*.
[110] Jan Wagenaar, "Verhaal van de overstroomingen heir te lande. En byzonderlyk van den Alblasserwaard in 1740 en 1741," in *Verzameling van historische en politieke tractaaten* (Amsterdam: Yntema en Tieboel, 1776), 4.
[111] Wagenaar, *Kort en Opregt Verhaal*, 8.
[112] Jan Wagenaar, *Verhaal van de wyze, op welke de penningen, ten behoeve der noodlydenden, in de Overstroomde Landen van Gelderland en Holland, in den aanvang des Jaars 1741, te Amsterdam en elders ingezameld, bestierd en besteed zijn* (Amsterdam: Isaak Tirion, 1742), 7.
[113] Ibid., 10.

Roughly a year after the flood, Wagenaar published a retrospective pamphlet that explained the methods and motivations of this small group of philanthropists. It described in detail the type and amount of relief and who it assisted. It was clear that Wagenaar intended this to be a model for future efforts. This was a new type of disaster literature. It was less interested in defending a causal story than expressing solidarity. According to Wagenaar, they "felt the need of their Countrymen from afar and provided the needy with immediate help."[114] The relief committee was almost certainly composed of Collegiants, a seventeenth- and eighteenth-century movement of freethinkers.[115] They couched their motivation in Christian language of moral obligation to fellow citizens and appealed to readers' "age-old, famous Dutch charity."[116] Their desire for social and moral improvement, their grounding in Christian values, and their appeal to historic ideals of the Golden Age all reflected the influence of the early Dutch Enlightenment.[117]

Appeals for aid likewise employed this rhetoric of enlightened solidarity. In spring 1741, François Kuypers traveled to major cities in Holland to solicit funds for his beleaguered community. He brought letters from victims and published sermons with pointed appeals for assistance. He argued that donations strengthened the bond of Christian unity fundamental to the Republic, "because every Christian is a member of our shared citizenry, generosity leads to the well-being of our State." He went on, saying that citizens blessed by God have a duty "not only under ordinary circumstances, but especially during extraordinary events and disasters to come to the aid of the country."[118] Kuypers argued that natural disasters demonstrated divine anger and confirmed the Republic's decline. Civic generosity, by contrast, redeemed the state.

This was neither the first time that flood victims appealed to Christian solidarity, nor the first time that groups of private citizens mobilized relief

---

[114] Wagenaar, "Verhaal van de overstroomingen," 31.
[115] Bosch, "Natuur en cultuur," 1–2, 42.
[116] *Reis naar Heusden of een kort doch waarachtig verhaal van ... de laatste overstrominge in 't Lant van Heusden op den 24 December 1740* (Rotterdam: Philippus Losel, 1741). Jan Wagenaar, *Verhaal van de wyze*, 3.
[117] Petra van Dam and Harm Pieters identify similar rhetoric of Enlightened solidarity and improvement in their study on nineteenth-century flood commemoration books. Petra J. E. M. van Dam and Harm Pieters, "Enlightened ideas in Commemoration Books of the 1825 Zuiderzee Flood in the Netherlands," in *Navigating History: Economy, Society, Science and Nature. Essays in Honor of Prof. Dr. C.A. Davids*, eds. Pepijn Brandon, Sabine Go, and Wybren Verstegen (Leiden: Brill, 2018), 275–97.
[118] Kuypers, *De vreeslyke overstrooming van Het Landt van Altena*, 87.

for distant communities. After the Christmas Flood of 1717, Groninger ministers had organized a collection to assist victims in Reformed communities in neighboring East Friesland. Wagenaar's group had already come to the assistance of business associates in Dantzig after a major flood in 1736.[119] Within the Republic, moralists often appealed to readers' shared identities to humanize the events and engender solidarity. Writers usually targeted regional or provincial identities, but appeals to the "Fatherland" had grown increasingly common during the eighteenth century, especially following large-scale disasters like cattle plague, the Christmas Flood, and shipworm epidemic. Most appeals, however, focused on shared responsibility for moral recovery rather than requests for material relief.

The floods of 1740–1 provoked a different response. This was the first time that Dutch citizens effectively mobilized proto-nationalist rhetoric to generate a significant flood-relief response. It reflected an altered view of disaster solidarity that expanded upon, rather than replaced the earlier emphasis on, shared moral responsibility. Popular, private philanthropy provided moral and material relief. Private relief also expanded the capacity of response. Their work certainly critiqued the inadequacy of decentralized state response. Wagenaar even called out the "cities and Estates of the suffering provinces" that failed to offer aid "either quickly or completely enough."[120] Victims, however, saw public and private streams of support as complementary. In a "Sermon of Thanks," the minister Anthonie van Hardeveldt expressed gratitude to private donors as well as the "Fathers of the Fatherland" – the Estates of Holland – for their assistance to his congregation. They had weathered "a long period of dearth, and a hard winter followed by flooding," thanks largely to the "Christian hearts and compassionate souls" of their countrymen.[121] In an

---

[119] Bosch, "Natuur en cultuur," 1–2, 43. Erica Boersma has recently argued that urban conflagrations also generated significant state-led expressions of national solidarity, which she traces to the mid-seventeenth century. These efforts rarely relied on expressions of humanitarianism but rather promoted the interests of the state. Erica Boersma, "Noodhulpbeleid bij stads-en dorpsrampen in de Republiek," in *Crisis en catastrofe. De Nederlandse omgang met rampen in de lange negentiende eeuw*, ed. Lotte Jensen (Amsterdam: Amsterdam University Press, 2020), 187–206.

[120] Wagenaar, *Verzameling*, 31.

[121] Anthonie van Hardeveldt, *De verbaasde en erbarmlyke vlucht der inwoonderen des Landts van Heusden, wegens de laatste verschrikkelyke overstroominge, vertoont in eene predikatie, over Matth. XXIIII: vers 20. Met een Dank-Addres aan de Chasidim of Weldadigen*... (Dordrecht: Fredrik Oudman, 1741), 4–5.

era of cascading and increasingly severe disasters, victims welcomed the expanded interpretation of solidarity and diversified source of assistance.

5.8 CONCLUSION

It is easy to see why Jacob Pierlinck singled out the floods of 1740–1 as an important inflection point in the history of Dutch river flooding. By 1757, when he published his pamphlet, many of the fears voiced by Holland's hydraulic experts in 1741 seemed to have materialized. Flooding had grown more frequent and severe. The floods he witnessed in 1757 were the third major inundation of the 1750s. In 1751, flooding along the Nederrijn and IJssel broke records set in 1740, and flooding hit Overijssel and Gelderland in 1753.[122] The mid- to late 1740s had been no less traumatic. River flooding returned to the Alblasserwaard in 1744, prompting another round of panic focused on the northern Lek dike.[123] In 1747, the northern Lek dike actually breached. Water inundated much of the territory between Utrecht and Leiden. Just as they did in 1726, the three major water boards of Holland, Rijnland, Schieland, and Delfland scrambled to reinforce inland dikes along their borders. Luckily, the wind favored their efforts. In a letter to the Estates of Holland several months later, they acknowledged how close they had come to the doomsday scenario described six years earlier, "the water came up to our lips, and if heaven had not turned the wind around, we feared the total ruin of the water boards."[124] The dike was quickly repaired, but fear of high water along the Lek persisted through the end of the eighteenth century.

The floods of 1740–1 reframed the stakes of river management. From the perspective of water managers, nobody doubted that flooding would worsen. After 1741, anxieties about Holland and Utrecht's increasing flood risk and moral decay grew increasingly integrated into proto-national, declensionist rhetoric. In 1749, Velsen published his magnum opus, *Treatise on River Hydraulics* (*Rivierkundige Verhandelingen*) drawing on historical lessons learned from floods such as 1740–1. To Velsen, the changing nature of the river system posed an

---

[122] Buisman, *Duizend jaar weer, wind en water in de Lage Landen. Deel 6: 1751–1800*, Vol. 6 (Franeker: Uitgeverij Van Wijnen, 2015), 38–44, 85–90.
[123] Buisman, *Duizend jaar*, Vol. 5, 747–9.
[124] Quote from A. A. B. van Bemmel, *De Lekdijk van Amerongen naar Vreeswijk: Negen eeuwen bescherming van Utrecht en Holland* (Hilversum: Verloren, 2009), 70.

existential threat. His anxieties echoed fears voiced during the shipworm epidemic that flooding would turn Holland into "its own sea." Velsen framed flooding in clear declensionist terms. He warned "the people" (*Volk*) that if they did not enact necessary changes, it would "result in their own ruin."[125] Velsen's vision of Dutch peace, commerce, and culture depended upon proper management of rivers and protection against disaster. Those elements of Dutch prosperity that Velsen underlined were not coincidentally the same subjects that generated the greatest anxiety about decline.

> What advantage does a people, or nobility, or government gain by protecting itself against the assaults of foreign enemies; or to prosper in commerce; or to nurture the arts and sciences; or to bring the finances of the country into a good state; or to relieve taxes; or to preserve the common peace and ensure the happiness and prosperity if the inhabitants must in the meantime remain subject sooner or later to flooding from their rivers and thus lose unexpectedly, suddenly, and forever their most wished for things (for some their lives)?[126]

Stabilizing the river system, he believed, was prerequisite to the economic, social, and political recovery of the Republic.

Hydraulic experts like Cruquius, Velsen, and Bolstra responded to this challenge by reevaluating flood causation. They treated river management as a systemic, interprovincial challenge that required careful observation, analysis, and visualization. They produced maps (often copying each other) to memorialize historical floods and visualize increasing risk. Every flood after 1741 reinforced the fear that total inundation was imminent. Velsen's successor and the first inspector general of Dutch Rivers – Leiden professor Johan Lulofs – warned after the 1751 floods that "the water would chase thousands of people ... countless wealthy and poor would lose hearth and home, the county's finances would receive such a hit, that for a long series of years, there would be no hope or means to recover."[127] State dependence on hydraulic expertise, the refinement of river cartography, and provincial intervention in the river system increased in response. Centralized river governance remained

---

[125] Cornelis Velsen, *Rivierkundige verhandeling, afgeleid uit waterwigt en waterbeweegkundige grondbeginselen, en toepasselyk gemaakt op de rivieren den Rhyn, de Maas, de Waal, de Merwede en de Lek* (Amsterdam: Isaak Tirion, 1749), 183.
[126] Ibid., 2.
[127] J. Lulofs and M. Bolstra, *Kortbondig Vertoog van het eminente gevaar, in het welk zig de Provincie van Holland bevindt, wegens de gesteldheid van de Lek*. From Paul van den Brink, "*In een opslag*," 89–91.

FIGURE 5.12. *Source*: Jan Caspar Philips, Title page from *Nederlands waternood van den jaare MDCCXL en MDCCXLI*, Print, 1741. Rijksmuseum, Amsterdam, http://hdl.handle.net/10934/RM0001.COLLECT.479263.

decades away, but after 1741, the stakes assumed interprovincial and existential significance.[128]

---

[128] Van de Ven, *Man-made Lowlands*, 347–67.

FIGURE 5.13. Later copy of Philips image included five disaster medallions along its top border. *Source*: Jan Caspar Philips, *Nederlands Water-Nood en verscheide bezoekingen*, Print, 1751, (detail) Rijksmuseum, Amsterdam, http://hdl.handle.net/10934/RM0001.COLLECT.389835.

Disaster relief likewise evolved in response to the 1740–1 floods as new charitable efforts compensated for overwhelmed systems of local management. Change arrived slowly, but relief efforts in 1741 pioneered a model that could be repeated and improved.[129] The growth of philanthropic relief efforts responded to two fundamental anxieties about the future of the Republic. On the one hand, flooding seemed to be growing worse. This expanded demand for relief. Flood *memoria* such as sermons, maps, chronicles, pamphlets, prints, and poetry documented this need. They reported the devastation and appealed to readers' shared identities. Private disaster-relief efforts reflected both the growing awareness of environmental change and expanding boundaries of social solidarity.

Private philanthropy also addressed the providential connections between past and future disasters. Relief efforts redeemed the moral failures seen as causally responsible. Each new disaster in the 1740s and 1750s reinforced this interpretation. To the printmaker Jan Caspar Philips, the connection he saw between disasters spanning decades prompted him to reprint, with small but important changes, an iconic image of the 1740–1 floods. The original depicted a *schuit* sailing across an inundated landscape

---

[129] In 1753, inundated regions in Overijssel received donations of hay from the Alblasserwaard in direct reciprocation of the aid received in 1741. This reflected a growing interprovincial solidarity that would be refined following floods in 1764, 1770–1, 1781, and 1784. Driessen, "Hulpverlening."

The River Floods of 1740–1741

(Figure 5.12). For his new image, which he published after the floods of 1751, he added five medallions along its top edge (Figure 5.13). Each depicted a disaster – the shipworm epidemic, the winter of 1739–40, hail and windstorms, and a plague of mice that ravaged Dutch harvests in late 1741. The fifth medallion depicted a familiar foe – cattle plague. This may have reminded viewers of the cattle plague episode two decades earlier, but it also reflected currents events. An epizootic had reemerged in the Dutch Republic leading to the longest, costliest, and deadliest disaster of the eighteenth century.

6

"From a Love of Humanity and Comfort for the Fatherland"

The Second Cattle Plague, 1744–1764

The year 1744 began under an ominous star in the Dutch Republic. A comet with six tails blazed across the night sky[1] (Figure 6.1). From his vantage in the Utrecht town of Woerden, the priest Godefridus Ram marveled at the celestial display. He noted his impressions about its appearance and visibility in the margins of his baptismal book.[2] To Ram and others, prodigies like comets were providential omens. Some believed comets foretold dramatic – even cataclysmic – events. Their interpretations drew from a venerable tradition of meteorological observation that saw human fate inscribed in the stars. Others considered the "Great Comet" an open page in the Book of Nature. This assurgent physico-theological interpretation read environmental phenomena like comets as a divine script that revealed the majesty of creation. Onlookers agreed that close observation and a detailed record might yield insight into divine agency. Between late 1743 and spring 1744, stargazers across Europe – from Swiss astronomers to Dutch cattle

---

[1] Two Dutch astronomers, Jan de Munck and Dirk Klinkenberg, and Swiss astronomer Jean-Philippe Loys de Chéseaux independently discovered the "Great Comet of 1744," called the Comet Klinkenberg-Chéseaux. De Munck and Chéseaux each published their findings. Jan de Munck, *Sterrekundige waarneemingen op de comeet of staart-sterre sedert den 29 November ... 1743 tot op den l Maart ... 1744* (Amsterdam: Isaak Tirion, 1744). Jean-Philippe Loys de Cheseaux, *Traité de la comète qui a paru en décembre 1743 & en janvier, fevrier & mars 1744: contenant outre les observations de l'auteur, celles qui ont ÉtÉ faites À Paris par Mr. Cassini & à Geneve par Mr. Calandrini : on y a joint diverses observations & dissertations astronomiques, le tout accompagné de figures en taille douce* (Lausanne: Chez Marc-Michel Bousquet & Compagnie, 1744).

[2] Ram, "Doopboek," fol. 59.

The Second Cattle Plague, 1744–1764

FIGURE 6.1. Early modern observers often interpreted meteors as divine, sometimes ominous, portents. This print depicts the meteors that appeared in 1742 and 1744. *Source*: Jan de Groot, *Afbeelding van de staartsterren, verscheenen in de jaaren 1742 en 1744*, Print, 1744, Rijksmuseum, Amsterdam, http://hdl.handle.net/10934/RM0001.COLLECT.43953.

farmers – followed the astronomical event with a mixture of curiosity, wonder, and fear.³

The recent history of natural disasters in the Dutch Republic encouraged particularly grim interpretations of the Great Comet. In his 1744 physico-theological sermon, *The Inhabitants of the Earth encouraged to a proper fear of Gods Signs ... in particular of the Comets*, the minister Hendrik van Barn-in 't-Loo argued that the comet revealed God's anger. It was a powerful reminder that earlier providential warnings had gone unheeded. Years of cattle plagues, coastal and river floods, and the shipworm epidemic had produced little lasting moral improvement. Yet he held out hope that the comet might provoke a different, more proactive response. "Surely," he argued, "it is a sign of God's holy displeasure over a land and a people when He begins to burn his fire everywhere."⁴ This type of interpretation persisted long after the comet disappeared from the night sky. Indeed, later events seemed to justify dire readings. Looking back on the most notable events of his life, the miller Simon Jacobszoon Kraamer included his impressions of the comet in his handwritten chronicle. The comet was "a woeful sign of the plague that God Almighty brought thereafter."⁵

Between 1740 and 1748, the Dutch Republic experienced its most tumultuous period since the Disaster Year of 1672. The peace it had enjoyed since the Treaty of Utrecht in 1713 proved a temporary interlude before the Republic found itself reluctantly embroiled in yet another European conflict. The War of the Austrian Succession (1740–8) cast the United Provinces in the familiar role of French antagonist, and like the *Rampjaar* of 1672, the war began disastrously. In 1744, the French

---

³ Jorink, *Reading the Book of Nature*, 109–79. Andrew Fix, "Comets in the Early Dutch Enlightenment," in *The Early Enlightenment in the Dutch Republic, 1650–1750*, ed. Wiep van Bunge (Leiden: Brill, 2003), 157–72. Vladimir Janković, *Reading the Skies: A Cultural History of English Weather, 1650–1820* (Chicago: University of Chicago Press, 2000). Danish farmers also connected cattle plague and the comet. Karl Peder Pedersen, "Als Gott sein strafendes Schwert über dem dänischen Sahnestück Fünen schwang. Über Verlauf und Bekämpfung der Viehseuche auf Fünen 1745–1770 unter besonderer Berücksichtigung des Bauernschreibebuchs von Peder Madsen auf Munkgaarde," in *Beten, Impfen, Sammeln-Zur Viehseuchen-und Schädlingsbekämpfung in der Frühen Neuzeit Graduiertenkolleg Interdisziplinäre Umweltgeschichte*, eds. Katharina Engelken, Dominik Hünniger, and Steffi Windelen (Göttingen: Universitätsverlag Göttingen, 2007), 59–60.

⁴ Hendrik van Barn in 't Loo, *De inwoonders der aarde aangespoort tot een betamelyke vreze voor Gods tekenen, in en aan den hemel: Handelende van de Hemel-Ligten in 't gemeen, als luidbare verkondigers van Gods volmaaktheden, maar in 't byzonder van de comeeten* (Leiden: Johannes Hasebroek, 1744), 4–5.

⁵ Kraamer, *Eenijge Merckwaerdige Gebeurttenisse*.

overran Dutch fortresses across the Austrian Netherlands (now much of Belgium). By 1747, they occupied towns in the Generality Lands, and French troops threatened the province of Zeeland.[6] Fear of another French invasion provoked a series of events that eerily echoed the *Rampjaar*. Lacking faith in their regent-led government, a popular revolution spread from Zeeland across the Republic, once again upending the political status quo in favor of the House of Orange. Much of the Dutch Republic had operated without a stadhouder since the death of William III in 1702. This "Second Orangist Revolution" of 1747 unseated the governing regent's party and installed William IV as stadhouder.[7] The cultural memory of 1672 and the fear of invasion prompted political revolution.

Economic troubles had also deepened since 1713 and contributed to the political turmoil of the 1740s. Domestic industry had declined precipitously, especially for export-oriented goods like linen. Once-vibrant cities such as Leiden, Haarlem, and Delft withered as their industry atrophied. The herring fishery, a "golden mountain" of the Dutch economy until the 1650s, continued to contract.[8] The Dutch overseas trading system largely survived the upheavals of the *Rampjaar* but gradually lost its primacy after 1720.[9] Much of the rural Netherlands, particularly in the Maritime Provinces, continued to suffer from a century-long agricultural recession. Extraordinary wartime taxation continued decades after the cessation of hostilities in 1713. Cattle plague, river floods, and the shipworm epidemic amplified these financial burdens.[10] Against this backdrop, popular dissatisfaction with rural tax collection led to mob violence against tax farmers in Friesland in 1748. This "tenant revolt" (*pachtersoproer*) was in part an expression of economic discontent and it spread quickly to Groningen and Holland.[11] Optimists could take

---

[6] The military inundation of the West Brabant water line near the town of Bergen op Zoom compounded the disastrous consequences of the invasion for inhabitants. Erica Boersma argues that this resulted in the first appeal for national, state-directed flood relief in 1749. Boersma, "Noodhulpbelied," 203.

[7] William III died childless. William IV was a close relative of William III and head of the Frisian branch of the House of Orange-Nassau. Israel, *Dutch Republic*, 1067–78.

[8] Ibid., 1006–12. The Dutch herring industry had been contracting since the 1650s because of warfare, mercantilism, and structural changes in consumption. Bo Poulsen, *Dutch Herring*. Van Bochove, "Golden Mountain," 209–44.

[9] Israel, *Dutch Primacy*, 377–98.

[10] J. Aalbers, "Het machtverval van de Republiek der Verenigde Nederland 1713–1741," in *Machtverval in de internationale context*, eds. J. Aalbers and A. P. Goudoever (Groningen: Wolters-Noordhoff/Forsten, 1986), 8–9.

[11] Israel, *Dutch Republic*, 1066–78.

comfort that some sectors of the economy, such as international finance and trade in colonial commodities, remained stable or even expanded. Per capita income remained among the highest in Europe.[12] The Republic's economic prognosis, in other words, was far from terminal. By the end of the 1740s, however, its diminished vitality was difficult to ignore.

Natural disasters peppered this period of political uncertainty and economic stagnation. The winter of 1739–40 had been one of the harshest of the century and likely the deadliest. The 1740s featured multiple coastal and river floods, including the disastrous Rhine and Meuse floods of 1740–1.[13] In 1742 and 1743, farmers in Utrecht and Holland complained of mice plagues that devoured their harvests. Flooding returned to the riverlands in 1744 and again in 1747. Malaria was endemic in communities along the North Sea, and outbreaks followed floods, high summer temperatures, and droughts.[14] Koenraad Blom, a Reformed minister from Groningen, summed up the hardships they had experienced in recent decades: "war, terrible storms that destroy many ships, floods, deadly infectious fevers, boring worms that threaten to turn us into a sea, droughts, ... hard frosts, gnawing mice, [and] the lessening of navigation and commerce." One anonymous chronicler wryly noted that the Treaty of Utrecht had inaugurated "a peace full of turmoil that for many was perhaps worse than open war."[15] Simon Jacobszoon Kraamer likewise recorded these disasters in his journal, yet the "plague" he connected to the Great Comet referred to none of these events, nor the political or economic trials of the 1740s. It heralded the return of cattle plague.

This chapter explores the shifting character of disaster origins and outcomes between the first (1713–20) and second cattle plague episodes (1744–64). Dutch responses exhibited significant continuities across both

---

[12] Christiaan van Bochove, *The Economic Consequences of the Dutch: Economic Integration around the North Sea, 1500–1800* (Amsterdam: Aksant, 2008), 154. De Vries and Van der Woude, *First Modern*, 699–710.

[13] See Chapter 5.

[14] Otto S. Knottnerus, "Malaria around the North Sea: A Survey," in *Climatic Development and History of the North Atlantic Realm: Hanse Conference Report*, eds. Gerold Wefer, Wolfgang H. Berger, Karl-Ernst Behre, and Eynstein Jansen (Berlin: Springer-Verlag, 2002), 339–53. Van Tielhof and van Dam, *Waterstaat in Stedenland*, 212–17.

[15] Koenraad Blom, *Hiskia's dank- en lof-offer voor Gods wonderbare genesing, den Heere toegebragt in syn dank-schrift Jesaia 38 vs. 16–20* (Amsterdam: Adrianus Douci 1746). *Een historiesch verhaal van veele en nooit meer gehoorde voorvallen die geschiet zyn in verscheide harde winters inzonderheid van den jaare 1709 en 1740: van tyd tot tyd aangetekend door een liefhebber en dus 't ligt gegeeven met een fraaye kopere plaat* (Amsterdam: Steeve van Esveldt, 1740), 185.

epizootics. Beef and dairy production remained central elements of the Dutch rural economy. The symbolic association of cattle with prosperity and fertility likewise retained its cultural resonance. Disease management displayed remarkable similarities as well. Barely a generation removed from the last outbreak, provincial and local authorities in the 1740s capitalized on their recent experience. Prior to 1744, every province had enacted policies to contain or mitigate the impacts of cattle plague and they quickly reintroduced them. Folk medicine and other popular remedies likewise lost none of their appeal.

Historians have tended to emphasize these continuities if the plagues are treated to separate consideration at all. The chief distinction most acknowledge between the two episodes relates to their widely varying scope and severity. Whereas mortality estimates for the entire country during the first cattle plague totaled at most 300,000 cattle, the second may have reached one million. The first plague episode lasted seven years, the second lasted two decades.[16] The consequences of the second cattle plague dwarfed the first. This clear difference in severity was an important indication that conditions had changed since the early eighteenth century.

Just as it had in the first decades of the eighteenth century, the "fatal synergy" of warfare, social dislocation, and climatic extremes again brought cattle plague to the Republic's doorstep and encouraged its spread. Only this time, the cascading consequences of environmental and social disasters leading into and running concurrent with the epizootic intensified its impacts. The moral and social context of disaster response likewise shifted. The turbulent economic and political conditions of the 1740s colored moral interpretations of providence, encouraged patriotic appeals to adopt crisis management, and lent urgency to declensionist anxieties. The conjuncture of so many economic, political, and natural disasters in the 1740s and 1750s seemed convincing evidence to some contemporaries that "the Netherlands appears to have already reached its peak."[17]

The most striking change since the earlier outbreak was the increasing interest and emphasis on medical innovation. Physicians expanded their interest in animal medicine, and provinces increasingly relied on their expertise. In the mid-1750s, citizen-scientists performed the first cattle inoculation experiments on the continent. The first Learned Societies emerged in the Dutch Republic and they immediately began investigating

---

[16] Koolmees, "Epizootic Diseases," 23.   [17] Blom, *Hiskia's Dank- en lof-offer*, 3–4.

cattle plague as an integrated medical, moral, and social problem. By the end of the cattle plague in the 1760s, the disease had emerged as a central project of the early Dutch Enlightenment. Cattle plague, more than any disaster of the era, reflected the manifold meanings of Dutch decline.

6.1 MORTALITY AND SCOPE OF THE SECOND CATTLE PLAGUE

The Republic had been plague-free for nearly a generation when cattle plague reappeared in 1744.[18] Some older farmers likely remembered the earlier disaster, but direct experience was not a prerequisite for nervousness. The Delft farmer Paulus van der Spek had not yet been born when the first outbreak emerged, yet he monitored its reappearance with trepidation. After the first Dutch reports appeared, rumors of sick and dying cattle spread far in advance of illness.[19] "I heard say from Pieter Melief who lives between The Hague and Scheveningen that one animal was already dead," he noted apprehensively in his journal in November 1744.[20] By the following spring, disease was widespread. Media coverage expanded enormously after disease appeared inside Dutch borders. "Newspapers and state proclamations are full of the ongoing death of cattle," one pamphleteer declared in 1745.[21] Reports acknowledged its similarity to the first epizootic, yet already during this first year, news coverage and private documents tended to emphasize its unprecedented scope and scale. "It was a wonder to find one farmer who had not lost his animals," Van der Spek reflected.[22] Cattle plague had returned to the Dutch Republic and it appeared far deadlier than ever.

Few documents captured these early days of the outbreak as powerfully as Jan Smit's print *God's Striking Hand over the Netherlands* (*Gods Slaandehand over Nederland*)[23] (Figure 6.2). It is easily the most well-

---

[18] A less virulent disease (likely hoof and mouth) affected parts of the Netherlands in the 1730s. W. H. Dingeldein, "Iets over mond en klauwzeer in 1732 en vroeger," *Tijdschrift voor Diergeneeskunde* 36 (1933): 389–92. Rommes, "Geen vrolyk," 89.

[19] Dutch newspapers and news periodicals scarcely mentioned the disease before it appeared in the Republic. One exception: the *Amsterdamse Courant* noted cattle plague near Frankfurt in 1742. November 3 (1742) *Amsterdamse Courant*.

[20] Van der Spek, "Het Boek," fol. 12.

[21] Jan Marchant, *Naagalm over de vee-ziekte, met een jaarlijst der voorgaande vee-sterftes, sédert de plagen van Egipten: alsmede de waare oorzaak der koeje-ziekte, en de middelen om die voor te koomen* (Haarlem: J. Marshoorn en Iz. vander Vinne, 1745), Introduction.

[22] Van der Spek, "Het Boek," fol. 13.

[23] Jan Smit, *Gods slaandehand over Nederland, door de pest-siekte onder het rund vee naar het leeven getekent, en gegraveert door Jan Smit* (Amsterdam: Steven van Esveldt, 1745).

FIGURE 6.2. Jan Smit's print depicted the inversion of the Golden Age landscape. The fertile polder landscapes of the back- and middle ground merge into a landscape of diseased and dead bodies strewn across the foreground. Source: Jan Smit, Gods slaandehand over Nederland door de pest-siekte onder het rund vee, Print, 1744, Rijksmuseum, Amsterdam, http://hdl.handle.net/10934/RM0001.COLLECT.479868.

known image of eighteenth-century European cattle plague. Both allegory and reportage, its vivid depiction of cattle mortality and human misery mirrors the lived experience of many rural communities. A pile of diseased cattle bodies dominates the foreground, while dead or dying cattle lie stricken in the middle and background. Its moral commentary is clear. The depiction of corpses and burial transform an otherwise bucolic polder landscape into a grotesque vision of inverted fertility. As reportage, it tells a more complicated story. The multitude of cattle bodies in the foreground signifies the severity and scope of the outbreak. The polders in the background – some depicting the outbreak, others populated with healthy livestock – showcase its asymmetry. Both stories are accurate. The epizootic was both unevenly experienced at a local level and exceptionally severe in the aggregate.

Contemporaries widely remarked upon the extreme mortality of the cattle plague. They underscored this point by comparing it to past disasters. The text accompanying Smit's print noted that cattle plagues had appeared in 1651, 1682, and 1713. These "visitations were terrible, but they cannot be compared to the present."[24] Mortality in 1744 far exceeded historic analogues, but its virulence did not mean the disease was unique. "The feeling is that this is the same sickness that befell our country thirty-three years ago," remarked Godefridus Ram in September 1744. Despite this similarity, he admitted that "there is such a great death of cattle in the Netherlands that it can never be recalled from before."[25] Memory of the earlier cattle plagues contextualized interpretations in 1744, but observers apprehended its exceptional severity almost immediately.

Observers emphasized the scope of the disaster as well. Simon Jacobszoon Kraamer noted in his chronicle how widely the disease spread after appearing near his home in North Holland in 1744. It appeared to encompass "the entire country, every province, yes one may say the entire Christian world."[26] It is difficult to discern whether this universalization of disease impacts was merely rhetorical strategy or an attempt to describe the events. On the one hand, the first years of the outbreak may have appeared truly universal to farmers like Van der Spek or millers like Kraamer. They likely read troubling reports arriving from across the Republic and across Europe. On the other hand, asserting its general nature performed a useful interpretive task. Like the Christmas Flood of 1717, expanding the scope of the disaster separated it from similar events in the past. The disaster was not novel, but its severity and scope set the disaster apart and encouraged greater appreciation of its meaning and implications.

The reality of the outbreak's differential impacts reinforces both interpretations. Cattle mortality was both widely distributed and sharply disparate. As had likely been the case three decades earlier, Holland, Utrecht, and Friesland experienced the greatest losses. Their large beef and dairy herds and integration into a market economy that required cattle movement exposed farmers to larger potential losses.[27] Utrecht lost over 33,000 cattle between 1744 and 1746, and Northern Holland lost 77,000 cattle between October 1744 and April 1745. Friesland suffered even greater mortality. Approximately 135,000 cattle died in Friesland between November 1744 and August 1745.[28] On this provincial scale,

---

[24] Ibid. [25] Ram, "Doopboek," fol. 60–3.
[26] Kraamer, *Eenijge Merckwaerdige Gebeurttenisse*, fol. 8.
[27] Record keeping had improved since the previous epidemic, but it was not comprehensive. Reliable figures exist for specific towns during a specific set of years. Fewer regional assessments exist, and most current historical estimates compile these scattered sources. Priester, *Geschiedenis*, 252. Bieleman, *Boeren in Nederland*, 150.
[28] Van der Woude, *Noorderkwartier*, 592. Faber, "Cattle Plague," 2. Bieleman, *Boeren Op Het Drentse*, 320.

The Second Cattle Plague, 1744–1764

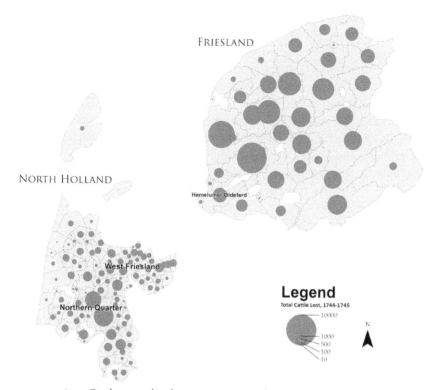

FIGURE 6.3. Cattle mortality by *gemeente* (North Holland) and *grietenij* (Friesland) between 1744 and 1745. The peat and marine clay regions in both territories experienced the highest mortalities.
Source: *Nederlandsche Jaerboeken*, Amsterdam: Erven van F. Houttuyn, 1757. Boundaries modified from Boonstra, NLGis shapefiles.

the relative impact depended on the extent to which the rural economy depended upon cattle production.

Regional differences reflected this pattern as well. Within Friesland, cattle mortality was highest in its western and central grasslands, which specialized in dairying and hay production[29] (Figure 6.3). Areas dominated by arable farming suffered fewer losses. The chance of infection was significantly higher in the grasslands because of larger herd sizes and the greater influence of the cattle trade.[30] Farmers in arable regions, by

---

[29] Faber, "Cattle Plague," 3.
[30] Van Roosbroeck, "To Cure Is to Kill?," Van Roosbroeck and Sundberg, "Culling the Herds," 31–55. Ronald Rommes identified similar regional differences between east and west Utrecht. Rommes, "Geen vrolyk," 88–136.

contrast, depended upon cattle for manure production but had fewer, less mobile animals.[31] The regional differences in mortality were stark. Whereas total provincial losses reached as high as 70 percent in the first year of the outbreak, mortality in the grasslands was exceptionally high. The municipality (*grietenij*) of Hemelumer Oldeferd lost 90 percent of its cattle in less than five months between 1744 and 1745.[32] Northern Holland displayed a similar trend. The reclaimed lake districts of the Northern Quarter and the peatlands of West Friesland saw the highest mortalities. Cattle plague's impacts varied by province and region. Larger herds and greater specialization in dairying, breeding, or beef production exposed farmers to elevated levels of impact.

While the worst cattle losses clustered in the western provinces, no province or region escaped completely. Cattle mortality remained high outside Utrecht, Holland, and Friesland. In the *Westhoek* region of the Generality Lands, 84 percent of cattle died between 1744 and 1745; in the northern portion of Zeeland, over 20,000 cattle died by 1749; and in the *Landschap* of Drenthe in the east of the Netherlands, rough estimates place cattle losses at 50 percent or more.[33] In arable regions, cattle grew so scarce that farmers were forced to use their own waste instead of manure.[34] If farmers were able to keep cattle healthy, skyrocketing prices dampened their overall losses. Farmers could also sell *gebeterde* cattle that had recovered from plague at higher prices because they understood that survival conferred immunity.[35] From a regional perspective, farmers in the marginal pasturelands of Drenthe may have benefited at the expense of foreign suppliers by intensifying their production of cattle to compensate for import bans. Most farmers were not so lucky, and high prices offered no benefits to farmers who repeatedly lost herds and sometimes even their farms.[36] No province and few regions escaped the outbreak completely.[37] In this context, it is easy to see why contemporaries described disease using universalizing language. Every

---

[31] Faber, *Drie Eeuwen*, 162–5.   [32] Ibid., 166.
[33] Priester, *Geschiedenis*, 252. Bieleman, *Boeren in Nederland*, 150.
[34] Van Schaik, *Overlangbroek*, 15.
[35] H. K. Roessingh, "Landbouw in de Noordelijke Nederlanden, 1650–1815," *Algemene geschiedenis der Nederlanden* 8 (1979): 18–19, 21. Dutch newspapers featured advertisements for cattle from "healthy and fresh meadows" and cattle that "have already had the sickness." Advertisement February 18 (1746), *'s Gravenhaegse Courant*. Advertisement November 20 (1748), *'s Gravenhaegse Courant*.
[36] J. N. H. Elerie, "De Rundveehouderij van Schoonebeek in de 17e en 18e eeuw," *Historia Agriculturae* 18 (1988): 97.
[37] One exception seems to be the western portion of Zeelandic Flanders, which, by virtue of isolation, escaped the epizootic. Farmers raised cattle to restock herds in the North. Van Cruyningen, "Behoudend," 153–4, 204.

cow seemed susceptible and considering the exceptional severity of the plague, the potential consequences were enormous.

6.2 DISASTER CASCADES AND DISEASE MANAGEMENT

The timing and unprecedented scale, duration, and severity of this epizootic outbreak were heavily influenced by social and environmental changes throughout the 1740s.[38] Just as it had four decades earlier, cattle plague emerged in 1740 in the context of extreme weather. The winter of 1739–40 was the harshest in a generation and marked the return of "Little Ice Age–type impacts" that had been absent much of the warm previous decade.[39] Exceptionally cool weather extended into the subsequent year. Indeed, historical climatologists identify the cool temperatures of 1740 as a climatic anomaly unmatched over the past 300 years.[40] The cold weather would have had direct effects on animal and human health, but subsequent seasonal precipitation extremes compounded its consequences for pasturage and the harvest. Springs between 1740 and 1742 were typically cool across Western Europe, and drought prevailed in the summers. These conditions prompted widespread harvest shortfalls, and in the absence of effective markets and relief institutions, the resulting food and fodder shortages increased the likelihood of famine and disease outbreaks. Historians have linked excess human and animal mortality across Europe in the early 1740s in part to this cluster of extreme events.[41]

The Dutch experience of this period of anomalous weather largely corresponded to these trends. The average temperature of 1740 was among the coolest in recorded observation. Winter and spring temperatures remained remarkably low until 1744, with the exception of a mild winter in 1741.[42] Notwithstanding the wet conditions that produced

---

[38] The natural variability of the disease may have contributed as well, but its varied symptomology and sporadic documentation limit retrospective evaluation.

[39] Christian Pfister and Rudolf Brázdil, "Social Vulnerability to Climate in the 'Little Ice Age': An Example from Central Europe in the Early 1770s," *Climate of the Past* 2, no. 2 (2006): 120. The decade between 1733 and 1742 was likely the second warmest since 1500. Luterbacher et al., "European Seasonal and Annual Temperature," 1502.

[40] Jones and Briffa, "Unusual Climate in Northwest Europe," 364–5.

[41] Effective famine relief strategies and grain storehouses buffered the worst impacts in the Dutch Republic. John D. Post, *Food Shortage, Climatic Variability, and Epidemic Disease in Preindustrial Europe: The Mortality Peak in the Early 1740s* (Ithaca, NY: Cornell University Press, 1985), 224.

[42] Post, "Climatic Variability," 8. Surface air temperature at De Bilt. Labrijn Series. KNMI Climate Explorer and KNMI. http://climexp.knmi.nl/data/ilabrijn.dat

widespread flooding in 1740, average rainfall dipped in the period between 1741 and 1743, especially in the spring and summer.[43] This combination of cool, wet weather in 1740, followed by cold and dry weather in subsequent years, created significant difficulties for farmers who depended upon timely rains and mild temperatures. Reports of the harsh winter of 1739–40 noted their connection to the 12,000 cattle deaths in Friesland and 17,000 deaths in Groningen alone.[44] The cool, wet spring that followed was likely more detrimental for cattle than extreme cold because it undermined grass growth and reduced nutrient content.[45] In Utrecht, the city administrator Thadeus François Quint observed that these cool, wet conditions lasted into late May. Even the oldest people could not believe that "after the strength of the winter, the spring remained incredibly cold with … lengthy snow, hail, and rainstorms."[46] According to eyewitnesses, "feed for the animals was in such short supply that thousands of cattle and sheep died of starvation."[47] Lacking sufficient grass production, Friesland banned the export of hay to prevent further losses.[48] The dry conditions in later years may have affected grain production, but the persistent cool weather likely had more impact on cattle.[49] Milk, butter, and cheese production declined as dairy cattle starved. Oxen that should have been fattened in Dutch meadows instead weakened or died.[50]

Cattle plague also reemerged and spread in the context of continental conflict and transnational commerce. The disease likely reemerged in Hungary at the outset of the War of the Austrian Succession in 1740 and followed Fredrick II's Prussian army into Bohemia and Bavaria. Eighteenth-century armies were constantly on the march, driven in part by their voracious appetites. If troops encamped too long, they quickly depleted local resources, including livestock. Provisioning these mobile armies and restoring herds after they departed likely contributed to the spread of the disease.[51] The international oxen trade likewise

---

[43] Labrijn, "Het Klimaat van Nederland," 95–8.
[44] *Historische Beschrijving Van Duure Tijden*, 195.
[45] B. H. Slicher van Bath, "Die europaischen Agrarverhältnisse im 17. und der ersten Hälfte des 18. Jahrhunderts," *A.A.G. Bijdragen* 13 (1965): 149. Pfister, "Weeping in the Snow," 64.
[46] Quint, *Nieuwe Tijdingen*, 67.   [47] Kraamer, *Eenijge Merckwaerdige Gebeurttenisse*.
[48] Bieleman, *Boeren in Nederland*, 217.   [49] Pfister, "Weeping in the Snow," 62–4.
[50] De Vries, "Measuring the Impact," 610. Post, *Food Shortage*, 103.
[51] Spinage, *Cattle Plague*, 117. Martin Van Creveld, *Supplying War: Logistics from Wallenstein to Patton* (Cambridge: Cambridge University Press, 2004), 26–39.

created ideal conditions for a panzootic outbreak. Long-distance travel by foot or ship not only moved animal bodies harboring disease to new locations and susceptible populations but also increased the likelihood that animals succumbed to infection. Environmental historian Sam White has argued that the confluence of climatic factors and the bodily stress of movement may have also compromised cattle immune systems.[52] Travel increased animal stress and malnutrition. In ideal conditions, oxen already lost considerable weight traveling overland due to the metabolic demands of long-distance walking. While not as taxing as extreme heat, extreme cold weather also increased energy needs.[53] Sea transport was arguably more strenuous because it required cattle to endure the stress of loading and unloading and remain standing during the three-week voyages from Denmark to Holland.[54] The plague entered the Republic from two directions: overland and by sea. The invading French army carried the disease overland into the Austrian Netherlands during summer 1744 and it moved slowly north. A second outbreak erupted in northern Holland in fall 1744, spreading from the cattle market in Hoorn. Echoing events in 1713, this outbreak likely emerged as a result of the overseas oxen trade from Denmark.[55]

Cascading disasters between 1740 and 1744 further stressed cattle immune systems and the cattle economy. In addition to cattle lost during the bitter winter of 1739–40 and subsequent seasons of poor grass production, river flooding in South Holland, Gelderland, and Utrecht during the winter of 1740–1 killed thousands of cattle, which required restocking from other territories.[56] The relatively warm winter of 1741–2 and subsequent dry spring fostered mice infestations that denuded

---

[52] White, "A Model Disaster," 107–11.

[53] Wilhelmina Maria Gijsbers and P. A. Koolmees, "Food on Foot: Long Distance Trade in Slaughter Oxen between Denmark and the Netherlands (14th–18th Century)," *Historia Medicinae Veterinariae* 26 (2001): 121–3. Gijsbers, *Kapitale ossen*, 16–17. John C. Malechek and Benton M. Smith, "Behavior of Range Cows in Response to Winter Weather," *Rangeland Ecology & Management/Journal of Range Management Archives* 29, no. 1 (1976): 9, 11–12. Vijai P. Maurya, Veerasamy Sejian, Kamal Kumar et al., "Walking Stress Influence on Livestock Production," in *Environmental Stress and Amelioration in Livestock Production*, eds. Veerasamy Sejian, S. M. K. Naqvi, Thaddeus Ezeji, Jeffrey Lakritz, and Rattan Lal (Berlin, Heidelberg: Springer Berlin Heidelberg, 2012), 76–8.

[54] Wilma Gijsbers and Bert Lambooij, "Oxen for the Axe. A Contemporary View on Historical Long-Distance Live Stock Transport," in *By, Marsk og Geest* 17, eds. Morten Søvsø and Jakob Kieffer-Olsen (Ribe: Ribe Byhistoriske Arkiv & Den antikvariske Samling I Ribe Forlaget Liljeberget 2005), 62–4, 69, 71–2.

[55] Rommes, "Twee eeuwen," 33–4.   [56] Gijsbers, *Kapitale ossen*, 93. See Chapter 5.

pastures. "I saw with my own eyes that most of the lands had been affected by mice so that one could see almost nothing green," the Holland farmer Jacob Pos proclaimed in 1742.[57] Mice could be just as problematic for grass growth as cool, wet summers. During "mice years" (*muizenjaren*), lack of fodder meant that cattle produced less milk and farmers could sell fewer animals.[58] The return of harsh winter weather in 1743 impacted yields across Southern Holland and Utrecht, and cold weather continued into the spring. Godefridus Ram lamented in 1743 that "this spring has been very cold and meager again, and because everything was destroyed by the mice in 1742, there is a lack of hay and straw, many animals are dead, and by the end of May many animals could scarcely find enough food to eat."[59] The cascading consequences of extreme or unseasonable weather, flooding, and agricultural pests had already exacted a large toll on rural society before the cattle plague arrived.

Facing a "scarcity of cattle," Holland enacted policy changes. It restricted the export of hay and cattle and forbade the slaughter of calves between 1741 and 1743 to encourage herd recovery.[60] They also reduced tariffs on imported cattle and reached an agreement with Danish producers to expand the oxen trade.[61] Replacing lost cattle exposed farmers to worse impacts abroad, however. Dutch dependence upon Danish beef supplies had decreased since the last cattle plague due to protectionist tariffs on imported cattle. Threats to Dutch domestic stock reenergized the trade.[62] Denmark, however, was suffering from what historian Thorkild Kjærgaard refers to as an "ecological crisis" prompted by deforestation and overgrazing. Danish cattle production had almost doubled between the 1680s and 1740s, unsettling a fragile ecological balance in production regions. The combination of flooding, drifting sand, and the harsh winter of 1740 killed or weakened Danish stock in

---

[57] J. F. C. Schlimme, ed., *Het dagboek van Jacob Pos* (Hilversum, 1992), 6. Buisman, *Duizend jaar*, Vol. 5, 728–42. Stephen A. Davis, Herwig Leirs, Roger Pech, Zhibin Zhang, and Nils Chr. Stenseth, "On the Economic Benefit of Predicting Rodent Outbreaks in Agricultural Systems," *Crop Protection* 23, no. 4 (2004): 305–14.
[58] Van der Poel, "Noord Hollandse Weidebedrijf," 148–9.
[59] "Ram, "Doopboek," fol. 56.
[60] *Groot placaet-boeck, vervattende de placaten, ordonnantien ende edicten van de ... Staten Generael der Vereenighde Nederlanden, ende van de ... Staten van Hollandt en West-Vrieslandt, mitsgaders vande ... Staten van Zeelandt* ('s Gravenhage: Isaac Scheltus, 1770), 1599, 1602.
[61] Proclamation. November 29 (1741). *NHA*. Ambachtsbestuur van Velsen. 3701.490. Gijsbers, *Kapitale ossen*, 93, 390, footnote 412.
[62] Gijsbers, *Kapitale ossen*, 87, 93.

advance of the plague.⁶³ A century of ecological change had transformed Danish pastures into a perfect harbor for disease. When Dutch provinces lowered trade restrictions on Danish oxen in the 1740s, they widened pathways for disease reintroduction. These cascading political, economic, and environmental crises were self-reinforcing and sowed the seeds of the worst epizootic outbreak of the eighteenth century.

6.3 CHANGING INSTITUTIONAL RESPONSE, 1744–1755

Disease management fell largely to provincial and local authorities, just as it had during the first cattle plague. Provinces dictated policies, often performed by local officials, that delineated diseased spaces, managed the movement of animals into and out of their jurisdictions, and policed adherence. Most policies built upon precedents established decades earlier. Like the earlier outbreaks, rural communities contended with multiple concurrent or related disasters. Crisis management thus remained flexible, adapting to the evolving needs of economically distressed communities. Although outwardly similar, notable differences separated state response between the two disasters. Less than a generation removed from the prior outbreak, state actors enacted policies more swiftly than in the past. Disaster management policies also remained in place far longer. The duration of the cattle plague, its immense economic costs, and the unevenness of its impacts encouraged provincial authorities to accommodate the often-conflicting imperatives to prevent outbreaks and mitigate economic losses.

Previous experience with cattle plague heavily influenced state response. Relying upon institutional precedent and memory afforded authorities two advantages. States had ready-made strategies they could enact. The first disease proclamation issued by Holland, for instance, restored many of the regulations and procedures enacted during the first plague episode. It prescribed a laundry list of rules, including certification for travel, quarantine, the killing of stray dogs, the cleaning of stalls, and burial.⁶⁴ These responses drew from both environmental and contagionist interpretations of disease propagation and did not veer far, if at all, from earlier orders.

---

⁶³ Thorkild Kjærgaard, *The Danish Revolution, 1500–1800: An Ecohistorical Interpretation* (Cambridge: Cambridge University Press, 2006), 26–7. Between 1740 and 1747, more than half of Danish stock died from the disease.
⁶⁴ *Groot placaet-boeck*, 1605–7.

Familiarity also encouraged rapid reinstatement. The epizootic likely appeared in Holland, Friesland, Zeeland, and the Generality Lands in late fall 1744, and provinces immediately enacted prophylactic measures to prevent its further spread. The Estates General banned imported cattle from the Austrian Netherlands into the Generality Lands in early November, and Zeeland, Holland, and Gelderland followed suit over the next two weeks. By December, the Prince of Orange forbade cattle importation into his territory in Utrecht. Gelderland and Friesland followed suit quickly thereafter.[65] Similar regulations during the first epizootic had required half a year to enact, and it took even longer to develop the full range of regulations immediately available during the second outbreak.

None of the challenges that authorities faced during the first cattle plague had been resolved by 1744, however. The same fundamental conundrum confronted them: cattle plague threatened cattle economies, but so did provincial and local efforts to manage the disease. States needed to protect existing herds, but also generate revenue; the former restricted mobility, the latter demanded unrestrained cattle movement. This dual burden to limit the impact of disease and disease management encouraged provinces to return to strategies of continuous crisis management between 1744 and 1765.

Provincial regulations reflected this dual imperative. Holland enacted its first import ban on November 21, 1744. Although intended to last until the following May, the desperate need to restore herds forced Holland to reopen the trade in February 1745. They instead imposed an alternative suite of regulations that "could be practiced with peace of mind." Certification and environmentalist precautions like cleaning and burial ensured a steady supply of cattle. "On the one hand," they argued, these policies would "prevent the progress and communication of [cattle plague] as much as possible, and on the other hand ... facilitate and encourage the importation of healthy cattle from outside this Province with confidence."[66] Friesland followed a similar policy progression. They imposed their first import ban on cattle on December 11, 1744 and later added a second ban on the slaughter and export of calves and dairy cattle.

[65] Ibid., 1603–8. Proclamation. November 18 (1744). UvA. Estates of Gelderland and Zutphen, O 63-3903. Gijsbers, *Kapitale ossen*, 93. J. Van de Water, C. W. Moorrees, and P. J. Vermeulen, Groot plakkaatboek 's lands van Utrecht: aangevuld en vervolgd tot het jaar 1810 (Utrecht: Kemink en Zoon, 1856), 234. Proclamation. December 12 (1744) GrA. Staten van Stad en Lande, 1594–1798. 1.478.

[66] *Groot placaet-boeck*, 1605–7, 1610–13, 1618.

By March 1745, Friesland replaced outright import bans with two-week-long quarantines. This decision was risky because it increased the likelihood of a new outbreak, but it was deemed necessary to restore some semblance of normalcy to the cattle economy.[67]

Local and provincial disease management required widespread public participation. As they had done during the first outbreak, state authorities policed adherence to regulations and occasionally prosecuted offenders. Friesland, for instance, attempted to mitigate the danger of removing import bans by increasing the penalties for circumventing quarantines in 1746.[68] State authorities had learned during the previous outbreak that complete control of cattle movement was functionally impossible, however, and generally demonstrated greater willingness to accommodate affected communities than previously.[69]

Previous experience and memory of the disease encouraged accommodation. People exhibited greater confidence about what likely did (and did not) transmit the disease. In early 1745, Johan Vegelin van Claerbergen, then mayor (*grietman*) of the Frisian town Doniawerstal, asked for "a concession to be made for the use of the fat of the beasts that have been killed by the ongoing and spreading sickness of cattle ... all the more because certainly the consumption ... [of] the fat and even the meat of dead animals has not had the smallest ill effects."[70] In Utrecht, Godefridus Ram noted a similar shift in policy. "When these deaths of the cattle first appeared, no one was permitted to skin them, but were required to bury them in the earth. This requirement was revoked, and they are now allowed to be skinned, the fat was sold to the candle makers, and the meat was eaten by many people."[71] The sale of diseased cattle products had been a contentious issue during the first outbreak, but greater leniency seems to have been afforded their use by 1745.

Farmers also received structural assistance, although this varied by province and locality. Tax remissions mitigated land abandonment,

---

[67] Resolution. December 11 (1744), March 17 (1745). *TRE. Gewestelijke bestuursinstellingen van Friesland 1580–1795*. 5.152.

[68] Resolution. March 17 (1746). *TRE. Gewestelijke bestuursinstellingen van Friesland 1580–1795*. 5.152

[69] In Groningen, only one case went as far as the sentencing court in Groningen, a case against the butcher Daniel Cuiter in 1745 for the slaughter of a calf. "De Slager Daniel Cuiter Wegens Het Slachten Van Wrak Vee" (1745) *GrA. Volle Gerecht van de stad Groningen, 1475–1811*. 1534.2074.

[70] Resolution. January 8 (1745). *TRE. Gewestelijke bestuursinstellingen van Friesland 1580–1795*. 5.152.

[71] Ram, "Doopboek," fol. 63.

which authorities considered the greatest threat posed by the epizootic. Friesland provided remissions for farmers whose cattle died in the first six months of the plague and enacted additional relief policies in subsequent months and years. In other regions, states and landowners offered tax remissions and rent reduction upon request.[72] In many cases, these measures were not enough. Land abandonment in hard-hit North Holland increased dramatically by midcentury.[73] Landholders abandoned 125 hectares of land in the town of Assendelft between 1700 and 1755.[74] In the town of Langedijk, land abandonment had been problematic throughout the eighteenth century as well. Abandonment spiked after natural disasters such as the winter of 1708–9, the first cattle plague, the shipworm epidemic, and the winter of 1739–40. The problem reached new heights in 1745 when farmers abandoned thirty-one properties. Facing an unprecedented crisis, local authorities warned that "everything descends into the bitterest confusion and desperate of situations ... Give us redress from taxation to prevent this general abandonment." They requested and received remission for the next six years.[75] Cattle plague management required dutiful adherence to regulation but also acknowledged the extreme state of distress in rural regions.

Visual and material culture reveal how this management strategy filtered into the public sphere. Jan Smit's *Gods Slaandehand over Nederland*, for instance, promoted public participation in disease management, but also encouraged empathy for distressed farmers struggling amidst crisis. In the foreground, a landlord raises his right hand pointing to heaven. His rich clothing and carriage contrast with the plainly dressed farmers (likely tenants) pleading for respite before a mound of dead cattle (Figure 6.4a and b). The landlord "neither sneers nor snarls at the poor farmers, nor threatens him with a penalty if he cannot pay his rent," the accompanying text reads, but instead beseeches them to seek "comfort and piety" from God.[76] Smit used providential rhetoric to promote

---

[72] Brusse, *Overleven*, 171.
[73] Van der Woude, *Noorderkwartier*, 599–600. This was also the case in other areas of Holland. H. P. Moelker, "De Diemerdijk: de gevolgen van paalwormvraat in de 18e Eeuw," *Tijdschrift voor Waterstaatsgeschiedenis* 6 (1997): 48. Fransen, *Dijk onder spanning*, 393.
[74] Bert Koene, *Goede luiden en gemene onderzaten: Assendelft vanaf zijn ontstaan tot de nadagen van de Gouden Eeuw* (Hilversum: Verloren, 2010), 302.
[75] Schilstra, *In de ban*, 92. Egbert Barten,"... *en rukte overal alles weg, waer zij trof ...*": *De runderpest in Nederland, met nadruk op de Langedijk, in de 18de Eeuw (1713-1784)* (Noord-Scharwoude: De Schrijfcaemer, 1983), 48–9.
[76] Smit, *Gods slaandehand over Nederland*.

The Second Cattle Plague, 1744–1764

FIGURE 6.4. (a) Detail of farmers pleading with landlord. (b) Detail of sheriff directing cattle burials. *Source*: Jan Smit, *Gods slaandehand over Nederland door de pest-siekte onder het rund vee*, Print, 1744, Rijksmuseum, Amsterdam, http://hdl.handle.net/10934/RM0001.COLLECT.479868.

leniency and debt forgiveness. In the left foreground, a sheriff directs the burial of dead cattle. Farmers dutifully strip their hides and bury the bodies to the required depth (usually six feet) (Figure 6.4a and b). "Learn from this, and behave obediently to what your government orders," Smit went on, as "they bear the responsibility of the welfare of all those in common."[77] Material and visual culture thus wielded providence to promote social solidarity while simultaneously reinforcing administrative authority.

Continuous crisis management accommodated the diverse and oftentimes contradictory demands of communities dependent on cattle economies. In combination with recurrent extreme weather events and associated disasters, it likely extended the duration of the plague as well. The mild winters followed by dry springs characteristic of the 1740s encouraged repeated mice plagues, for instance.[78] Farmers might flood their fields to kill the mice, but this only worked if water levels were high enough.[79] Winters that featured high water and wet windy weather (combined with structural issues in water management) produced deadly flooding in 1744, 1747, 1751, and 1753. Flooding also occurred during the dry, cold winter of 1757 when frozen rivers thawed and produced ice dams.[80] All these conditions affected cattle. During floods, farmers might save their cattle by moving them to higher ground, into elevated attic spaces of farmhouses (*waterzolders*) or "flood barns" (*waterschuren*), but extreme river flooding sometimes overwhelmed these adaptations.[81] Catastrophic loss of hay production following mice plagues, extreme or unseasonable weather, or flooding reduced domestic stock and required importation of new animals from Denmark and other regions suffering from cattle plague.[82] In ordinary conditions, poor hay harvests and flood

---

[77] Ibid.
[78] Contemporaries noted mice plagues in 1742, 1747, 1749, and 1750. Van der Spek, "Het Boek", 11, 17. Kraamer, *Eenijge Merckwaerdige Gebeurttenisse*, 7, 9, 18. Ram, "Doopboek," fol. 53–6. Meihuizen, "Sociaal-economische geschiedenis," 315.
[79] Van der Poel, "Noord Hollandse Weidebedrijf," 149. In addition to the drought years of 1740–3, the period between 1747 and 1750 was also dry. Buisman, *Duizend jaar*, Vol. 5, 939. Jones and Briffa, "Unusual Climate in Northwest Europe."
[80] Buisman, *Duizend jaar*, Vol. 5, 747, 792. Jan Buisman, *Duizend jaar*, Vol. 6, 38–42, 85–90, 151–68.
[81] Judith Toebast, "Voor als de dijken doorgingen: maatregelen tegen rivieroverstromingen bij boerderijen, zeventiende-negentiende eeuw," *Tijdschrift voor Waterstaatsgeschiedenis* 21 (2012): 11–22. Pierlinck, *Verschrikkelyke watersnood*, 49–50.
[82] Jacob Pos noted cattle in his small village from Hungary, England, Scotland, Hanover, Prussia, Munsterland, and Jever. Schlimme, *Dagboek*, 10.

The Second Cattle Plague, 1744–1764

disasters remained local or at most regional affairs. Cattle plague and continuous crisis management linked them together. The steady pulse of recurrent environmental shocks and continual reintroduction of animals from foreign pastures extended the duration of the epizootic beyond any in living memory.

Although mortality in general declined after the first and deadliest wave abated in 1746, the epizootic remained a threat to rural livelihoods for two decades. Marijtje Cornelisdr Westgeest and her nephew Leendert Jans Buytendijk noted that cattle plague hit their farm in South Holland three separate times between 1745 and 1748. They ultimately lost their entire herd of fifty-five cattle despite (and perhaps because of) restocking.[83] Cattle plague continued to impact Jan Barte Fokkert's farm in Utrecht between 1744 and 1753, and his financial struggles lasted into the 1760s.[84] In 1757, thirteen years after its first appearance, the biannual periodical *Nederlandsche Jaerboeken* reported that the plague persisted in parts of Friesland and Holland. In the city of "Zwolle [in the province of Overijssel] and the surrounding areas, the disastrous infection rages as strongly as ever, and in Gelderland, the city Hattem and surrounding areas [where] a miserable share are still plagued."[85] Other sources confirm the disaster's longevity. Van der Spek noted cattle plague in the area surrounding The Hague and Delft until 1760, and in *Rijnland*, between Leiden and Amsterdam, "one still hears a little bit, here and there, of cattle plague in some places."[86] Cattle plague appeared to be waning after 1754, but the threat of renewed outbreaks never truly disappeared. In this context, medical responses to the disease underwent significant reevaluation.

---

[83] A. G. van der Steur, "Een Warmondse boerenboekhouding uit de tijd van de veepest 1742–1749," *Leids Jaarboekje* 62 (1970): 161–7.

[84] Fokkert's debt ballooned to 1,400 guilders in 1753 due to flooding and cattle plague. In 1764, his debt remained 1,200 guilders. Jan Barte Fokkert, *Gedachtenis van de wonderbare wegen der goddelijke Voorzienigheid gehouden in de weg der bekering* (Middelburg: Stichting de Gihonbron, 2017), 30–2.

[85] *Nederlandsche jaerboeken, inhoudende een verhael van de merkwaerdigste geschiedenissen, die voorgevallen zyn binnen den omtrek der Vereenigde Provintiën, sederd het begin van 't jaer* (Amsterdam: F. Houttuyn, 1757), 692. The *Nederlandsche jaerboeken* provides the most comprehensive account of the second outbreak after 1747. Between its first publication in 1747 and 1754, every issue included a segment entitled "Letters Concerning the Death of Cattle." From 1755 to 1764, cattle plague appeared in a separate recurring segment called "blessings and disasters."

[86] Van der Spek, "Het Boek", fol. 22.

6.4 CHANGING MEDICAL RESPONSE TO CATTLE PLAGUE

The reemergence of cattle plague triggered renewed attention to animal healing, much of which took place outside the bounds of formal medicine. Farmers and dilletante medical practitioners published and purchased medical texts. Folk remedies were copied and likely shared through family or informal networks.[87] Provinces distributed promising treatments, and news journals collected and published them.[88] Pamphlets and periodicals translated foreign remedies for their Dutch readership, and newspapers announced new editions of medical books. Remedies arrived from far afield, including Brussels, Ghent, Frankfort, Aalst, Raamsdonk, and Normandy.[89] Medical therapies, especially medicinal remedies, enjoyed a deepening of public interest during the 1740s.

The return of the disease also prompted physicians to expand their engagement with animal medicine. This challenged traditional hierarchies of medical authority, which tended to exclude animals from analysis and care, but the virulence and economic consequences of the outbreak encouraged states to provide new incentives.[90] In its first cattle plague proclamation in 1744, the Estates General offered a reward of 1,000 florins for a cure.[91] Holland promoted a similar contest that November, which offered "a premium of 1000 guilders for anyone who could find a new and innovative remedy for the cattle plague, which upon being tested is found effective."[92] In light of these inducements, "there came, one after another, more cow remedy books and other things to light." "Everyone want[s] to play cow-doctor," the pamphleteer Jan Marchant proclaimed

---

[87] "Rapporten over Waarnemingen van Ziekte onder het Rundvee te Westbroek en Achttienhoven, met Mededelingen over Symptomen van deze Ziekte en Middelen tot Bestrijding, 1744–1774," UA. Huis Zuilen. 76.145. Van Deinse, "Uit het dagboek," 540.

[88] December 15 (1744). GrA. Staten van Stad en Lande, 1594–1798. 1.484. *Nederlandsche jaerboeken* (Amsterdam: F. Houttuyn, 1749), 90.

[89] Marchant, *Naagalm*, 2.

[90] Louise Hill Curth, *The Care of Brute Beasts: A Social and Cultural Study of Veterinary Medicine in Early Modern England* (Leiden: Brill, 2010). Dominik Hünniger, *Die Viehseuche von 1744–52: Deutungen und Herrschaftspraxis in Krisenzeiten* (Neumünster: Wacholtz Verlag, 2011).

[91] Gijsbers, *Kapitale ossen*, 387.

[92] Republished in *Nederladsch Gedenkboek of Europische Mercurius* (Amsterdam: J. Ratelband en Compagnie, 1744), 298–9. Friesland and Utrecht also offered prizes. P. C. M. Hoppenbrouwers, "Geen heer, geen boer: de bemoeienissen van Johannes le Francq van Berkhey met de Hollandse landbouw," *Holland: Regionaal-historisch Tijdschrift* 18, no. 3 (1986): 143.

in 1745.⁹³ State authorities sought medical expertise to evaluate this flood of remedies. Just as they had during the shipworm episode and during the river floods of the 1740s, provinces turned to the University of Leiden for guidance. The resulting report produced by Leiden's medical faculty cautioned that many proposed remedies were "very old and inconsistent."⁹⁴ Ironically, the faculty's "provisional precautions" were not substantively different from popular remedies.⁹⁵

Two months later, the Estates of Utrecht requested a similar study from Utrecht University. Although their recommendations were nearly identical to Leiden's, the Utrecht faculty's methods differed. They tasked four surgeons to travel throughout the countryside in teams of two to observe the effects of cattle plague and interview affected rural populations. These interviews provided practical information about the symptoms and history of the disease. The Utrecht study also created an exception to a provincial law that banned the dissection of diseased cattle. The governing council of Utrecht suspended this rule so that "four experienced surgeons" could "open as many beasts as needed, and to precisely examine [the disease] and to note the condition of all parts, in the head, breast and belly, and everything else."⁹⁶ No similar exception was made in 1714. In view of the severity of the 1744 outbreak, however, medical accommodation seemed necessary. People "have more reasons to fear, that the deaths today are much more serious than in earlier times."⁹⁷

Despite over a decade of efforts, no solution appeared. Nervous observers hoped that the disease would disappear as it had two decades earlier. By the early 1750s, this optimism was wearing thin. "How we hold out hope, month after month, and how ardent is our desire to be able to share something good, about the lessening of the disaster," reported the *Nederlandsche Jaerboeken* in 1752, "but it remains impossible for us."⁹⁸

---

⁹³ Marchant, *Naagalm*, introduction.
⁹⁴ "Aanwysinge Der Teekenen," *GrA*. 1.485. fol. 301-2.
⁹⁵ *Nederlandsch Gedenkboek of Europische Mercurius* (1744), 302-4. Van Roosbroeck reports similar findings during the third outbreak in Flanders. Van Roosbroeck, "To Cure Is to Kill?"
⁹⁶ J. V. Baerle, *Gedagten van de heeren professoren uytmakende de medicynsche faculteyt op de Academie van Utrecht over de ziekte onder het rund-vee thans woedende: waar by nog gevoegd zyn eenige der observatien, uit welke de voornoemde gedagten zyn opgemaakt* (Utrecht: Jacob van Poolsum, 1745), 5.
⁹⁷ Ibid., 6. The Estates of Groningen also commissioned the physician Egbert Wiardi to conduct similar experiments in 1745. *GrA*. Register Feith-stukken Stadsarchief Groningen, 1594-1815. 2041.1344.
⁹⁸ *Nederlandsche jaerboeken* (Amsterdam: F. Houttuyn, 1752), 551.

In the meantime, the economic costs of plague mounted and other natural disasters compounded its effects in the countryside. Mice plagues returned in 1749 and 1750, shipworms reappeared near Hoorn in 1751, and rivers again flooded in 1751 and 1753.[99] The legacy of social unrest in 1747–8 and the continued economic recession in the countryside amplified this turmoil. The farmer Jacob Pos aptly termed this period a "chain of punishments, one after another."[100]

The intractability of the problem amidst economic decline and social volatility demanded a novel approach. Inoculation presented one promising option to simultaneously stem the spread of cattle plague and heal the diseased body politic. Inoculation (also called variolation) is the transfer of live virus material via incision into a healthy body. The resulting illness was often less severe and, in the case of smallpox, conferred lifelong immunity to the disease.[101] The first plague outbreak had demonstrated that cattle who survived infection also enjoyed immunity. Some medical authorities thus considered it analogous to smallpox.[102] Inoculation might present a solution for animal disease as well. Beginning in 1755, the Dutch Republic became the continental staging ground for an inoculation movement that spread across Europe. It emerged out of changes in the sociability of animal medicine, the providential meaning of disaster, and the enlightened conviction that science could be wielded to promote national recovery.

6.5 INOCULATION AND ENLIGHTENMENT CITIZEN-SCIENCE

One of the earliest inoculation experiments in the Republic took place in Beverwijk, a small town north of the city of Haarlem in 1755.[103] Three

---

[99] *Nederlandsche jaerboeken* (1749), 832; (1750), 775, 853, 928; (1751), 255–8; (1754), 129.

[100] Israel, *Dutch Republic*, 994–7. Schlimme, *Dagboek*, 14.

[101] Genevieve Miller, *The Adoption of Inoculation for Smallpox in England and France* (Philadelphia: University of Pennsylvania Press, 1957). Uta Janssens, "Matthieu Maty and the Adoption of Inoculation for Smallpox in Holland," *Bulletin of the History of Medicine* 55, no. 2 (1981): 246–56.

[102] Hervé Bazin, *Vaccinations: A History: From Lady Montagu to Jenner and Genetic Engineering* (Esher: John Libbey Eurotext, 2011), 126. Joana Swabe, *Animals, Disease and Human Society: Human–Animal Relations and the Rise of Veterinary Medicine* (London: Routledge, 2002), 63.

[103] The *Leydse Courant* noted an earlier inoculation experiment in the Heer Hugowaard on February 18, 1755. Ph. H. Breuker, "De achttjinde-ieuske yninting tsjin de Feepest yn West-Europa," in *Freonen om ds. J.J. Kalma hinne: Stúdzjes, meast oer Fryslân, foar syn Fiifensantichste Jierdei*, eds. Ph. H. Breuker and M. Zeeman (Leeuwarden: Tille, 1982), 325–50.

men – Cornelis Nozeman, Agge Roskam Kool, and Jan Tak – conducted the trial. Using donated funds, they purchased seventeen cattle and transferred disease material to incisions on the cows' bodies. They monitored changes in cattle appearance, pulse, and stool; documented environmental conditions; and controlled food and water intake. Only three of the original seventeen animals survived.[104] Undeterred, they published a pamphlet that described their methods and reasserted their "chief hope to make this experiment successful."[105] Their pamphlet was less an intervention in medical theory than a response to the practical challenge of protecting vulnerable animals.[106] Nozeman, Kool, and Tak understood the social and economic consequences of the plague. They had watched "with great attentiveness" while hundreds of cattle died in Holland between 1744 and 1746.[107] Their inoculation experiment addressed an acute medical challenge as well as a civic imperative that "took its origins from a love of humanity and comfort for the Fatherland."[108]

The Beverwijk study's emphasis on the practical implications of inoculation reflected the evolving social importance of medicine during the Dutch Enlightenment. The perception of medical science and practices were changing across Europe in the mid-eighteenth century, showcasing a new optimism about the human capacity to govern environments and improve society.[109] Dutch anxieties about decline, which intensified during the turmoil of the 1740s, lent unique urgency to these medical goals. Cattle plague was a scourge that intensified economic hardship and revealed the declining moral condition of the Republic. It was also a disease that might be mastered. Inoculation presented one potential

---

[104] Tom Barrett, Paul-Pierre Pastoret, and William P. Taylor, *Rinderpest and Peste Des Petits Ruminants: Virus Plagues of Large and Small Ruminants* (London: Elsevier Science, 2005), 92.

[105] Cornelis Nozeman, Agge Roskam Kool, and Jan Tak, Eerste proefneeming over de uitwerkingen van de inentinge der besmettende ziekte in het rundvee, gedaan in de Beverwijk (Amsterdam: K. van der Sys and K. de Veer, 1755), 88.

[106] Margaret DeLacy argues that the experience of the first cattle plague outbreak increased the popularity of contagionism. The Beverwijk experimenters remained agnostic about the subject. Margaret DeLacy, *The Germ of an Idea: Contagionism, Religion, and Society in Britain, 1660–1730* (New York: Palgrave Macmillan, 2016), 76–7.

[107] Nozeman et al., *Eerste proefneeming*, 18.   [108] Ibid., preface.

[109] Mary Lindemann, "Medical Practice and Public Health," *A Companion to Eighteenth Century Europe*, ed. Peter H. Wilson (Oxford: Blackwell, 2008), 165. Roy Porter, "The Eighteenth Century," in *The Western Medical Tradition: 800BC–AD1800*, Vol. 1, eds. Lawrence I. Conrad, Michael Neve, Vivian Nutton, Roy Porter, and Andrew Wear (Cambridge: Cambridge University Press, 1995), 474.

pathway to solve two intractable problems: cattle plague and the broader issue of decline.

It was likely no coincidence that Beverwijk was the site of the first inoculation experiments. Situated near the once-flourishing industrial center of Haarlem, both town and city were microcosms of the challenges facing many urban and rural communities in the 1750s. Haarlem had been the site of intense social and political unrest between 1747 and 1750, partly prompted by economic and demographic changes.[110] The city was experiencing rapid depopulation and deindustrialization due to the decline of its textile and brewing industries.[111] The epizootic intensified stress on its linen bleaching industry, which relied on buttermilk and whey. The rural town of Beverwijk was also stagnating. After experiencing a doubling of its population between the sixteenth and the seventeenth century, it flatlined during the eighteenth. Cattle plague hit Beverwijk as well. Already by November 1744, town officials alerted the Estates of Holland that the disease was spreading in their territory. Although its rural economy centered on horticulture, farms invariably relied upon cattle, and the region likely experienced severe mortality during the second outbreak.[112] Experimenters could have found few more suitable locations to test the medical and social promise of inoculation.

The Beverwijk experiments also reflected the new sociability of Enlightenment medicine in the 1750s. Dutch Enlightenment medicine was a bourgeois phenomenon that mobilized physicians and citizen-scientists largely drawn from the middle class and the ministry. The movement was relatively nondenominational and embraced a wide number of professions. Nozeman was a Remonstrant (the Arminian tradition of Calvinism), and Kool and Tak were Mennonites. Tak was a Leiden physician, Kool was a Beverwijk merchant, and Nozeman was

---

[110] J. A. F. de Jongste, *Onrust aan het Spaarne: Haarlem in de jaren 1747–1751* (Dieren: Bataafsche Leeuw, 1984).

[111] Richard W. Unger, *A History of Brewing in Holland, 900–1900: Economy, Technology, and the State* (Leiden: Brill, 2001), 223.

[112] November 21 (1744). NHA. Stads- en Gemeentebestuur van Beverwijk, 1298–1817. 3769.266. Van der Woude, *Noorderkwartier*, 619. De Vries and Van der Woude, *First Modern*, 225. There are no records of cattle losses in the 1740s and 1750s. Beverwijk lost over 30 percent of its cattle between 1768 and 1770 during this (less severe) epizootic. Comparative tables of the number of cattle between 1768 and 1770. NA. Collectie Goldberg, 2.21.006.51.

The Second Cattle Plague, 1744–1764 239

minister and well-regarded naturalist from Alkmaar.[113] What united them was their belief that experimental science could be wielded in service to national improvement. This conviction was broadly shared by citizen-scientists and it grew in popularity over the second half of the eighteenth century.[114]

Much of the growth and development of these sciences were channeled through Learned Societies.[115] The first great Learned Society in the Netherlands unsurprisingly appeared in Haarlem in 1752. Cornelis Nozeman was one of its founders, and Agge Roskam Kool was a member. The *Holland Society of Sciences* was a prototypical Dutch Enlightenment institution – as focused on the challenges of Dutch decline, as it was optimistic about recovery. "The splendor and prosperity of the commonwealth or *Volk* [is] very connected to the flowering or neglect of loyalty and virtue, and the need for the arts and sciences," the inaugural issue of the *Society*'s journal declared.[116] Improvement could thus take many forms, including civic and moral education, as well as practical interventions into the integrated social, economic, and environmental challenges that threatened the Republic. The Society sponsored prize competitions to foster these goals and published the results in their journal. Several early prize competitions, including their first in 1753, tackled river flooding, which they argued had grown to "dreadful importance" over the previous two decades and thus "attracted the attention and concern of righteous Dutchmen."[117] River flooding reflected the declining state of Dutch rivers and, more broadly, threatened Dutch society and prosperity. The virtuous application of scientific knowledge promised greater ability to master nature and promote recovery.

Cattle plague, despite its obvious differences from inundation, offered a similar set of challenges and opportunities. Like river flooding, it

---

[113] Erik Jan Tjalsma and Ronald Rommes, "Dominees en de runderpest in de tweede helft van de 18e eeuw," *Argos* 59 (2018): 358–9. Nienhuis, *Environmental History of the Rhine-Meuse Delta*, 151.

[114] Frank Huisman, "Medicine and Health Care in the Netherlands, 1500–1800," in *A History of Science in the Netherlands: Survey, Themes and Reference*, eds. Klaas van Berkel, Albert van Helden, and Lodewijk Palm (Leiden: Brill, 1999), 278.

[115] Wijnand Mijnhardt, *Tot Heil van 't menschdom: Culturele genootschappen in Nederland, 1750–1815* (Utrecht: Rodopi, 1987). Koen Stapelbroek and Jani Marjanen, eds. *The Rise of Economic Societies in the Eighteenth Century: Patriotic Reform in Europe and North America* (New York: Palgrave Macmillan, 2012).

[116] *Verhandelingen uitgegeven door de Hollandse Maatschappy der Weetenschappen te Haarlem* (Haarlem: J. Bosch, 1754), 3–4.

[117] Ibid., foreword.

presented a clear and present danger to Dutch social and moral vitality, yet it also appeared amenable to scientific assessment and intervention. The *Society* thus underwrote the Beverwijk trials. Undaunted by this initial experiment's failure, they continued to invest in cattle plague studies throughout the 1750s. In 1759, for instance, it hosted a competition and offered a prize for the researcher best able to explain "the natural origins" of the disease. They also sought insight into why it is "lasting so much longer than in earlier times" and called for "strategies to prevent cattle plague when it begins to spread in our stalls and in our neighborhoods."[118] These competitions reflected the belief that scientific study might yield practical insight into environmental vulnerabilities and strategies to mitigate the disease. The Beverwijk experimenters joined an expanding cohort of citizen-scientists who believed that the virtuous application of rational knowledge could address the problems of the Fatherland.[119]

Enlightened sociability also connected citizen-scientists in the Republic to their counterparts abroad through networks of print and correspondence. The Beverwijk study was part of a decades-long international effort to inoculate against cattle plague.[120] The first confirmed experiment took place in England in 1754. Its results were disseminated in the *Gentleman's Magazine*, an important English periodical for the transmission of early Enlightenment medical knowledge to the lay public.[121] Journals of learned societies in England, Germany, and Holland translated and republished subsequent attempts. The well-known anatomist and surgeon in The Hague, Thomas Schwenke, published his work in both the *Gentleman's Magazine* and the *Bremisches Magazin zur Ausbreitung*

---

[118] Melchior Bolstra won the first prize. The 1757 floods also prompted a prize competition on dike breaches, which Jacob Pierlinck won in 1760. J. A. B. de Haan, *De Hollandsche Maatschappij der Wetenschappen, 1752–1952* (Haarlem: J. Enschedé en Zonen, 1952), 202, 209. J. G. de Bruijn, *Inventaris van de prijsvragen uitgeschreven door de Hollandsche Maatschappij der Wetenschappen 1753–1917* (Haarlem: H. D. Tjeenk Willink, 1977), 34–5.

[119] Mijnhardt, "The Dutch Enlightenment: Humanism, Nationalism, and Decline," 216–21. N. C. F. van Sas, "De vaderlandse imperatief. Begripsverandering en politieke conjunctuur, 1763–1813," in *Vaderland: Een gescheidenis van de vijftiende eeuw tot 1940*, ed. N. C. F. Van Sas (Amsterdam: Amsterdam University Press, 1999), 279.

[120] Huygelen, "The Immunization of Cattle against Rinderpest," 182–96.

[121] Roy Porter, "Lay Medical Knowledge in the Eighteenth Century: The Evidence of the Gentleman's Magazine," *Medical History* 29, no. 2 (1985): 138–68. Trials possibly began as early as 1747 in Burgundy. Breuker, "De achttjinde-ieuske yninting," 340.

der Wissenschaften in Germany in 1755 and 1756.[122] Schwenke admitted his experiment had "very bad effects," but he remained optimistic.[123] The Beverwijk study was reported in Dutch newspapers in 1755 and the *Gentleman's Magazine* picked up the story as well. They reprinted a letter from Amsterdam reporting that "the ill success of the above instance had, however, so effectually deterr'd others, but that an owner near the same place, who has suffer'd much from this fatal distemper among his cattle, is determined to make a still further trial."[124] Daniel Peter Layard, a physician and Fellow of the Royal Society in London, published inoculation results in 1757, and a Dutch physician from Hoorn named J. Grashuis conducted his own study and published it in 1758.[125] Grashuis reported on the continuing inoculation efforts of his countrymen and noted the improvements he had made to the methods laid out at Beverwijk. Consistent, verifiable success remained elusive, however. The promise of inoculation was easy to justify in principle but difficult to prove in practice. Experiments nevertheless continued and expanded across Europe in the 1760s.[126]

6.6 INOCULATION, PROVIDENCE, AND PHYSICO-THEOLOGY

This unyielding optimism in the face of persistent failure reflected important changes in the Dutch perception of medicine, disease, and their

---

[122] Thomas Schwenke, "Schreiben des Herrn Doct. Swenke der Anatomie und Chirurgie Professors im Haag die Einpropfung des Hornviehs betreffend," *Bremisches Magazin zur Ausbreitung der Wissenschaften, Künste und Tugend* (1756).
[123] "A Letter from Dr. Swencke, Professor of Anatomy and Surgery at the Hague, on the Subject of Inoculating Horned Cattle," *Gentleman's Magazine* (1755), 464.
[124] March 22 (1755) *Leeuwarder Courant*. "Mr. Urban, Amsterdam, April 1755," *Gentleman's Magazine* (1755), 160.
[125] Daniel Peter Layard, "A Discourse on the Usefulness of Inoculation of the Horned Cattle to Prevent the Contagious Distemper among Them. In a Letter to the Right Hon. George Earl of Macclesfield, P. R. S. From Daniel Peter Layard, M. D. F. R. S.," *Philosophical Transactions* (1757), 528. J. Grashuis, "Berigt aangaande de in-enting der besmetlyke ziekte in het rundvee in een aanmerkelyk getal beesten te werk gesteld, en tot verdere beproeving onze landgenooten voorgehouden door den heer J.G.," *Uitgezogte verhandelingen uit de nieuwste werken van de societeiten der wetenschappen in Europa en van andere geleerde mannen* (1758), 248.
[126] Most experiments took place in the Dutch Republic, Denmark, and Germany. Kai F. Hünemörder, "Zwischen 'abergläubischem Abwehrzauber' und der 'Inokulation der Hornviehseuche.' Entwicklungslinien der Rinderpestbekämpfung im 18. Jahrhundert," in *Beten, Impfen, Sammeln: Zur Viehseuchen- und Schädlingsbekämpfung in der Frühen Neuzeit. Graduiertenkolleg Interdisziplinäre Umweltgeschichte*, eds. Dominik Hünniger, Katharina Engelken, and Steffi Windelen (Göttingen: Universitätsverlag Göttingen, 2007), 35–43.

relationship to decline. By the mid-eighteenth century, a wide array of documents increasingly conflated these themes, which had appeared largely independent during the prior outbreak. Medical treatises emphasized the "natural" origins of disease, yet most acknowledged that divine agency remained the "first cause." Religious books and sermons incorporated and reflected upon the secondary, natural origins of disaster. Dutch decline had long remained a concern largely confined to sermons and other moralizing literature but gradually found purchase in broader discussions of practical challenges facing Dutch society, including natural disasters. By the mid-eighteenth century, the rigid boundaries between those concerns had eroded. In the context of cattle plague, decline became an integrated moral, economic, and medical challenge.

Three developments smoothed this transition. By the 1740s, moralists found it increasingly difficult to distinguish the causal origins of decline. The many social, political, and natural disasters of the 1740s contributed to this shift. Pietist disaster sermons in particular emphasized the hybrid, cascading consequences of disasters.[127] "We have in truth still not recovered from the burdens of the last lengthy war," Johan van den Honert proclaimed in 1747, "and the troubles arising from cattle plagues, through the many terrible floods, through the worms that attack our ships and the lands piles, through the mice that eat our harvests, through the decline on trade, commerce, and shipping." Every land is being tested, but "ours in particular" suffers.[128] Disasters pointed to the multidimensional, uniquely Dutch character of decline.

Second, punishment sermons shifted their emphasis away from concerns about doctrinal unity and assumed a more civic orientation. Sermons no longer demanded repentance for breaches in orthodoxy but rather expected moral behavior as a civic obligation to specific challenges facing the Republic.[129] This change had been building for decades, and every new disaster incentivized greater attention to this national civic discourse. By 1760, Joachim Mobachius's plea that "people, as citizens, and as Christians [must] ... acknowledge the ways of God's providence and to pray!" became commonplace.[130] He preached, "[I]t should be

---

[127] Van Eijnatten, *God, Nederland en Oranje*, 63.
[128] Johan van den Honert, *De kerk in Nederland beschouwd, en tot bekeering vermaand* (Leiden: Samuel Luchtmans en Soon, 1747), 55.
[129] Edwards, "Varieties," 40–1.
[130] Joachimus Mobachius, *Den Vrezelyken Zamenloop Van veele gedugte Oordelen Gods, Als van Alverwoestende Oorlogen, swaere Overstromingen, ontzachelyke Aerdbevingen, aenhoudende, en meer en meer woedende Runder-pest, enz. enz. Als*

your duty, not only to criticize the decay of the Church, the sins of the Land ... but first and foremost to address your own decline, your own sins, and deviations from God in deep discontent. To confess and to seek reconciliation first of all."[131] In her providential poem on flood disasters in 1754, Maria van den Berge asked God to "[b]less ... the state, round the entire Netherlands, clad them with holy strength, and grasp with righteous hand, strengthen our governance, in the work that she has been presented."[132] These entreaties for national moral improvement were conducive to the program of Enlightened recovery.[133] They blunted the sharp edge of doctrinal difference and encouraged religious leaders to participate, as at Beverwijk, alongside physicians and merchants in the practical task of national recovery.

Finally, the rise of physico-theology encouraged this transition. Physico-theology was an Enlightenment methodology that employed reason and the systematic observation of nature to demonstrate God's providence. This international movement had been popularized in the Republic through Bernard Nieuwentijdt's 1715 publication, *The Right Use of the Contemplation of the World for the Convincing of Atheists and Unbelievers*. Physico-theology bridged the gap between natural theology, providence, and scientific inquiry in the early eighteenth century.[134] It reinforced the conviction that God created a just world in the

---

zoo veele voorboden van eenen geheelen ondergang, zoo het niet door eene spoedige bekeringe werde voorgekomen ... (Gorinchem: Teunis Horneer, 1760), 33.

[131] Ibid., 48.

[132] Maria van den Berge, *Klaag-liet, Uitgeboesemt over de inbreuke en overstroming der vreeslyke wateren ontrent Zwolle, Kampen en Zutphen ... voorgevallen in 't laatste ... desvoorgaanden jaars 1753. en in het begin ... deses jaars 1754* ... (1754), 8.

[133] Ernestine van der Wall, "The Religious Context of the Early Dutch Enlightenment: Moral Religion and Society," in *The Early Enlightenment in the Dutch Republic, 1650–1750*, ed. Wiep van Bunge (Leiden: Brill, 2003), 39–57. This did not extend to Catholic practice. In a repeat of the phenomenon that occurred in Heiloo in 1713, Catholic pilgrims flocked to the healing waters of Culemborg in 1745. J. G. M. Boon, "Historische aantekeningen van een pastoor," *Heemtydinghen: orgaan voor de streekgeschiedenis van het Stichts-Hollandse grensgebied* 8, no. 15 (1965): 6–9. Protestant ministers condemned this practice. Johannes Barueth, *Davids wyze keuze uit drie plagen en ware boetvaerdigheid onder gekoze pestilentie: in deze dagen van de pest-ziekte onder het rundvee* (Rotterdam: Jacobus Bosch, 1745), 31. D. van Adrichem, "Een nieuwe devotie door de Vee-Ziekte van 1744," *Archief voor de geschiedenis van het Aartsbisdom Utrecht* 42 (1916): 373–83.

[134] Bernard Nieuwentijdt, *Het regt gebruik der werelt beschouwingen ter overtuigingen van ongodisten en ongelovigen* (Amsterdam: Joannes Pauli, 1714). J. Bots, *Tussen Descartes en Darwin: Geloof en natuurwetenschap in de achttiende eeuw in Nederland* (Assen: Van Gorcum, 1972). Jorink, *Reading the Book of Nature*. In the Netherlands, it targeted

service of man and governed nature to achieve harmonious ends. Physico-theology linked scientific reason and revealed that religion and its popularity transcended confessional beliefs. It was especially influential with dissenters from Reformed orthodoxy. Nozeman, Kool, and Tak were all dissenters, and many later figures in the Dutch inoculation program were also Mennonites or Remonstrants, who adhered to the principles of physico-theology.[135] Physico-theology created the intellectual space for broad participation in animal medicine. It encouraged awe of nature's power as exhibited in natural disasters, just as it stimulated scientific and technological intervention. Like the rest of creation, epizootic disease was a sign of God's providence and subject to human understanding and control.

6.7 EXCLUSIVE PROVIDENTIALISM

Inoculation appealed to a segment of Dutch society invested in the improvement of the human condition and the recovery of the Republic. That optimism favored inoculation experimentation but did not necessitate it. Even within enlightened circles, inoculation faced resistance from medical authorities. The physician Jan Engelman, an early member of the *Holland Society of Sciences*, was critical of animal inoculation. He wrote several articles about the disease in its *Transactions* that denied the analogy between smallpox and cattle plague. He (correctly) argued it more closely resembled measles.[136] In England, anti-contagionist

---

unbelief and Spinozism. Jonathan Israel, *Enlightenment Contested: Philosophy, Modernity, and the Emancipation of Man 1670–1752* (Oxford: Oxford University Press, 2006), 385.

[135] Mijnhardt, *Tot Heil van 't menschdom*, 53. Van der Wall, "The Religious Context of the Early Dutch Enlightenment," 43. Huisman, *Neerlands Israël*, 36. Wiep van Bunge, "Introduction: The Early Enlightenment in the Dutch Republic, 1650–1750," in *The Early Enlightenment in the Dutch Republic, 1650–1750*, ed. Wiep van Bunge (Leiden: Brill, 2003), 3. Tjalsma and Rommes, "Dominees," 356–63.

[136] Jan Engelman, "Waarnemingen in de rundveesterfte in 1756 en 1759, dienende tot een voorloper ter nader verhandeling over dezelfde stoffe," *Hollandsche Maatschappye der Weetenschappen, te Haarlem* 6 (1760). Jan Engelman, "Nader verhandeling over de rundveesterfte, betreklyk tot de waarneemingen, vervat in 't VIde deels IIde stuk," *Verhandelingen van de Hollandsche Maatschappye der Wetenschappen* 7 (1762): 247–318. W. J. Paimans, "De Veeartsenijkunde in Nederland vóór de stichting Der Veeartsenijschool de Utrecht," in *Een eeuw veeartsenijkundig onderwijs. 's Rijks-Veeeartsenijschool, Veeartsenijkundige Hoogeschool, 1821–1921*, eds. H. M. Kroon, W. J. Paimans, and J. E. W. Ihle (Utrecht: Senaat der Veeartsenijkundig Hoogeschool te Utrecht, 1921), 9.

physicians opposed the procedure because it did not address the physiological origins of the disease and because it exposed people to quackery and "wild chemists."[137] This type of resistance may have been evidence of physicians policing medical boundaries. It also pointed to a familiar problem, which had limited acceptance of stamping out during the first outbreak. Inoculation presumed acceptance of contagionism. In an era when causal theories of disease transmission still operated along a spectrum from contagionism to environmentalism, inoculation seemed at best a partial solution.

Inoculation also presented economic and moral dilemmas that worked against widespread acceptance in the medical professions and society at large. Inoculation was expensive and risky. Experimenters sometimes used their own animals, but large trials like the Beverwijk study required investment from wealthy benefactors or learned societies. Other critics questioned the wisdom of intentionally infecting animals when infection might accidentally spread to healthy cattle. Every inoculation contained the seeds of a new outbreak. Still others argued the practice was immoral.[138] According to this view, cattle plague was divine punishment and any intervention undermined God's intentions. In the preface to the Beverwijk study, the authors noted opposition driven by "prejudice," spurred on by this "so-called religion and conscience."[139] Proponents frequently singled out this exclusive providential opposition to inoculation.

Eelko Alta, a Reformed minister and citizen-scientist from Friesland, was perhaps the most strident voice in favor of inoculation and against exclusive providentialism. Alta performed several inoculation experiments in 1755 and 1759 and published his findings in 1765.[140] Like many others, he had been motivated by a *Holland Society of Science*'s prize competition.[141] Alta began by apologizing for having kept his findings secret until publication. He did so, not out of "willfulness, but because a report about inoculation for this sickness was such a tender

---

[137] DeLacy, *Germ of an Idea*, 136–7.   [138] Barrett et al., *Rinderpest*, 92.
[139] Nozeman et al., *Eerste proefneeming*, preface, 5.
[140] Eelko Alta, *Verhandelinge over de natuurlyke oorzaaken der ziekte onder het rund-vee en derzelver langere duuringe als te vooren, waar in ... de in-enting derzelve ... wordt aangepreezen: gaande vooraf twee vertoogen, gezonden aan de Hollandse Maatschappy der Weetenschappen, ter beantwoording eener vraag over deeze stoffe in 1759 voorgesteld en in 1760 herhaald* (Leeuwarden: Wigerus Wigeri, 1765).
[141] Alta, *Verhandelinge*, 4. Alta also competed for a similar prize sponsored by the King of Prussia, Frederick the Great. He won neither. Tjalsma and Rommes, "Dominees," 359.

point, of which almost nobody dared to speak, where those that understood it and those that did not were very much against it, and most out of an incorrect understanding of the disease."[142] According to Alta, resistance to inoculation was widespread but grounded in faulty logic about the nature of providence and potential for improvement.

Alta's work is notable because he engaged his opposition. In his book *Reports on the Natural Origins of Cattle Plague*, he attributed opposition to several misunderstandings. "Some err," he argued, out of "simplicity or oversight, some from a wrong understanding of God's providence, or out of conceitedness."[143] According to Alta, antagonism sometimes extended to outright denial of the "natural origins" of the epizootic. Opponents believed that disease "was wrought by God's immediate, supernatural, and exclusive *slaande hand*."[144] Alta distanced himself from exclusive providentialism by pointing to his training and social position. "Although most knowledgeable people agree with me, namely scientists and physicians," he explained, "there are still far too many people that think that the cattle plague is an unavoidable judgment of God."[145] They included farmers who, by virtue of their "plodding, prejudices, and adherence to their ancient and ancestral habits, and their resulting small-mindedness," opposed the idea. These same tendencies sometimes applied to "the Learned among all the Sciences," who, he argued in a later pamphlet, were "guilty of the same sickness."[146] Alta placed himself squarely on the side of an enlightened science that resisted the impulse to rely on tradition and worked to the improvement of society.

According to Alta, cattle plague was merely a disease, no different from other calamities. "One has no more reason to think this sickness is an unavoidable punishment from God or providence than any other sickness, disaster or uncommon occurrence," he argued.[147] He criticized the notion that failure of inoculation was evidence of the affliction's supernatural origin, calling this notion "absurd and ridiculous."[148] The disease was natural because it behaved as a contagion. It moved between animals in a predictable manner that mimicked the pattern of the previous

---

[142] Alta, *Verhandelinge*, 3.   [143] Ibid., 31.   [144] Ibid., 32.   [145] Ibid., 31.
[146] Eelko Alta, *Nodige raadgeevingen aan overheden en ingezetenen, dewelke in het bijzonder voor den boer in deeze akelige omstandigheden, waarin het vaderland zig door de ziekte van het rundvee thans bevind, van zeer veel nut kunnen zijn* (Leeuwarden: H. A. de Chalmot, 1769), 4–5.
[147] Alta, *Verhandelinge*, 35.   [148] Ibid., 39.

epidemic.[149] Opponents of inoculation, particularly those who drew upon an exclusivist providential rationale, characterized the plague as unique, unknowable, and unstoppable. Alta, by contrast, marveled at the power of science to illuminate the "great secrets of nature."[150]

Alta offered an optimistic appraisal of the power of reason to solve the social and medical challenge of the disaster. His optimism did not preclude providence. Alta believed that God steered the world via general providence but denied its exclusivity. In a manner consistent with the physico-theological underpinnings of the early Dutch Enlightenment, he viewed the disease as an expression of God's natural order. This interpretation demanded intervention. Alta likened the plague to other natural disasters and questioned why they provoked less public opposition. Why should people work to limit wars, famine, and floods if they were also divinely ordained? He then quoted the Beverwijk study, which argued that it would be foolish to avoid "paying taxes, stocking food, strengthening dikes, and practicing medicine" for reasons of "religion or conscience."[151] To do so would deny the God-given power to understand the Book of Nature and undermine the virtuous application of reason. Perhaps most importantly, improvement benefited the Republic. "I have considered it my duty to inform society of my thinking and to enlighten with this work," he explained. Inoculation "is now of the utmost importance for Volk and Land, for every citizen in particular, as well as for the State in general."[152]

6.8 CONCLUSION

Cattle plague had virtually disappeared by the time Alta published his book in 1765. After twenty years of repeated outbreaks, cattle plague finally receded from the Dutch countryside. In the meantime, the Republic had weathered a suite of social, political, and natural disasters that deepened public fears of decline. By the 1760s, decline was broadly recognized as an integrated social, moral, and economic challenge. State, scientific, and religious responses had evolved in the context of these declensionist fears as well as the optimistic influence of the Dutch Enlightenment. The inoculation trials were an enlightened rejoinder to the severity and intractability of epizootics and the problem of decline more broadly.

[149] Ibid., 52–6. [150] Ibid., 46. [151] Ibid., 163. [152] Ibid., 149.

Cattle plague would return four years later to prompt a third round of responses. Physicians and citizen-scientists renewed experimentation with inoculation. Public fear of Dutch decline intensified, revitalizing discussion among Enlightenment thinkers worried about its political, economic, and moral origins. In this context, secular interpretations of both disaster and decline slowly displaced exclusivist interpretations of divine agency. Historians have identified the 1740s and 1750s as a watershed moment in this transition. "No Reformed minister ever doubted that God ruled the world," historian Joris van Eijnatten argues, "but after 1746, he would be hard pressed to demonstrate how, exactly, and when God governed it."[153] Causal stories of disaster and decline rarely precluded providence, but they reserved increasing attention for material concerns once considered unrelated or subordinate.

Changes in providential thinking and the rise of physico-theology challenged the "self-evidentness" of punishment theories. They transformed phenomena like cattle plague, flooding, shipworms, and comets from ominous portents into "edifying appendices."[154] When the Great Comet had streaked across the night sky in 1744, it sparked widespread wonder and apprehension, and many later connected it to cattle plague. By the time Halley's comet reappeared in 1759, skeptical interpretations of prodigies were ascendant.[155] Some observers dismissively noted that "[Halley's comet] resulted in none of the disasters that ignorant common peoples had feared." "And no wonder that this heavenly body shows no sign of disaster, since, like others connected to a fixed course, they stretch to fulfill the infinite intentions of the all-wise Creator, as time and diligence has discovered." The comet's predicted appearance was both triumph of science and evidence of the majesty of creation. That same year, the *Holland Society* held its first contest to discover the "natural origins" of cattle plague.[156] These events were related. Both were opportunities to

---

[153] Van Eijnatten, *God, Nederland en Oranje*, 67.
[154] Buisman, *Tussen Vroomheid*, 143. Huisman, *Neerlands Israël*, 38. Jorink, *Reading the Book of Nature*, 109–79. Fix, "Comets," 161.
[155] Some skeptics in 1744 already bemoaned the "old superstition" of the "common people" who connected cattle plague to the Great Comet. Ludvig Holberg, "Korte Betænkning over den nu regierende Qvæg-Syge med nogle œconomiske Anmerkninger," *Skrifter som udi Det Kiøbenhavnske Selskab af Lærdoms og Viderskabers Elskere ere fremlagte og oplæste i Aaret 1745* 2 (1745), 396.
[156] *Nederlandsche Jaerboeken* (1759), 683, 716. The *Nederlandsche Jaerboeken* promoted enlightened interpretations of disaster. Donald Haks, *Journalistiek in crisistijd: De (Nieuwe) Nederlandsche Jaarboeken 1747–1822* (Hilversum: Verloren, 2017), 37–59.

subject environmental change to rational explanation and optimistic intervention.

In light of these broad changes, it is easy to overstate the degree to which they replaced traditional interpretations of disaster. Alta and other Enlightenment-era thinkers defined themselves in contrast to an opposition they often characterized as ignorant, superstitious, or fatalistic. Response to the second cattle plague exposed the limits of this view. The epizootic prompted neither broad acceptance of inoculation nor an eclipse of providential reasoning. Inoculation trials never produced results convincing enough to tempt farmers or state authorities to back the endeavor. They were too expensive and risked infection of healthy herds. To physicians convinced of the internal, physiological origins of disease, inoculation would only increase the likelihood of an epizootic outbreak. For moralists convinced of its exclusive providential origins, inoculation undermined moral improvement. Despite efforts by proponents of inoculation to mischaracterize opposition as irrational, little evidence supports this assertion.

The cattle plague that descended upon the Dutch Republic in 1744 sparked a disaster unsurpassed in the eighteenth century. Appearing amidst social and political turmoil, economic dislocation, and repeated environmental shocks, cattle plague signaled a new nadir in a century long bedeviled by calamity. In fact, these social, economic, and environmental disasters of the 1740s and 1750s were interrelated and self-reinforcing. They contributed to a cascading epizootic and economic catastrophe, at once unevenly distributed and nationally significant.

At the same time, cattle plague presented new opportunities to adapt to ongoing challenges and promote strategies to reclaim Golden Age virtue and prosperity. As they had during previous disasters, state authorities, moralists, and the public relied on diverse interpretations and strategies to cope with disease. Memory of the previous outbreak was central, both as a rubric against which exceptionality could be measured and a source of management strategies. The disaster was not novel, but its severity, duration, and coupling to concurrent crises prompted people to adapt their time-tested responses. Punishment sermons remained powerful vehicles of moral improvement by emphasizing civic virtue. Provinces enacted flexible crisis-management strategies that accommodated economic demands and epizootic threats. Traditional animal medicine expanded in popularity, and inoculation experimentation revealed newfound confidence in the potential of science and medicine to control nature and improve the Republic. It also showcased a further elaboration of the improvement and innovation discourses that had prevailed across much of the early eighteenth century. Both change and continuity defined this period of crisis. Cattle plague provoked an optimistic perspective about recovery and simultaneously reached back to the golden achievements of the past.

7

The Twin Faces of Calamity

Lessons of Decline and Disaster

Disasters marked the end of the Dutch Golden Age. Between the *Rampjaar* of 1672 and the early 1760s, the Republic weathered the deadliest North Sea storm in its history, reoccurring and increasingly severe river floods, two devastating cattle plagues, and an unprecedented molluscan outbreak. These disasters coincided with profound changes in Dutch society and culture. By the end of the War of the Spanish Succession, the Republic had lost its position as a Great Power in Europe. Important sectors of its economy were deteriorating. A century-long recession tested rural and urban communities, and natural disasters exacerbated its impacts. After 1720, export-oriented industry collapsed. By the 1740s, industrial and fishing towns appeared dull facsimiles of their seventeenth-century counterparts. With a few notable exceptions, most cities shank as the Republic deurbanized.[1] The troubling reality of these urban, economic, and geopolitical changes became virtually inescapable. This same period witnessed widespread social unrest and two political revolutions.[2] Cultural and moral critiques of Dutch character, virtue, and piety linked economic decay, political dissatisfaction, and environmental deterioration to growing fears of decline. By the second half of the eighteenth century, decline appeared nearly universal,

---

[1] Roessingh, "Landbouw," 16–72, 451–3. Israel, *Dutch Republic*, 998–1008. Populations in Amsterdam, Rotterdam, and The Hague remained fairly static.

[2] The period between 1671 and 1750 was a high-water mark for popular uprisings. Rudolf Dekker, *Holland in beroering: oproeren in de 17de en 18de eeuw* (Baarn: Ambo, 1982), 37.

provoking moral, economic, social, cultural, and political anxieties.[3] Natural disasters thrust these evolving fears into public view.

Disasters were also trials of faith and opportunities to improve the physical and moral security of the Republic. They showcased the intersection of declensionist anxieties with evolving water and disease management strategies, science and medicine, and the shifting moral and civic meaning of environmental change. Disasters could be deadly and divisive, but they could also promote solidarity and innovation. The universalizing vision of total decline does little justice to the richness of Dutch response to eighteenth-century adversity. Disasters exposed real and accumulating fear, but they also revealed a society creatively refashioning itself.

This final chapter steps back to assess the changing perception of decline between 1672 and the second half of the eighteenth century. Thus far, this book has treated disasters as inflection points in a long trajectory of social and environmental change, prompting reconsideration of the meaning of decline. Declensionist anxieties did not emerge fully formed in 1672, nor any other date. The meaning of decline evolved, and flooding, shipworms, and cattle plague marked its transformation. This chapter asks what natural disasters revealed about the changing perception of decline in the long term. Disasters showcased the growing influence of proto-national declensionist narratives, the emergence of economic and especially commercial anxieties, and the influence of underrepresented, rural interpretations of decline.

It also considers what this extraordinary period of economic, social, environmental transition teaches us about disasters more broadly. Disasters, like the perception of decline, emerged in the long term. They were events of immediate and uneven consequence, but also processes born out of the slow transformations of Dutch society and its environment. Dutch perception of these conditions affected their responses in varied, sometimes conflicting, ways, just as they do today. This era of Dutch history reminds us that "learning from disasters" – a subject of intense interest for historical disaster studies – was never a straightforward process. Environmental history presents useful interpretive tools to understand disaster in the past and in the current era of novel and increasing risks.

---

[3] Kossmann, "Dutch Republic," 28.

7.1 DISASTERS AND THE CHANGING PERCEPTION OF DECLINE

During the 1670s, observers marveled at the miracle of Dutch prosperity. The Republic, in the words of English ambassador William Temple, was "the Envy of some, the Fear of others, and the Wonder of all their Neighbours."[4] The political, economic, and military disaster of the *Rampjaar* of 1672, coupled with floods and windstorms in subsequent years, tested Dutch confidence about their geopolitical and environmental security. As De Hooghe's 1675 print *Ellenden klacht* illustrated, however, the "restoration and hope of peace" was eminent. William III's consolidation of political power and Dutch victories on land and sea secured peace by 1678. Economic recovery was halting, but few in the Republic believed that seventeenth-century prosperity was irrevocably lost. The *Rampjaar* was a trial of faith that provoked moral and political self-reflection. At the dawn of the eighteenth century, decline was far from a foregone conclusion.

Declensionist narratives expanded and diversified during the first quarter of the eighteenth century. In 1702, William III died, and the War of the Spanish Succession began. This was the Dutch Republic's last European conflict as a Great Power.[5] After the war ended in 1713, declensionist anxieties about the Republic's evolving geopolitical and moral condition began to coalesce, and natural disasters forced these issues center stage. By cruel coincidence, the signing of the Treaty of Utrecht (1713) barely preceded the emergence of a new threat. A panzootic of cattle plague spread from the steppes across central and eastern Europe and entered the Republic. Its scale and severity deeply unsettled Dutch farmers, moralists, and state authorities. Mortality varied, but every province suffered. The disease threatened dairy cattle and oxen – both integral to the rural economy and powerful symbols of the state and Dutch cultural identity. Between 1713 and 1720, plague transformed productive pastures into burial grounds. The cattle bodies that once represented Arcadian fertility and the vast extent of Dutch economic hegemony withered and died. Cattle plague heightened apprehensions about the Republic's changing position.

Amidst the cattle plague, the catastrophic Christmas Flood of 1717 struck the North Sea coast. It was likely the deadliest flood in North Sea history, but its impacts remained remarkably limited within the

---

[4] Temple, *Observations*, preface.   [5] Israel, *Dutch Republic*, 960.

Netherlands. Most Dutch casualties lived in socially and geographically vulnerable communities along the Wadden Sea coastline of Groningen. The Christmas Flood demonstrated how localized impacts translated to provincial and pan-provincial dialogues of responsibility. The city of Groningen and the *Ommelanden* debated financial responsibility for reconstruction, and punishment sermons refashioned this Groninger disaster into a broader condemnation of Dutch moral decline. The cultural memory of flooding was key to this transfiguration of scale. No natural disaster was as culturally resonant or as central to coastal identity. Flood *memoria* commemorated its deep history in the Netherlands. These memories operated as coping mechanisms that reminded victims of their own resilience. At the same time, the exceptional severity of the flood separated it from past disasters. Historic resilience was no guarantee of future security.

Cattle plague and the Christmas Flood presented monumental, but ultimately manageable challenges. Response to both hazards drew upon causal stories of disaster and management strategies based in cultural memory and previous experience. Trade networks could be regulated, polder landscapes and cattle stalls cleansed, and dikes and water administration could be improved. The disasters appeared exceptional in severity, but longstanding experience and memory provided useful frameworks for response. Far from inspiring fatalistic acceptance or dependence on passive and ineffectual management strategies, responses grounded in memory and tradition proved malleable and capable of accommodating diverse needs. The deep structural changes in the urban economy and the overseas trading system that would eat away at the Republic's economic vitality after 1720 had not yet crested.[6] In this context, declensionist rhetoric overwhelmingly focused on moral critiques.

The shipworm epidemic of the 1730s presented an altogether different problem. The novelty of this animal threat and the apocalyptic fear of total inundation forced coastal communities, moralists, and state officials to adapt their playbooks. Shipworms required new designs for water infrastructure and new moral meanings. By 1732, the failure of water authorities to resolve the problem pushed shipworm paranoia into the public sphere. Dike breaches threatened not only the safety of the affected coastal communities but also endangered the economic heart of the Republic. Shipworms presented a flood risk that demanded novel

---

[6] Ibid., 993.

adaptations and new interpretations of their meaning. Shipworms transformed coastal infrastructure and moral critiques of the Republic. Religious authorities had long linked calamity with sin, but providential castigations rarely targeted specific crimes. The shipworm epidemic nearly coincided with a wave of sodomy persecutions between 1730 and 1732. Pietist ministers and enlightened spectatorial essayists alike connected shipworms to sodomy, refining their declensionist rhetoric. For the first time, a natural hazard appeared existentially significant for the Republic, and its insidious, unnatural, and unexpected arrival clearly pointed to moral decline.

Moral declensionism grew increasingly tied to economic and environmental anxieties throughout the 1740s and 1750s. Repeated river flooding convinced hydraulic experts that the Rhine–Meuse River System was deteriorating. Water boards and the province of Holland tasked a group of Leiden-trained surveyors and cartographers to evaluate the declining state of Dutch rivers and present solutions. They produced a series of risk maps and hydraulic studies that framed the disaster in interprovincial, systemic terms. The declining environmental stability of the river system increased economic and social vulnerabilities, and experts like Cornelis Velsen argued that adapting river management was prerequisite to addressing Dutch economic decline. Ministers and other moralizing authors responded differently. They connected the cascading environmental and economic disasters of the 1740s to longstanding anxieties about moral decline. An unprecedented confluence of disastrous events in the 1740s provoked new relief efforts. Private relief committees addressed the social and moral vulnerability of the riverlands. They reinforced civic Christian solidarity while at the same time offered material assistance. Their efforts addressed victims' economic dislocation and redeemed the moral vulnerabilities of the Republic. Between the floods of the 1740s and 1750s, disasters became increasingly interprovincial affairs, and declensionist rhetoric merged economic, moral, and environmental fears.

The last major disaster of the early eighteenth century began in 1744 when cattle plague returned to the Dutch Republic. The plague was far deadlier and longer in duration than the earlier outbreaks. Severe frosts and floods in the Netherlands, concurrent ecological crises in foreign territories, and the outbreak of the War of the Austrian Succession conditioned both the severity and extraordinary duration of the epizootic. These environmental traumas coincided with civic unrest, political revolution, and continued economic deterioration. In the face of a disaster cascade that echoed the darkest moments of the *Rampjaar* and the failure of management strategies to contain the disease, people sought new meanings and novel strategies to understand and solve the crisis. Cattle plague emerged as an integrated social, economic, and moral challenge that required input from ministers, enlightened citizen-scientists, and formally trained physicians. Traditional regimes of interpretation and responses persisted, but the first inoculation trials in the 1750s presented an optimistic rejoinder to this multidimensional problem. By the 1760s, cattle plague encapsulated the multiple meanings of decline and became a central project of the early Dutch Enlightenment. Decline was widely recognized and assumed broad social, economic, and moral valence.

Observing the long trajectory of decline through the lens of natural disasters highlights three important changes in the perception of decline during the first half of the eighteenth century. First, disasters track the expanding influence of proto-national decline narratives. In the first decades of the eighteenth century, declensionist discourses that focused on the "Fatherland" were nascent. The Republic was highly decentralized, and Dutch identities remain tied to local, provincial, and confessional concerns. On top of this, natural disasters produced uneven, often localized impacts. Even the cattle plague outbreaks, which touched every province, remained regionally and locally variable. Disasters informed decline narratives, provided they could be convincingly scaled up.

Ministers and moralizing pamphleteers pioneered approaches that translated local and provincial concerns to a broader scale. They used the language of exceptionality to widen disasters' moral meaning. Over the course of the eighteenth century, public participation in disaster dialogues expanded as state authorities came to terms with the limits of prior experience. After the 1730s, growing fears of economic and social decline encouraged a wider range of responses. Provincial authorities increasingly sought formal and informal expertise in hydrology, natural history, and animal medicine. This reflected widening public familiarity with water management and medicine, as well as growing civic engagement with the problem of decline. State authorities, water managers, and ministers argued that shipworms and river floods presented novel, existential, and increasing threats to the vulnerable heart of the Republic.

By the second cattle plague, little translation was necessary. The growing emphasis on improvement and innovation (not to mention intense interest in the roots and meaning of decline more generally) reflected the enlightened project of social and environmental perfectibility. These ambitions cut across social, confessional, and geographic distance, at least rhetorically. Although the notion of the Dutch "nation" would remain anachronistic until the end of the century, these changes fed from, and likely contributed to, national identity formation – a subject of recent and increasing interest in disaster studies in the Netherlands.[7] By the 1750s, moralists and enlightened citizen-scientists alike framed decline as an unambiguous threat to *Volk* and state and an opportunity for redemption and improvement.

Natural disasters also highlighted the increasing influence of economic perspectives on decline. These concerns had already emerged at the end of the seventeenth century.[8] The decay of commerce (*neeringloosheid*) had featured in De Hooghe's print *Ellenden klacht* in 1672, yet few worried these economic problems would endure. In subsequent decades, the fear of commercial stagnation grew more pronounced and appeared in disaster sermons, prints, and state reports. Cultural and moral perspectives on decline continued to dominate early eighteenth-century disaster literature, but economic concerns steadily gained influence and attention. This "economic turn" in decline literature was part of a broader European phenomenon that sought human improvement and national prosperity in the enlightened study of political economy.[9] Disasters illuminated this transition. In contrast to commercial diagnoses that framed decline as a relative change in the European theater of power, disaster literature promoted inward-looking perspectives that highlighted moral, environmental, and economic vulnerabilities internal to the Republic.

Finally, the natural disasters of the early eighteenth century uncover a distinctive rural decline paradigm that has thus far captured little attention. The official reports and widely disseminated autopsies of Dutch decline that monopolize scholarly attention today tend to reflect the anxieties of urban elites in the second half of the eighteenth century.[10] These concerns were largely commercial and geopolitical in nature. Easily

---

[7] N. C. F. van Sas, "The Netherlands: A Historical Phenomenon," in *Dutch Culture in a European Perspective: Accounting for the Past, 1650–2000*, eds. Douwe Fokkema and Frans Grijzenhout (New York: Palgrave Macmillan, 2004), 52–2. Jensen, *Wij tegen*.

[8] Koen Stapelbroek, "Dutch Decline as a European Phenomenon," *History of European Ideas* 36, no. 2 (2010): 139–52. Nijenhuis, "Shining Comet."

[9] Sophus Reinert and Steven Kaplan, *The Economic Turn: Recasting Political Economy in Enlightenment Europe* (London: Anthem Press, 2019). Erik S. Reinert, "Emulating Success: Contemporary Views of the Dutch Economy before 1800," in *The Political Economy of the Dutch Republic*, ed. Oscar Gelderblom (Surrey: Ashgate, 2013), 35–7.

[10] Nijenhuis, "Shining Comet."

the most famous report was stadhouder Willem IV's 1751 "Proposition for the Restoration and Improvement of Commerce in the Republic."[11] Historians consider it a seminal document in decline scholarship. His plan to improve the navy and lower import levies to revitalize trade and restore financial stability exemplified a commercial decline paradigm and drew from ideas circulating in the merchant class of Amsterdam and Rotterdam.[12] It reflected an elite interpretation of Dutch power rooted in maritime trade. It also signaled an important shift in baseline assumptions about the roots of Dutch prosperity and decline. Rather than emphasizing the Arcadian fertility and productivity of Dutch rural landscapes, commercial theories after the 1750s framed the Dutch environment as a sterile liability only compensated through industry and foreign trade. Eighteenth-century commercial stagnation, in this later view, undermined the roots of Golden Age prosperity.

Rural concerns in the first half of the eighteenth century operated under different assumptions. Fear of agricultural recession, total inundation of the countryside, and moral degeneration of Dutch society emerged out of a rural worldview that tied moral and civic virtue to agricultural productivity, fertility, and mastery of water. Disasters inverted these associations, forced people to confront the limits of experience and memory, and thrust their anxieties into public view. Although largely absent from later elite debates about Europe's shifting balance of power, natural disasters reinforced, transformed, and challenged Dutch perceptions of decline.

7.2 THE LESSONS OF DISASTER

From his farmhouse in the rural town of Oud Loosdrecht, a few kilometers north of Utrecht, Jacob Jacobsz. Pos (1703–79) bore witness to many of the political and economic changes that transformed the Dutch Republic over the course of the eighteenth century. We know this because Pos kept a journal that recounted the "most wondrous" and consequential events of his life.[13] It included his impressions of important shifts in domestic politics, trade, the consequences of war and peace, and the

---

[11] *Propositie van syne hoogheid.*
[12] Ida Nijenhuis, "Captured by the Commercial Paradigm: Physiocracy Going Dutch," in *The Economic Turn: Recasting Political Economy in Enlightenment Europe*, eds. Sophus A. Reinert and Steven L. Kaplan (New York: Anthem Press, 2019), 640–1.
[13] Schlimme, *Dagboek*.

subtle and complex ways these events affected his life. He described wartime taxation and trade embargoes, encounters with mob violence, and their consequences for "common folk" beset by years of trauma and uncertainty.[14] Pos was a cattle farmer and merchant, and perhaps unsurprisingly, he reserved most of his journal for extended descriptions of natural disasters.

Beginning with the shipworm epidemic, his narrative reads like a compendium of calamity. In the 1740s alone, he weathered the bitter winter of 1739–40, flooding in 1741, mice plagues in 1742 and 1743, and the resulting years of failed hay harvests. He watched in awe as the Great Comet of 1744 passed overhead and listened to Amsterdam street preachers exhorting citizens of the Republic to repent of their sins. In 1745, the "plague years" began in Oud Loosdrecht; the epizootic continued its unwelcome, itinerant visitations for two decades.[15] Pos interspersed his narrative with accounts of the financial and social strain of these disasters, their intersections with other social changes, and outlined with unusual clarity their cascading economic and environmental consequences. Pos's journal is a window into a rural worldview that translated the tumultuous first decades of the eighteenth century to the human scale.

Pos left this record for posterity – and also to impart important lessons about the meaning of the disasters that beset this period of his life. Addressing his children, he implored them to be patient and humble in the face of adversity. "Under God's Striking Hand," he advised, "pray to the Lord and He will be your guide."[16] Pos believed that his experience of cattle plague was particularly instructive. "Thousands of animals have been dragged away by this pestilence," robbing Pos and his neighbors of their wealth and security. This reversal of fortune offered another lesson. "Do not exalt in prosperity, for the Lord may strike." Pos interpreted the arrival of disease as consequence of pride – and not merely his own. The punishing decades of the early eighteenth century were broad in scope and seemingly worse than those in the "days of our fathers."[17]

Pos lived in an age of increasing risk, yet he remained on the whole optimistic. Looking back on "God's judgments" years later, Pos admitted that "at its beginning, people thought it impossible to overcome. With hindsight, it appears differently because the land is already filled once more with livestock, and most farmers retained their homes."[18] He and his neighbors had suffered years of tragedy and uncertainty, but with

---

[14] Ibid., 13. [15] Ibid., 10. [16] Ibid., 30. [17] Ibid., 24. [18] Ibid.

patience and faith, they endured. Pos directed these lessons to his children, but they were also a prescription to redeem the moral character of Fatherland. Certainly, Pos's characterization of rural resilience was not all encompassing. Many farmers would not escape these disasters with their farms or their lives intact. However, Pos's underlying message – that these disasters were trials of faith – was broadly shared across the Republic and deeply influenced disaster perception and response throughout the eighteenth century. At its core, it connected disasters to decline, but like so many accounts of this tumultuous era, it also promised hope and renewal.

Pos's search for lessons drawn from his own experience speaks to a common impulse in the study of historical disasters to derive insights that inform the present and the future. Of course, any efforts to glean lessons from the past are necessarily limited by the chasm of difference that separates these societies from our own. How instructive could Pos and his contemporaries, who lived in a world governed by different social demands, cultural imperatives, and environmental pressures and opportunities, possibly be? The disasters described in this book were hybrid events that emerged at the nexus of social and environmental change. They were dramatic, even catastrophic events, but also long-term processes whose consequences far outlasted these relatively brief moments of acute danger. Disaster impacts likewise varied immensely and their uneven outcomes were contingent upon preexisting vulnerabilities. Finally, the past was a malleable and necessary tool to construct and convey meaning. Pos's disaster experiences reflected these insights, and his decision to write a journal and conclude it with a set of retrospective lessons underscores their enduring value. These broad interpretative frameworks simultaneously emphasize and transcend the specific conditions and contingencies that birth and define calamity. Disasters show that much separates our lives and concerns from Jacob Pos and his contemporaries, but interpreting disasters in this manner also uncovers undeniable throughlines that connect their experiences to our own.

Evaluating the *perception* of disasters in the eighteenth-century Dutch Republic also encourages us to reconsider these frameworks. The values, traditions, and beliefs of Jacob Pos and his contemporaries defined the range of their potential responses.[19] They encouraged and constrained adaptation, set the bounds of their causal stories, and enabled or occluded

---

[19] Adger et al., "Are There Social Limits," 345.

interpretations that connected calamity to social or physical environmental change. These perceptions were fluid, and every disaster supplied an opening to reevaluate and refine their meaning. The Dutch eighteenth century provides the historian a wealth of opportunities to assess not only the conditions that produced calamity but also the diversity of ways that perception molded response. Despite their very different worldviews, the patterns of interpretation that emerge from eighteenth-century stories provide important insights for the present as well.

7.3 DISASTERS AS HYBRID PHENOMENA

Between the late seventeenth and the first half of the eighteenth century, Dutch communities weathered a series of devastating phenomena. Today, we might refer to them as natural disasters, although closer consideration reveals a variety of interrelated social and environmental conditions that both catalyzed events and influenced their outcomes. Cattle plagues emerged, spread, and declined based in part on shifting patterns of war, trade, and disease management efforts, and also seasonal and interannual changes in weather that influenced grass growth and flooding. Shipworm populations waxed and waned based on the availability of habitats provided by dike infrastructure, including changing precipitation patterns, widening coastal inlets, and shrinking river mouths. The likelihood of river flooding increased or decreased based in part on climatic trends, and also river management. Shifting economic relationships, evolving institutional governance, and changing technologies influenced disaster emergence and outcomes, but acknowledging their importance should not diminish the significance of environmental conditions, nor should physical hazards be relegated to the role of "triggering events." Seeking disaster origins and tracking their consequences in the social and physical environment discourages us from seeing either as static or independent.

Disaster hybridity also encourages us to examine trends that linked seemingly disconnected disasters together. It was no coincidence that the eighteenth century featured a series of devastating natural disasters, nor that rural communities bore the brunt of their impacts. Rural recession, prompted by a century-long decrease in agricultural prices, was the clearest common denominator. The associated reductions in income undermined the capacity of communities to weather and rebound from natural disasters. Disasters also amplified their impacts and ensured their unequal distribution. Communities less dependent upon specialized production were generally less susceptible in the short term, and some enjoyed relative prosperity. People who experienced the scissor-like consequences of decreasing income and increasing expenditures – whether due to wartime taxation, increasing water management duties, or the costs of restoring cattle herds – experienced the greatest economic dislocation. Jacob Pos and his neighbors in Oud Loosdrecht, for instance, experienced this full suite of adverse conditions. The region was repeatedly flooded by the Lek, it helped finance the reconstruction of the *Diemerdijk* along the Southern Sea following the shipworm outbreak, and its inhabitants endured high taxes in times of war. Its specialization in capital-intensive dairying exposed farmers to high mortality during cattle plague outbreaks and the repeated expense of replenishing diminished herds. A disastrous event on its own may hint at these underlying social and economic vulnerabilities, but examining multiple, seemingly disconnected events in sequence casts their outlines in sharper relief.

The disasters of early eighteenth century were also the unintended legacy of social and technological systems to control water. River flooding increased in frequency and severity throughout the eighteenth century due in part to centuries of interventions in the Rhine–Meuse River System. Shipworms colonized wooden dike technologies developed to stabilize the coast. Embankments, sluice and dam building, and pumping technologies encouraged reclamation along coastlines and rivers and transformed low-lying peat districts into cultural landscapes suitable for grazing cattle and epizootic spread. Drainage often simultaneously encouraged agricultural development and elevated flood risk, especially in communities where social marginalization undermined their capacity to maintain water infrastructure. Maintaining the balance between land and water was a constant struggle, which in the best of times brought significant rewards. By the seventeenth century, the social and technological mastery of water produced a degree of security and fostered a level of agricultural productivity unparalleled in Europe. Many of the environmental interventions that fostered prosperity during the Golden Age also lay the groundwork for disasters to come. Eighteenth-century disasters were often a consequence, not only of water management failures in short term but also their successes in the long term.

The interactions between climate change, extreme weather, and calamity were the quintessential expressions of disaster hybridity. These connections were internally complex and often ambiguous because they were contingent upon evolving social, technological, and economic conditions. Climate and extreme weather were neither the lone nor necessarily the most significant factors that contributed to disastrous outcomes, yet shifts in temperature, wind patterns, and precipitation undoubtedly influenced every disaster covered in this book. This history began during one of the coldest periods of the Little Ice Age and concluded in decades of relative warmth. Historians have traditionally associated Little Ice Age disasters with its periods of coolest temperature, and, indeed, this warming trend certainly reduced the likelihood of some climate and weather-related impacts. Yet this remains at best a partial story. Warmer conditions actually fostered some disasters. The shipworm epidemic was far more likely in the warmer, drier 1730s than it had been even two decades earlier. Other times, the decreasing likelihood of disasters in warmer temperatures were offset by social, economic, or technological shifts. Ice dams were less common in warmer decades, for instance, yet the increasing inadequacy of river management ensured that flooding increased in frequency and severity. Hybrid approaches encourage us to identify underlying connections between disasters, to recognize them as the consequences of failure and success, and to weigh their influence in light of multiple, sometimes conflicting environmental and social trends.

Evaluating eighteenth-century disaster perception reinforces the value of hybrid frameworks and also encourages their reevaluation. Eighteenth-century observers were well aware that catastrophes resulted from inter-related natural and cultural origins. This perception deeply informed their causal stories, and providence supplied the most important link. Providential readings did not preclude secondary environmental or social explanations, but neither were they dominated by them. Hydraulic experts, moralists, and state authorities understood that the frequency and severity of river floods resulted from weaknesses in the river system, many of which resulted from their own intervention. They also reflected moral decline. They tracked environmental changes such as subsidence, high water, and seasonal weather conditions, which informed their perceptions of increasing social, environmental, and moral risk. The weight of these causal stories tilted on very different axes than modern interpretations – yet they nevertheless incentivized new interpretations of disaster origins that demanded improved relief strategies.

These stories were powerful because they were not limited to the disasters themselves. Victims and observers of calamity yoked their causal narratives to deep-seated cultural imperatives such as the mastery of water, civic identities, and anxieties about the future of the Republic. Hybrid interpretations of disaster origins are also causal stories. They offer robust, new interpretations that place disaster at the nexus of social and environmental change. Early-modern examples suggest that hybrid narratives are incomplete absent engagement with cultural concerns. Disaster responses today are often refracted through the prism of cultural ideals, symbols and imagery, and anxieties either directly or indirectly related to the disasters themselves.[20] If hybridity promises deeper understanding and perhaps improved response, the disasters themselves must be neither the beginning nor the end of the appeal.

7.4 DISASTERS AS EVENTS AND CUMULATIVE PROCESSES

The disasters that struck the Republic were both events and cumulative processes. They manifested as singular moments of violence, but also emerged out of slower environmental and social changes that interacted in complex ways, affecting their frequency and severity. Each framework presents unique and important insights for historians. Analyzing disasters as events foregrounds the conditions that shaped their immediate impacts, as well as historically contingent variables that bounded perception and response. Without this event-centered approach, it would be difficult to account for the emergence of private relief efforts during the river floods of 1740–1 – a unique inflection in the long history of Dutch response. It would be hard to account for the extreme mortality of the Christmas Flood of 1717 without acknowledging the role of the North Sea storm that catalyzed it. It would likewise be challenging to explain paths not taken, whether the refusal to enact the policy of stamping out during the epizootic outbreaks or failed efforts by the *Ommelanden* to communalize dikes in the aftermath of the Christmas Flood. For historians, event-based interpretations ground disasters in the context and contradictions that emerge out of dramatic, if momentary, social and environmental upheaval.

Natural disasters were also cumulative processes. From this perspective, seemingly discrete disaster events appear as nonrandom outcomes in an extended negotiation between nature and culture. Slow, process-driven interpretations invite consideration of new, hybrid causal stories that account for long-term environmental changes such as subsidence,

---

[20] W. Neil Adger, Jon Barnett, Katrina Brown, et al., "Cultural Dimensions of Climate Change Impacts and Adaptation," *Nature Climate Change* 3, no. 2 (2013): 112–17. Saffron O'Neill and Sophie Nicholson-Cole, "'Fear Won't Do It': Promoting Positive Engagement with Climate Change through Visual and Iconic Representations," *Science Communication* 30, no. 3 (2009): 355–79. Anna Lukasiewicz, "The Emerging Imperative of Disaster Justice," in *Natural Hazards and Disaster Justice*, eds. A. Lukasiewicz and C. Baldwin (Singapore: Palgrave Macmillan, 2020), 3–23.

sedimentation, or climatic change, as well as social changes that influenced vulnerability. The Christmas Flood storm affected the entire southern rim of the North Sea, yet it saved its deadliest consequences for marginalized communities that lay unprotected by *kwelders* from the full force of the storm surge. Cattle plague affected much of the Dutch countryside, but mortality was greatest in regions with the largest herds that specialized in beef and dairy production. These conditions resulted from processes – whether social marginalization, shifts in prevailing winds, coastal reclamation, or agricultural specialization – that resolved on decadal or even centennial scales and created conditions conducive to disaster. This process-based approach unloads the oftentimes unwieldy causal burdens from singular events and hazards to long-term transitions that influenced the frequency, severity, and likelihood of calamities. An integrative framework builds on the strengths of both approaches. Disasters were products of history that nevertheless reflected the contingencies and conjunctures of the moment.

The relationship between calamitous events and the longer-term conjuncture of social and environmental changes also held important implications for disaster perception. Without exception, early-modern victims and observers focused on disastrous events, but rarely in a narrow sense that precluded consideration of longer-term social and environmental changes. The historical reoccurrence of floods, plagues, and extreme weather events was one important measure of that change. Victims and observers were intimately aware they lived in a region of risk. Disasters were frequent life experiences, which fostered cultures of coping. Disaster *memoria* such as flood markers, disaster chronicles, and maps preserved references from the deeper past. Previous experience and cultural memory informed both their causal stories and the range of potential responses.

Perhaps the most pervasive framework of early-modern perception emphasized the cumulative, and often cascading, relationships between disastrous events. Rather than treating them in isolation, people placed them in the context of chronic adversity, including but not limited to natural disasters. Jacob Pos's description of the tumultuous 1740s tracked a sequence of events precipitated by seasonal extremes in precipitation and temperature. Harvest failures followed bitter winters and cool, wet springs and summers. Later years brought winter rains and river flooding or spring droughts and plagues of mice. The cumulative impacts of repeated harvest failure resulted in emaciated cattle, food shortages, and price spikes in grain and dairy products. Pos described the cascading social impacts as well, including rural hardship and public unrest in cities across Holland. In this climate of anxiety, Pos interpreted the arrival of the Great Comet as an ominous portent, which the cattle plague made manifest when it reemerged that November and quickly made its way to his village.[21] The trajectory of his story was neither linear nor uniformly dire. Unlike many of his neighbors who lost their entire herds, for instance, three of Pos's cattle survived the first wave of the disease. This would serve as a basis for his financial recovery. Others took advantage of decreasing rents or tilled their pastures to grow grain. The very markets that brought plague to his doorstep supplied the cattle that Pos and his neighbors required for recovery. The care Pos took to describe each event in a sequence that spanned decades, some multiplying his misery while others opening opportunities, reinforced how complex, significant, and long lasting these cascading processes appeared.

Disaster cascades were not limited to loss of lives or livelihoods either. Their cumulative outcomes filtered through perception to penetrate the social and cultural fabric of Dutch rural life. They transformed relationships between neighbors, between landlords and tenants, and between cities and their hinterlands. Their collective meanings were memorialized in print, enshrined in cultural memory, and embedded in folklore. In the process, the distinctions between individual disaster events blurred and intersected with underlying cultural imperatives and anxieties. Some consequences were direct and perhaps predictable. In the context of economic uncertainty and widespread land abandonment in northern Holland, the outbreak of shipworms understandably amplified deep-seated anxieties about flooding. This incentivized the redesign of coastal infrastructure. Other consequences were less straightforward. The connection between mollusks and sodomy persecution, for instance, was only explicable in the context of emerging declensionist anxieties that tied sodomy, sin, and shipworms together in a unique causal chain. River flooding, in the context of a disaster cascade and mounting declensionist anxieties, forged novel expressions of proto-national solidarity.

These early-modern disasters suggest we might find the enduring legacy of disasters in unexpected places – whether moral panics or more subtle shifts in communal identities seemingly disconnected from natural calamities. They also encourage us to reexamine disaster cascades as more than the collective consequence of their direct physical and economic impacts. They intersect with values, beliefs, and identities, producing

---

[21] Schlimme, *Dagboek*, 9.

feedbacks that might amplify, diminish, or transform response. Disaster scholar Susan Cutter has recently termed these phenomena "social cascades," which, she argues, remain an important, yet poorly understood dimension of disaster studies.[22] Environmental history seems uniquely suited to unravel these long-term processes of interaction. Disasters past and present are human experiences that ripple through diverse social, environmental, and cultural pathways provoking complex and often unexpected results. Disasters, these eighteenth-century stories suggest, do not just reflect social and cultural relations – they become inseparable from them.

7.5 DISASTERS AND THE CHALLENGE OF SCALE

The disasters of the early eighteenth century were experienced and perceived unevenly across scale. Their hybrid character and the cumulative nature of their impacts ensured a differential landscape of exposure. Storm surges and shipworms affected Holland, Zeeland, Friesland, and Groningen far more than inland provinces. Cattle plague visited every province, but it devastated the regions with the largest, most interconnected herds. Even within regions, the extent and severity of impacts mapped to local geographies of risk. The communities, regions, and provinces that tended to fair worst were those that weathered a variety of cascading calamities that compounded over time. By 1752, for instance, economic recession and recurring disasters had so overwhelmed the town of Grootebroek in North Holland that it could no longer support its churches, poorhouse, or orphanage. The value of taxable lands had plummeted. "The principal origins of this decline," they claimed, "[is] the decreasing value of agricultural products, the ruinous and persistent cattle plague, the shipworm epidemic, and the resulting unendurable dike taxes."[23] Farmers in rural communities such as Grootebroek were not without options, and their responses likewise varied across scale. On a local level, landlords might reduce their rents. In certain conditions, they might shift to different crops or livestock production. Regional water boards might accommodate communities by

---

[22] Cutter, "Compound, Cascading, or Complex," 23–4.
[23] "Akte waarbij de Staten van Holland en West-Friesland de stede Grootebroek t.b.v. haar kerken, armen en weeshuis kwijtschelding verlenen van achterstallige verponding," October 3 (1754). WFA. Stede en gemeente Grootebroek, 1364–1949. 1107.281.

expanding the sphere of responsibility for dike maintenance. Communities might also petition the province for tax remission or for assuming the costs of dike repair. Communities addressed the ultimate moral cause of disaster by turning to religious rituals mandated on local or provincial scales, or even the Netherlands as a whole. Disasters were invariably experienced as local phenomena, but as consequences compounded, communities often appealed to larger scales of response.

The perception of disaster asymmetry encouraged moralists, technocrats, citizen-scientists, state authorities, and rural communities to translate disasters across scale to suit their needs. Some, like Grootebroek, opted to scale up their appeals for tax remission to the provincial level by emphasizing the complex and compounding character of disaster. Others emphasized the perceived novelty or exceptionality of adversity. Ministers and spectatorial essayists, for instance, translated localized mollusk outbreaks into a parable of Dutch decline to promote moral improvement. River flood victims garnered relief from distant communities unaffected by inundation by appealing to their shared civic identity. Still other responses accommodated the diversity of disaster experiences. Disease management drew on multiple causal stories to enact regulations across local, provincial, and statewide scales. Not all attempts to scale up disaster response succeeded. The *Ommelanden* attempted to reframe the Christmas Flood as a provincial disaster to obtain financial assistance. These efforts ultimately failed because *Stad Groningen* wielded its power to enforce an alternate interpretation of responsibility, which no flood event superseded. The Christmas Flood may have seemed to victims like an exceptional moment out of time, yet response remained beholden to histories that entrenched social and political power in the city.

The uneven consequences of disasters today often limit our ability to translate disaster response to greater scales of responsibility and action. This disconnect has widened as entwined social and environmental challenges globalize. These stories suggest that novelty and exceptionality are powerful tools to bridge the difference, but their success is neither guaranteed nor without risk. Emphasizing novelty and exceptionality can position disasters as events outside of history and the long-term environmental and social processes that birthed them. This framing, which emphasizes acute rupture in everyday experience, can deny or obscure the "slow violence" that equally warrant response across scale.[24] They

---

[24] Rob Nixon, *Slow Violence and the Environmentalism of the Poor* (Cambridge: Harvard University Press, 2011).

have also been used as politically expedient tools to justify enacting "states of emergency" to transform the legal order or simply embolden efforts to restore the same conditions that fostered calamity in the first place.[25] Yet imagining disasters as outside the norm also presents undeniable opportunities to counteract or forestall these tendencies. The exceptionality of the 1740–1 river floods provoked novel responses that transcended localized experience precisely because the dawning recognition of historical trends pointed to increasingly vulnerable futures and a shared sense of responsibility. Neither were claims of urgency limited to elites.[26] A broad spectrum of Dutch society used novelty and exceptionality to demand improvement in security, prosperity, and virtue across scale. These stories reinforce the contingency and flexibility of these rhetorical tools, which at once express urgency *and* agency. While never without risk, they remain powerful appeals to convey tragedy, trauma, hope, and optimism.

7.6 THE TWIN FACES OF CALAMITY

These three frameworks of historical disaster interpretation help explain the origins and outcomes of eighteenth-century disaster, including their interpretation and response. Although grounded in history, they offer useful guidance for understanding disasters in the present. The study of disasters at the closing of the Dutch Golden Age presents one additional lesson about the value of history and memory in an era of increasing environmental risk.

The past loomed large in the perception of disasters in the Dutch Republic. Victims and observers of natural disasters reflexively looked backward to understand their predicaments. Memory, history, and tradition supplied their most valuable frameworks for interpretation, flexible templates for response, and baselines against which improvement could be measured. Disaster *memoria* – whether pamphlets, maps, monuments, dikes, or even the landscape itself – preserved and conveyed these meanings. They also reminded observers of their own continued risk.[27] Modern disasters rarely provoke this type of intense retrospection or

---

[25] Peter Adey, Ben Anderson, and Stephen Graham, "Introduction: Governing Emergencies: Beyond Exceptionality," *Theory, Culture, & Society* 32, no. 2 (2015): 6–7.
[26] Ben Anderson, "Emergency Futures: Exception, Urgency, Interval, Hope," *The Sociological Review* 65, no. 3 (2017): 465.
[27] Valérie November, Marion Penelas, and Pascal Viot, "When Flood Risk Transforms a Territory: The Lully Effect," *Geography* 94, no. 3 (2009): 189–97.

FIGURE 7.1. The "Stone Man" rests atop the Westerzeedijk, south of Harlingen. The Stone Man today sits several meters above the crown of the dike in 1570. Photo by author, 2016.

scrutiny of the past. Yet *memoria* can also mask the full meaning of disaster and convey instead triumphal stories of progress, improvement, and increasing mastery of nature. Environmental histories challenge these simple narratives by uncovering paths taken, abandoned, and forgotten over time.

Viewing disaster *memoria* through an environmental historical lens conveys this lesson. Less than a kilometer south of the city of Harlingen, a monument rests atop the *Westerzeedijk*, the westernmost sea defense protecting the province of Friesland. The dike is massive, rising several meters from base to crown. Standing at the base of its landward side, it completely obscures the Wadden Sea beyond. To view the marker, visitors must climb up a wide stairway to its crown (Figure 7.1). As you ascend, flat stone markers indicate the gradual elevation of the dike following important dates in its history. The markers begin in 1570 when the famous All Saints' Day Flood prompted coastal

FIGURE 7.2. The Janus face of Terminus, Roman god of borders atop the Stone Man. Photo by author, 2019.

communities to elevate the dikes and they end at the monument itself. Climbing the dike takes visitors across centuries of change marked by disaster and technological renewal.

Like the dike it rests upon, the monument is an imposing structure. It rises over 7 meters from its wide stone base to the top of its narrow, segmented pedestal. It stands alone, framed on one side by the sea and on the other by the cityscape of Harlingen. Atop the pedestal, a Janus-faced bronze sculpture stares north and south along the dike line.[28] The monument, informally referred to as the "Stone Man" (*Stenen Man*), once marked a border that divided zones of financial obligation for the two regions tasked with dike maintenance (Figure 7.2). The two-faced sculpture reminded viewers of this dual responsibility and hearkened back to the contentious period of dike management that preceded it.[29] Today, that boundary no longer exists, the water boards have been consolidated, and the dike has nearly tripled in size. Like the dike, the Stone Man has

---

[28] Ronald Stenvert, Chris Kolman, Sabine Broekhoven, Saskia van Ginkel-Meester, and Yme Kuiper, *Monumenten in Nederland. Fryslân. Rijksdienst voor de Monumentenzorg* (Zwolle: Zeist/Waanders Uitgevers, 2000), 160.

[29] Tolle, "Tweedracht Maakt Zacht," 41–58. Van Tielhof, *Consensus*, 225–9.

been restored and rebuilt over the centuries, and disasters marked its transformation.

The *Westerzeedijk* is a material embodiment of the learning process that fashioned and refashioned the Dutch coastal landscape for over 1,000 years. Disasters marked this history but in no way determined the later course of events. Learning from disaster is a cultural, political, and technological act – and thus subject to decisions as likely to elide or even "unlearn" importune lessons as they were to build from them.[30] Learning is also, and perhaps most of all, an act of continuous historical interpretation. Standing at the foot of the monument atop the *Westerzeedijk*, very little of its disaster-inflected history is apparent. The stones that line the dike's stairway tell a confident tale of gradual, near-inevitable progress. Upon reaching the crown of the dike, visitors read an inscription that dedicates the statue to Caspar de Robles, the Habsburg governor of Friesland who presided over the reconstruction of the dike at the dawn of the Golden Age. This text elides centuries of conflict about moral and social responsibility, obscures the push and pull of human and physical environmental changes, and ignores the multifaceted outcomes of those negotiations. The dike and the Stone Man are *memoria* that present a highly curated exhibit more intent on mythologizing Caspar de Robles and the triumph of Dutch engineering than commemorating its rich, disaster- and conflict-riven history.[31]

The *Westerzeedijk* hides this history so well because it appears safe. It is a "primary water defense" for the country, capable of withstanding a 4,000-year flood.[32] Standing atop the dike and before the Stone Man, this story is convincing. The coastal and river floods, shipworm outbreaks, and cattle plagues of the eighteenth century are recalled dimly (if at all), and the hazards appear resolved. Since the last great inundation of 1953, the Netherlands has suffered no coastal flood disasters. The last river flood disaster occurred in 1926. Shipworms no longer pose an existential threat, and rinderpest was globally eradicated in 2011.[33] These are unquestionably heartening and significant accomplishments, but they in

---

[30] Scott Gabriel Knowles, "Learning from Disaster? The History of Technology and the Future of Disaster Research," *Technology and Culture* 55, no. 4 (2014): 780–1.

[31] Kees Draaisma, "Een nieuwe kijk op Caspars dijk," *It Beaken* 79, no. 1/2 (2017): 27–105.

[32] J. M. Kind, "Economically Efficient Flood Protection Standards for the Netherlands," *Journal of Flood Risk Management* 7, no. 2 (2014): 116.

[33] Richard Tol and Andreas Langen, "A Concise History of Dutch River Floods," *Climatic Change* 46, no. 3 (2000): 366. F. Njeumi, W. Taylor, A. Diallo, et al., "The Long

no way erase the potential for future calamity. The loss of disaster memory presents its own risks.[34] Eighteenth-century observers were highly cognizant of that danger. Disaster *memoria* were testaments to the peril of forgetting lessons past.

Environmental histories of natural disaster temper the risk of forgetting, and also misremembering. They caution that social and technological arrangements intended to limit hazards are never completely separated from the natural processes they purport to control and thus must remain correspondingly fluid and mutable. The difficulties keeping pace with dynamic environments are manifold and undeniable, especially as those changes accelerate, populations grow, and cultural landscapes expand. Technologies can also foster path dependencies that further distort these asymmetries. Wetland drainage, for instance, promoted Dutch reliance upon sea and river dikes. This encouraged subsidence and limited sediment deposition, which necessitated greater reliance upon stronger and higher dikes. This centuries-long relationship locked communities into a narrowing range of responses, even as vulnerabilities increased. As the calamities of the eighteenth century amply demonstrate, cultures of coping that attempt to control nature can also inadvertently contribute to greater risk in the future. In a broad sense, these processes of environmental and social change and the tensions they elicit are not new. It is no coincidence that scientists today again warn of increasingly numerous and severe river floods along the Rhine, elevated risk of storm surges and coastal inundation, the reemergence of epizootics, and even the renewed spread of shipworms.[35] That these threats echo eighteenth-century adversity speaks to the enduring nature of our efforts to flourish even as those efforts compound or fashion additional challenges.

Journey: A Brief Review of the Eradication of Rinderpest," *Revue Scientifique et Technique* 31, no. 3 (2012): 729–46.

[34] Paul J. A. Baan and Frans Klijn, "Flood Risk Perception and Implications for Flood Risk Management in the Netherlands," *International Journal of River Basin Management* 2, no. 2 (2004): 114.

[35] J. van Minnen, W. Ligtvoet, L. van Bree, et al., *The Effects of Climate Change in the Netherlands: 2012* (The Hague: PBL Netherlands Environmental Assessment Agency, 2013). Sanne Muis, Martin Verlaan, Hessel C. Winsemius, et al., "A Global Reanalysis of Storm Surges and Extreme Sea Levels," *Nature Communications* 7 (2016): 11969. Delia Grace and John McDermott, "Livestock Epidemic," in *Routledge Handbook of Hazards and Disaster Risk Reduction*, eds. Ben Wisner, J. C. Gaillard, and Ilan Kelman (London: Routledge, 2011), 378. Peter Paalvast and Gerard van der Velde, "New Threats of an Old Enemy: The Distribution of the Shipworm *Teredo navalis* L. (Bivalvia: Teredinidae) Related to Climate Change in the Port of Rotterdam Area, the Netherlands," *Marine Pollution Bulletin* 62, no. 8 (2011): 1822–9.

Just as disasters provoked diverse reactions during the eighteenth century, however, modern responses continue to belie simplistic narratives of progress or decline. Dutch water authorities, for instance, are well aware of the need to revisit longstanding assumptions about the degree of their control. Following a series of near-flood disasters in 1993 and 1995, they began implementing "softer" approaches to water management that "build with nature" along the coast and widen river channels and floodplains to make "room for the rivers."[36] Proponents consider these changes a paradigm shift in flood protection, breaking free from centuries of reliance on "hard" engineering strategies in favor of approaches that mimic, restore, or accommodate natural processes.[37] Many of these plans were completed, although in practice the combination of high costs and divergent stakeholder interests encouraged continued reliance on hard flood defenses. Complicating these approaches still further is the fact that the very success of hard "solutions" at preventing disasters in recent decades has convinced many people living in flood-prone areas that they are not at risk, thus limiting support for alternatives.[38] This latter impediment is perhaps the most problematic because unlike subjects of economic or political disagreement, the absence of flood disasters is an issue few would wish to resolve. Yet here again the *Westerzeedijk* supplies a valuable message. If the *Westerzeedijk* evokes the technocratic dream of control, it is equally a monument to repeated failure. Read in this manner, the Dutch landscape is replete with disaster memories awaiting creative strategies to unearth them.

Environmental history is a useful tool in this project because it privileges readings that resist uncomplicated stories of failure or success. Disaster narratives all too often rest on assumptions that denote

---

[36] Huib J. de Vriend, Mark van Koningsveld, Stefan GJ Aarninkhof, et al., "Sustainable Hydraulic Engineering through Building with Nature," *Journal of Hydro-environment Research* 9, no. 2 (2015): 159–71. "Ruimte voor de rivieren," Room for the River Programme, Rijkswaterstaat, accessed December 31, 2020, www.rijkswaterstaat.nl/water/waterbeheer/bescherming-tegen-het-water/maatregelen-om-overstromingen-te-voorkomen/ruimte-voor-de-rivieren/index.aspx.

[37] Sander van Alphen, "Room for the River: Innovation, or Tradition? The Case of the Noordwaard," in *Adaptive Strategies for Water Heritage*, ed. Carola Hein (Cham: Springer, 2020), 308–23.

[38] H. P. Ritzema and J. M. van Loon-Steensma, "Coping with Climate Change in a Densely Populated Delta: A Paradigm Shift in Flood and Water Management in the Netherlands," *Irrigation and Drainage* 67 (2018): 58–9. Anna Wesselink, Jeroen Warner, M. Abu Syed et al., "Trends in Flood Risk Management in Deltas around the World: Are We Going 'Soft'?," *International Journal of Water Governance* 3, no. 4 (2015): 29–30.

powerlessness or control, each conjuring an aura of inevitability that richer retellings deny. At best, these simplistic stories leave us exposed and unprepared as hazards gather and vulnerabilities mount. At worst, their consequences may be magnified. The Stone Man's twin faces are a fitting reminder that past disasters need not be relics of history and memory. Janus was the Roman god of transitions; his two faces symbolize past and future. One points backward, the other warns of trials ahead. Our uncertain futures are sometimes characterized as no-analogue states due to climate change and other markers of increasing human influence in the Earth System.[39] The stories we tell about the nature and meaning of disasters, now or in the future, are no less powerful or malleable than those of the eighteenth century. These histories remind us that fears of unprecedented adversity can provoke useful and motivating responses, including the hope that novel challenges may produce extraordinary responses. Unlike the monument, these outcomes are not etched in stone.

---

[39] David Wallace-Wells, *The Uninhabitable Earth: Life after Warming* (New York: Crown/Archetype, 2019). Van Bavel et al., *Disasters and History*, 159.

Bibliography

Archives/Libraries

Gelders Archief (Arnhem, the Netherlands) – GA
Kaartenverzameling Rijksarchief in Gelderland

Gemeente Archief Delft (Delft, the Netherlands) – GaD

Gemeentearchief Zaanstad (Zaandam, the Netherlands) – GaZ
Doopsgezinde Gemeente Zaandam-West
Persoonlijk archief Honig

Groninger Archieven (Groningen, the Netherlands) – GrA
Bibliotheek RAG (Losse Boeken)
Doop, Trouw en Begraaf (DTB), Kerkelijke Gemeente Leens Doop en Trouwboek 1680–1748.
Register Feith-stukken Stadsarchief Groningen, 1594–1815
Staten van Stad en Lande, 1594–1798
Volle Gerecht van de Stad Groningen, 1475–1811

Hoogheemraadschap van Rijnland (Leiden, the Netherlands) – OAR
Oud Archief Rijnland

Koninklijke Bibliotheek Special Collections (The Hague, the Netherlands) – KB

Nationaal Archief (The Hague, the Netherlands) – NA
Collectie Goldberg
Gedeputeerden van Haarlem ter Dagvaart

Noord Hollands Archief (Haarlem, the Netherlands) – NHA
Ambachts-en Gemeentebestuur van Uitgeest en Markenbinnen
Ambachtsbestuur van Velsen
Archief Oudemannenhuis
Collectie van Losse Aanwinsten
Stads-en Gemeentebestuur van Beverwijk, 1298–1817

Regionaal Archief Rivierenland (Tiel, the Netherlands) – RaR
    Ambt en dijkstoel van de Overbetuwe, 1427–1838
Regionaal Historisch Centrum Rijnstreek en Lopikerwaard (Woerden, the Netherlands) – RRL
    DTB Registers Montfoort
Stadsarchief Amsterdam (Amsterdam, the Netherlands) – SA
    Archief van de Burgemeesters: Publicaties van de Staten-Generaal en van de Staten van Holland en West-Friesland
    Archief van de Burgemeesters: resoluties van de Staten van Holland en West-Friesland, van Gecommitteerde Raden van het Zuiderkwartier
    Archief van de Notarissen ter Standplaats Amsterdam
Streekarchief Langstraat Heusden Altena (Heusden, the Netherlands) – SLHA
    Waterschap Hoge Maasdijk van Stad en Lande van Heusden
TRESOAR – Frysk Histoarysk en Letterkundich Sintrum (Leeuwarden, the Netherlands) – TRE
    Gewestelijke bestuurinstellingen van Friesland 1580–1795
    Verzameling Handschriften, afkomstig van de Provinciale Bibliotheek
University of Amsterdam Library Special Collections (Amsterdam, the Netherlands) – UvA
Utrechts Archief (Utrecht, the Netherlands) – UA
    C. Berger, als burgemeester van Utrecht
    Huis Zuilen
Westfries Archief (Hoorn, the Netherlands) – WFA
    Ambacht van West-Friesland genaamd Drechterland en hoofdingelanden van West-Friesland
    Ambacht van Westfriesland genaamd de Vier Noorder Koggen
    Notarissen in West-Friesland tot 1843
    Stede en gemeente Grootebroek 1364–1949
Zeeuws Archief (Middelburg, the Netherlands) – ZA
    Polder Walcheren 1511–1870
    Verzameling Handschriften Gemeentearchief Veere

Periodicals

's Gravenhaegse Courant
Amsterdamse Courant
Gentleman's Magazine (London: D. Henry and R. Cave, 1755)
Justus van Effen, *De Hollandsche Spectator* (1732–33)
Leeuwarder Courant
Leydse Courant
Nederlandsche jaerboeken, inhoudende een verhael van de merkwaerdigste geschiedenissen, die voorgevallen zyn binnen den omtrek der Vereenigde Provintiën, sederd het begin van 't jaer (Amsterdam: F. Houttuyn, 1749–64).
Nederlansch Gedenkboek: Of, Europische Mercurius (Amsterdam: A. van Damme, 1713–44).

Nieuwe Nederlandsche jaerboeken, inhoudende een verhael van de merkwaerdigste geschiedenissen, die voorgevallen zyn binnen den omtrek der Vereenigde Provintiën, sederd het begin van 't jaer ... (Amsterdam: Erven van F. Houttuyn, 1766–76).
Opregte Haarlemse Courant

Printed Primary Sources

Alta, Eelko. *Nodige raadgeevingen aan overheden en ingezetenen, dewelke in het bijzonder voor den boer in deeze akelige omstandigheden, waarin het vaderland zig door de ziekte van het rundvee thans bevind, van zeer veel nut kunnen zijn.* Leeuwarden: H. A. de Chalmot, 1769.

Verhandelinge over de natuurlyke oorzaaken der ziekte onder het rund-vee en derzelver langere duuringe als te vooren, waar in ... de in-enting derzelve ... wordt aangepreezen: gaande vooraf twee vertoogen, gezonden aan de Hollandse Maatschappy der Weetenschappen, ter beantwoording eener vraag over deeze stoffe in 1759 voorgesteld en in 1760 herhaald. Leeuwarden: Wigerus Wigeri, 1765.

Baerle, J. V. *Gedagten van de heeren professoren uytmakende de medicynsche faculteyt op de Academie van Utrecht over de ziekte onder het rund-vee thans woedende: waar by nog gevoegd zyn eenige der observatien, uit welke de voornoemde gedagten zyn opgemaakt.* Utrecht: Jacob van Poolsum, 1745.

Barueth, Johannes. *Davids wyze keuze uit drie plagen en ware boetvaerdigheid onder gekoze pestilentie: in deze dagen van de pest-ziekte onder het rundvee.* Rotterdam: Jacobus Bosch, 1745.

Bedenkingen, en raad, noopende de tegenwoordige stervte onder het rundvee. 's Gravenhage: Pieter van Thol, 1714.

Beels, Leonard. *Sodoms zonde en straffe of streng wraakrecht over vervloekte boosheidt, en loths vrouw, verandert in een zoutpilaar.* Amsterdam: Adriaan Wor, en de erve G. Onder de Linden, 1730.

Beknopte beschryving, van de noodlydende en oversrtoomde landstreeken in Zuidholland, en een gedeelte van Gelderland. Gorinchem: Jacob van Wijk, 1741.

Belkmeer, Cornelis. *Naturkundige verhandeling of waarneminge, betreffende den hout-uytraspende en doorboorende zee-worm.* Amsterdam: J. Ratelband en Compagnie, 1733.

Beschryvinge, van de schade en raseringe aan de zee-dyken van Noort-Holland en West-friesland, door de worm in de palen, en de daar op gevolgde storm, en vervolgens: waar by komt een beschryving van een nieuwe water-machine. Hoorn: Jacob Duyn, 1732.

Biblia, dat is: de gantsche H. Schrifture, vervattende alle de Canonycke Boecken des Oude en des Niewen Testaments (Statenvertaling 1637). Leiden: Paulus Aertsz van Ravensteyn, 1637.

Bicker-Raye, Jacob. *Het dagboek van Jacob Bicker Raye 1732–1772*, edited by F. Beijerinck and M. G. de Boer. Amsterdam: H. J. Paris, 1935.

Blom, Koenraad. *Hiskia's dank- en lof-offer voor Gods wonderbare genesing, den Heere toegebragt in syn dank-schrift Jesaia 38 vs. 16–20.* Amsterdam: Adrianus Douci, 1746.

Bogaert, A. *De Kersvloedt van den jare mdccxvii, vermengt met de gedenkwaardigste vloeden sedert den algemeenen.* Amsterdam: Gerrit Bosch, 1719.

Centen, Is. *Gods oordelen over Nederland, in de sterfte van 't rundvee, den zwaren storm, en hogen watervloed.* Amsterdam: Johannes Oosterwijk, 1718.

C. J. *Korte beschryvinge van de schrikkelijke watervloed, veroorsaekt door een sterke stormvint/voorgevallen tusschen den 24. en 25. december 1717: soo in Holland/Friesland/Groninger-Embder en Oost-Friesland.* Leeuwarden: Johannes Thijssens, 1718.

Costerus, Bernard. *Historisch verhaal ... raakende het formeren van de Republique van Holland ende West-Vriesland.* Leiden: by Coenraad Wishoff, 1736.

Crous, A. E. *Opregt en nauwkeurig historis-verhaal van de verwonderenswaardige, droevige, schrikkelike en seer schaadelike waaters-vloed, voorgevallen in de provincie van Groningen en Ommelanden, op Kersdag den 25. december ao. 1717: daarom met regt genaamt de Kers-vloed of midwinters-vloed.: met al 't geene in deselve, aanmerkelik is voorgevallen ... : beneevens een byvoegsel van 't aangrensend Oost-Vriesland, en de daar bygeleegene.* Groningen: Seerp Bandsma, 1719.

De Hooghe, Romeyn. *Ellenden klacht van het bedroefde Nederland te sedert het jaer 1672 tot den Aller-heyligen Vloet van het jaer 1675.* Amsterdam: Romeyn de Hooge, 1675.

Deknatel, J. *Klaag en troost-dicht over den tegenwoordigen staat van Oostvriesland, door zwaare watervloeden van vyf achter een volgende jaaren in de uyterste ellende gebracht.* Amsterdam: s.n., 1722.

De Missy, Jean Rousset. *Aanmerkingen over den oorsprong, gesteltheit, en aard der zeewormen: die de schepen en paal-werken doorboren door den hr. Rousset.* Leiden: Gysbert Langerak, 1733.

Observations on the Sea- or Pile-Worms Which Have Been Lately Discover'd to Have Made Great Ravages in the Pile-or Wood-Works on the Coast of Holland, &C: Containing a Particular Account of Their Make and Nature, and of the Use of Their Several Parts in Boreing and Feeding; with a Particular Description of Their Cells or Lodgments in the Wood. London: J. Roberts, 1733.

De Munck, Jan. *Sterrekundige waarneemingen op de comeet of staart-sterre sedert den 29 November ... 1743 tot op den l Maart ... 1744.* Amsterdam: Isaac Tirion, 1744.

Dortse en Haagse woonsdag en saturdag, of nader opening van de bibliotheecq van mr. Jan de Witth, zijnde een samenspraak tusschen een Hagenaar en Dortenaar (s.n., 1672).

The Dutch Drawn to Life. London: Tho. Johnson, and H. Marsh, 1664.

Een historische beschrijving van duure tijden, en hongersnoden. Amsterdam: Arent van Huyssteen and Steeve van Esveldt, 1741.

Een historiesch verhaal van veele en nooit meer gehoorde voorvallen die geschiet zyn in verscheide harde winters inzonderheid van den jaare 1709 en 1740: van tyd tot tyd aangetekend door een liefhebber en dus in 't ligt gegeeven met een fraaye kopere plaat. Amsterdam: Steeve van Esveldt, 1740.

Engelhardt, Henricus. *Goede suffisante, Gods verleende, uitgevondene middelen, omme yzere, en steene zeemuuren, met paalen zonder metzelwerk, tegens het schadelyke zeegewormte te maken, dewelke in plaats van zeedyken, dammen, en in de zehavens zoude kunnen dienen.* 's Gravenhage: Laurens Berkoske, 1733.

Engelman, Jan. "Nader verhandeling over de rundveesterfte, betreklyk tot de waarneemingen, vervat in 't VIde deels IIde stuk." *Verhandelingen van de Hollandsche Maatschappye der Wetenschappen* 7 (1762): 247–318.

"Waarneemingen in de rundveesterfte in 1756 en 1759, dienende tot een voorloper ter nader verhandeling over dezelfde stoffe." *Verhandelingen van de Hollandsche Maatschappye der Wetenschappen* 6 (1760): 955–1015.

Fabricius, Isbrandus. *Nederlantse oordelen en rampen: na een schriftmatige verklaringe van den twist godts met Israël, in sijne vloeksprake over hunne sonden, by Hoseas cap. IV: vs. I, II, III ... betooght in de weeklaaglijke en nog durende runt-vee sterfte, en verschrikkelyke storm-winden en watervloeden. Toegepast met een ernstige opweckinge tot een Hoognodige bekeringe, voor Nederlants Inwoonders.* Alkmaar: Nikolaas Mol, 1718.

Fokkert, Jan Barte. *Gedachtenis van de wonderbare wegen der goddelijke Voorzienigheid gehouden in de weg der bekering.* Middelburg: Stichting de Gihonbron, 2017.

Francken, Aegidius. "Geestelyk houwelyk, dat is, een Verhandeling van de Ondertrouw der gelovigen met Kristus, door AEgidius Franken, Bedienaar van 't H. Evangely te Maas sluys. Te Dordregt hy J. van Braam, 1715. in 8. groot 144 Bladzyden, behalven de Opdragt en Voorreden; agter aan is ook bygevoegt, de kloppende Jesus aan de deure der Kerke van Laodicea, vertoond in een Schriftmatige verhandeling over Openb. III. 20. door den zelven Schryver groot 142 Bladzyden." *Boekzaal der geleerde wereld.* Amsterdam: Gerard onder de Linden, 1715.

Gabbema, Simon Abbes. *Nederlandse watervloeden, of naukeurige beschrijvinge van alle watervloeden voorgevallen in Holland ... en de naabuirige landen.* Gouda: Lucas Cloppenburg, 1703.

Grashuis, J. "Berigt aangaande de in-enting der besmetlyke ziekte in het rundvee in een aanmerkelyk getal beesten te werk gesteld, en tot verdere beproeving onze landgenooten voorgehouden door den heer J.G." In *Uitgezogte verhandelingen uit de nieuwste werken van de societeiten der wetenschappen in Europa en van andere geleerde mannen.* Amsterdam: F. Houttuyn, 1758.

Groot placaet-boeck, vervattende de placaten, ordonnantien ende edicten van de ... Staten Generael der Vereenighde Nederlanden, ende vande ... Staten van Hollandt en West-Vrieslandt, mitsgaders vande ... Staten van Zeelandt ... 's Gravenhage: Isaac en Jacobus Scheltus, 1725.

Groot placaet-boeck, vervattende de placaten, ordonnantien ende edicten van de ... Staten Generael der Vereenighde Nederlanden, ende van de ... Staten van Hollandt en West-Vrieslandt, mitsgaders vande ... Staten van Zeelandt. 's Gravenhage: Isaac Scheltus, 1770.

Halma, François. *Godts wraakzwaardt over Nederlandt, vertoont in de zwaare sterfte onder 't rundtvee.* Leeuwarden: François Halma, 1714.

Harkenroht, Jacobus Isebrandi. *Kerkreede over Oostfrieslands rundvees pest, gevreest, en helaas gekoomen, aangetoont uit Jes: VII. vs. 21. ... op een maandelijke bededag aan de gemeente te Larrelt in Oostfriesland.* Embden: Enoch Brantgum, 1716.

Hekelius, Johann Christian. *Ausführliche und ordentliche Beschreibung Derer beyden erschrecklichen und fast nie erhörten Wasserfluthen in Ost-Frießland.* Neue Buchhandlung: Halle im Magdeb, 1719.

Henriksz, Jan *Opregt Dog Droevig Verhael, Van De Groote Elende En Droefheyd, Veroorzaekt Door De Felle Kou En Sterk Vriesend Weer Van Deeze Winter.* UvA (1709).

H. F. *Treurdigt, Ter droeviger gedagtenisse Van de vreesselyken en verderffelyken Watervloed. Den 25sten van Wintermaand MDCCXVII en enige volgende Dagen, gebragt oover Holland, Friesland, Omlanden en derselver Nabuuren.* Amsterdam: Johannes Douci, 1718.

Holberg, Ludvig. "Korte Betænkning over den nu regierende Qvæg-Syge med nogle œconomiske Anmerkninger." In *Skrifter som udi Det Kiøbenhavnske Selskab af Lærdoms og Viderskabers Elskere ere fremlagte og oplæste i Aaret 1745* 2 (1745): 385–402.

Huysmans-praetje, Voorgestelt tot onderrechtingh, Hoe men sich in desen verwerden en murmurerige toestandt des tijdts behoorden te dragen: En met eenen Om tot beter verstant des oorsaecks van de tegenwoordige bedroefden toestant des tijdts te komen: Tusschen Jacob, Klaes en Symon. Amsterdam, 1672.

Kemner, Johannes. *De suchtende landtman in de provincie van stadt Groningen en Ommelanden over de hooge water-vloedt op kers-tydt den 25. van wintermaandt in het jaar onses heeren 1717.* Groningen: Johannes van Velsen, 1718.

"Kort Verhaal van de Hooge Watervloed, waar mede de Graafschap Culemborg, nevens andere nabuurige Landschappen, in December 1740, en January 1741, is bezogt geworden." In *Verzameling Van eenige geloofwaardige Berigten en Brieven Betreffende de Elende van de Opgezetenen der overstroomde landen in Nederland.* 23–31. Amsterdam: Isaak Tirion, 1741.

Korte schets van 'slands welwezen door de laatste vrede. Voor den Autheur, 1714.

Kuypers, François. *De vreeslyke overstrooming van het Landt van Altena, voorgevallen tusschen den 24 en 25 December van 't jaar 1740, en op den 29 April 1741.* Gorinchem: Nicolaas Goetzee, 1741.

Het Lant van Altena gewaarschouwt en ten laatste zwaarlyk gestraft. Gorinchem: Nicolaas Goetzee, 1741.

Lakenman, Seger. *Het wonder Oordeel Godts, ofte een kort verhaal van de ongehoorde bezoeking dezer Provintie door zekere plage van Zee-wormen,* 1732.

Layard, Daniel Peter. "A Discourse on the Usefulness of Inoculation of the Horned Cattle to Prevent the Contagious Distemper among Them. In a Letter to the Right Hon. George Earl of Macclesfield, P. R. S. From Daniel Peter Layard, M. D. F. R. S." *Philosophical Transactions* 50 (1757): 528–38.

L'Epie, Zacharias. *Onderzoek Over de oude en tegenwoordige Natuurlyke Gesteldheyd van Holland en West-Vriesland, Desselfs rivieren en landen; Aanwas, Ophooping, Zakking, Laagte en Dykagie; Mitsgaders eene Verhandeling over de ... Zee- of Kokerwormen ... Als mede de Middelen tot Verbetering en Versterking der Zee-weeringen ...* Amsterdam: Jacobus Hayman, 1753.

Linné, Carl. *Systema naturae, per regna tria naturae: secundum classes, ordines, genera, species cum characteribus, differentiis, synonymis, locis.* Holmiae: Salvius, 1758.

Linschoten, J. H. *Itinerario: voyage ofte schipvaert van Jan Huygen van Linschoten naer dost ofte Portugaels Indien ...* Amsterdam: Cornelis Claesz, 1596.

Loys de Cheseaux, Jean-Philippe. *Traité de la comète qui a paru en décembre 1743 & en janvier, fevrier & mars 1744: contenant outre les observations de l'auteur, celles qui ont été faites à Paris par Mr. Cassini & à Geneve par Mr. Calandrini: on y a joint diverses observations & dissertations astronomiques, le tout accompagné de figures en taille douce.* Lausanne: Chez Marc-Michel Bousquet & Compagnie, 1744.

L. T. and J. B., *Nederlant ten toneele, van Godts regtvaardige wormstraf gekozen.* Groningen: Laurens Groenewout, 1732.

Marchant, Jan. *Naagalm over de vee-ziekte, met een jaarlijst der voorgaande veesterftes, sédert de plagen van Egipten: alsmede de waare oorzaak der koejeziekte, en de middelen om die voor te koomen.* Haarlem: J. Marshoorn en Iz. vander Vinne, 1745.

Martinet, J. F. *Katechismus der natuur: met plaaten*, Vol. 2. Amsterdam: Allart, 1778.

Massuet, Pierre. *Wetenswaardig onderzoek over den oorsprongk, de voortteling, de ontzwachteling, het maaksel, de gedaante, de gesteltheit, den arbeidt, en de verbazende menigte der verscheide soorten van kokerwurmen die de dykpalen en schepen van enige der vereenigde Nederlandsche provintsien doorboren.* Amsterdam: Wor, 1733.

Mel, Koenraad. *Het gruwelijk sodom gestraft.* Middelburg: Stichting de Gihonbron, 2004.

Mills, John. *A Treatise on Cattle: Shewing the Most Approved Methods of Breeding, Rearing, and Fitting for Use, Horses, Asses, Mules, Horned Cattle, Sheep, Goats, and Swine; with Directions for the Proper Treatment of Them in Their Several Disorders; to Which Is Added, a Dissertation on Their Contagious Diseases.* Dublin: W. Whitestone, 1776.

Mobachius, Joachimus. *Den Vrezelyken Zamenloop Van veele gedugte Oordelen Gods, Als van Alverwoestende Oorlogen, swaere Overstromingen, ontzachelyke Aerdbevingen, aenhoudende, en meer en meer woedende Runder-pest, enz. enz. Als zoo veele voorboden van eenen geheelen ondergang, zoo het niet door eene spoedige bekeringe werde voorgekomen ....* Gorinchem: Teunis Horneer, 1760.

Mobachius, Johannes A. *De almagtige en regtveerdige slaande hand Gods, Ter Besoeking van het Land en deese Provincie met de verderfelyke Plage der Zeewormen, Vertoond en aangedrongen uit Nahum I: vs. 3. Op den plegtelyken Vast-en Bid-dag, gehouden in de provincie van Groningen ende Ommelanden, den 8 Octob. 1732 ... Alwaar met een ingevoegt is een Beschrijving van de Gedaante en Groote der Zeewormen, en hoe Die de Dykpalen doorboren ... nevens een Afbeelding daarvan ...* Groningen: Jurjen Spandaw, 1733.

De waare oorzaken van Neerlands plage, Wegens de groote Sterfte onder het Rundvee ... vertoond ... uit de klaagliederen van Jer. Cap. 3 vs. 38, 39, 40. Op den voorgaanden Bid-dag ... in 't Jaar 1715 ... Alwaar met een ingevoegt is, een Wederlegging van het gevoelen van F. Leenhof, aangaande den Hemel op Aarden. Groningen: Jurjen Spandaw, 1716.

Groningerlands zeer Hooge en Schrikkelyke Watervloed ... op Kerst-tyd den 25 Decemb. 1717. Verhandeld ... Op den eersten Biddag, gehouden in Groningerland, den 5 Jan. 1718. En daar na uitgebreid, en ter Gedagtenis in 't ligt gegeven. Groningen: Jurjen Spandaw, 1718.

Neerlands lang gewenschte Vreede, tot Roem van den Almagtigen ... Verkondigd en aangedrongen Uit Psalm CXLVII. 12, 13, 14. Op den ... 14. van de Maand Juny 1713. Groningen: Seerp Bansma, 1713.

Moubach, Abraham. *Schaadelijke Veepest, Ontstaan sedert den Jare 1711 in Italie , Duitsland, en Zwitserland; en die in deeze Nederlanden tot heden zoo jammerlyk onder 't Hoornvee heeft gewoed. Waar in de Oorzaak en Oorsprong deezer algemeene Landplaag word aangewezen ... Als mede wat noodige voorzorge deezer quaal dient gebruikt.* Amsterdam: Joannes Oosterwyk, 1719.

Mundy, Peter. *The Travels of Peter Mundy in Europe and Asia, 1608–1667*, Vol. IV, edited by Richard Carnac Temple. Cambridge: Hakluyt Society, 1907.

Nederlant aengemaent tot boetvaerdigheit. 's Gravenhage: G. Block, 1732.

Nieuwentijdt, Bernard. *Het regt gebruik der werelt beschouwingen ter overtuigingen van ongodisten en ongelovigen.* Amsterdam: Joannes Pauli, 1714.

Nozeman, Cornelis, Agge Roskam Kool, and Jan Tak, *Eerste proefneeming over de uitwerkingen van de inentinge der besmettende ziekte in het rundvee, gedaan in de Beverwijk.* Amsterdam: K. van der Sys and K. de Veer, 1755.

Omstandig verhaal van de schrikkelijke en nooid gehoorde watervloed, voorgevallen in Engeland, Zeeland, Holland, Frisland, Groeningen en Duytsland, &c. op den 25 December 1717. Middelburg: Johannes op Somer, 1718.

Outhof, Gerardus. *Damon. Ofte herdersklagte over de sterfte onder 't rundvee ... waar mede Godt Oostvrieslandt ... bezoekt.* Emden: E. Brantgum, 1716.

Gerhardus Outhofs Verhaal van alle hooge watervloeden, In meest alle plaatsen van Europa, van Noachs tydt af, tot op den tegenwoordigen tydt toe: Met een nieuw Kaertje van 't verdronken Landt in den Dollaart, afbeeldinge van Kosmas en Damianus en van den steenenman in Vrieslandt, Met eene breede beschryvinge van den zwaaren Kersvloedt van 1717, En bygevoegde Tafels der watervloeden... Embden: H. van Senden, 1720.

Pierlinck, Jacob. *De verschrikkelyke watersnood, langs de ... Waal en de Maas, voorgevallen in ... February des Jaars 1757 ...* Amsterdam: Isaak Tirion, 1757.

Poot, Hubert Korneliszoon. *Op den hoogen watervloet, omtrent het einde des Jaers MDCCXVII.* 1717.

Rampen van het vredejaer. s.l. 1713.
Propositie van syne hoogheid ter vergaderingen van haar Hoog Mogende en haar Edele Groot Mog. gedaan, tot redres en verbeeteringe van den koophandel in de Republicq. 's Gravenhage: Jacobus Scheltus, 1751.
R. B. *Nederlands klachte over Gods naekende oordeelen; Duidelyk te bespeuren in het knaegen der wormen aen de paelen der Nederlandze zeedyken; Ter gelegenheid van den dank, vast- en bededag, uitgeschreven by de Heeren Staeten van Vriesland, tegens woensdag den 3en van Herfstmaend des Jaers 1732*. Harlingen: Folkert Jansz van der Plaetz, 1732.
Reis naar Heusden of een kort doch waarachtig verhaal van ... de laatste overstrominge in't Lant van Heusden op den 24 December 1740. Rotterdam: Philippus Losel, 1741.
Rotterdam, Arnoldus. *Gods weg met Nederland, of vervolg op Blomherts geschiedenissen van het Vereenigde Nederland; ... onder 't verstandig beleid van Willem den IV*. Amsterdam: S. v. Esveldt, 1753.
Schenckel, H., L. Smids, and T. Brongersma, *Kort en bondige beschrijvinge van de schrickelijcke water-vloedt den 13. Novemb. 1686. over de provincie van Stadt en Lande ontstaen, behelsende een waerachtigh verhael van ... voorvallen ... alles door eygene ondervindinge ofte ... oor- en oogh-getuygen bevestight*. Groningen: Carel Pieman, 1687.
Schoock, Martin. *Martini Schoockii Tractatus de Inundationibus, Iis maxime, quae Belgium concernunt. Quatuor Disputationibus propositus In Academia Groningae et Ommelandiae*. Groningen: Johannis CollenI, 1652.
Schwenke, Thomas. "Schreiben des Herrn Doct. Swenke der Anatomie und Chirurgie Professors im Haag die Einpropfung des Hornviehs betreffend." In *Bremisches Magazin zur Ausbreitung der Wissenschaften, Künste und Tugend*. Hannover: Nicolai Forsters and Sohns Erben, 1756.
Sellius, Gottfried. *Natuurkundige histori van den zeehoutworm, ofte houtvreeter, zynde koker- en meerschelpigh, inzonderheit van den Nederlantschen ... : Eerste Deel*. Utrecht: H. Besseling, 1733.
Sluiter, Johannes. *Vernedert Nederlant, Of Klagte over Nederlants sonden, gestraft door de sterfte van het Rundervee*. Steenwyk: Pieter and Hendrik Stuyfzant, 1715.
Smids, Ludolf. *Diluviana of Daghwyser der Nederlandsche waternooden van het Jaar 793 Tot deesen laatsten van den 25 November des verledene Jaars 1717*. Amsterdam: Hendrik van de Gaete, 1718.
Smit, Jan. *Gods slaandehand over Nederland, door de pest-siekte onder het rund vee naar het leeven getekent, en gegraveert door Jan Smit*. Amsterdam: Steven van Esveldt, 1745.
Tweede Plaat der Overstroomingen vande Provincien Gelderland en Holland in den Jaare 1740 en 1741. Inzonderheyd vande Alblasserwaard, Het Land van Gorcum en het Land van Vianen, waar inde omgeleegene Landen, Revieren, Killen, Steeden, Dorpen, Doorbraaken, Hulpgaten en aanmerkelykste voorvallen, met de uytterste naauwkeurigheyd, naar het Leeven zyn afgebeeld. Amsterdam: A. van Huyssteen en S. van Esveldt, 1741.

Spinniker, Adriaan. *Gods gerichten op aarde, Vertoond in den Schrikkelyken Storm en hoogen Watervloed, op den 25 en 25sten van Wintermaand in 't 1717 de Jaar voorgevallen.* Groningen: Jurjen Spandaw, 1718.

Straat, Pieter and Pieter van der Deure, *Ontwerp tot een minst kostbaare zeekerste en schielykste herstelling van de zorgelyke toestand der Westfriesche zeedyken ... Met een nader ontwerp Hoe men de Dyken, daar de grootste dieptens zyn op de zekerste, minst kostbaarste en schielykste wyze kan herstellen ...* Amsterdam: J. Oosterwyk, 1733.

Temple, William. *Observations Upon the United Provinces of the Netherlands. By Sir William Temple of Shene, in the County of Surrey, Baronet, Ambassador at the Hague, and at Aix La Chapelle, in the Year 1668*, 2nd ed. London: A. Maxwell, 1673.

Tiele, Alardus. *Schouwburg der oordeelen en gerigten Gods, geoeffend en uitgevoerd door watervloeden, stormwinden, oorlogen, aardbevingen, pestilentien, siektens over menschen en vee, ...: in drie deelen.* Rotterdam: Nikolaas Topyn, 1736.

Van Barn in 't Loo, Hendrik. *De inwoonders der aarde aangespoort tot een betamelyke vreze voor Gods tekenen, in en aan den hemel: Handelende van de Hemel-Ligten in 't gemeen, als luidbare verkondigers van Gods volmaaktheden, maar in 't byzonder van de comeeten.* Leiden: Johannes Hasebroek, 1744.

Van Byler, Henricus Carolinus. *Helsche boosheit of grouwelyke zonde van sodomie, in haar afschouwelykheit, en welverdiende straffe uit Goddelyke, en menschelyke schriften tot een spiegel voor het tegenwoordige, en toekomende geslagte opentlyk ten toon gestelt.* Groningen: Jacobus Sipkes, 1731.

Historis-verhaal van de sterfte die in vorige eeuwen onder het rundvee, in deze en andere landen geweest is, en nog duurt. Groningen: Jurjen Spandaw, 1719.

Van de Water, J. *Groot Proclamationboek Vervattende Alle de Placaten, Ordonantien En Edicten, Der Edele Mogende Heeren Staten 's Lands Van Utrecht.* Utrecht: Van Poolsum, 1729.

Van de Water, J., C.W. Moorrees, and P.J. Vermeulen. *Groot plakkaatboek 's lands van Utrecht: aangevuld en vervolgd tot het jaar 1810.* Utrecht: Kemink en Zoon, 1856.

Van den Berge, Maria. *Klaag-liet, Uitgeboesemt over de inbreuke en overstroming der vreeslyke wateren ontrent Zwolle, Kampen en Zutphen ... voorgevallen in 't laatste ... desvoorgaanden jaars 1753. en in het begin ... deses jaars 1754 ....* 1754.

Van den Honert, Johan. *De kerk in Nederland beschouwd, en tot bekeering vermaand.* Leiden: Samuel Luchtmans en Soon, 1747.

Van Gouthoeven, Wouter. *D'oude chronijcke ende historien van Holland (met West-Vriesland) van Zeeland ende van Utrecht.* Dordrecht: Peeter Verhaghen, 1620.

Van Gyzen, Jan. *Klaagend Neederland, bezogt met sterfte onder het rund-vee en swaare stormwinden.* Amsterdam: Jacobus van Egmont, 1714.

Van Hardeveldt, Anthonie. *De verbaasde en erbarmlyke vlucht der inwoonderen des Landts van Heusden, wegens de laatste verschrikkelyke overstrominge, vertoont in eene predikatie, over Matth. XXIIII: vers 20. Met een Dank-Addres aan de Chasidim of Weldadigen ....* Dordrecht: Fredrik Oudman, 1741.

Velsen, Cornelis. *Rivierkundige verhandeling, afgeleid uit waterwigt en waterbeweegkundige grondbeginselen, en toepasselyk gemaakt op de rivieren den Rhyn, de Maas, de Waal, de Merwede en de Lek*. Amsterdam: Isaak Tirion, 1749.

Verhaal van de droevige waternood en inbreuken van een groot gedeelte van Gelderland en Holland, in 't laatst van den Jare 1740. en in den beginne van 't Jaar 1741. Amsterdam: Jan ten Houten, 1741.

Verhandelingen uitgegeven door de Hollandse Maatschappy der Weetenschappen te Haarlem. Haarlem: J. Bosch, 1754.

Vervolg der Beschryvinge van den zwaaren Ysgang, Dyk-breuken en Watervloeden voorgevallen in het begin van het Jaar 1726. In veele Landen en Steden van Europa. Haarlem: Aäron van Hulkenroy, 1726.

Verzameling van eenige geloofwaardige berigten en brieven betreffende de elende van de opgezetenen der overstroomde landen in Nederland. Amsterdam: Isaak Tirion, 1741.

Vierlingh, Andries. *Tractaet van dijckagie*. The Hague: Martinus Nijhoff, 1920. Available online at http://resources.huygens.knaw.nl/retroboeken/vierlingh/

Wagenaar, Jan. *Kort en Opregt Verhaal van de Elende der Opgezetenen van de overstroomde Landstreeken in Nederland en byzonderlyk van den Alblasserwaard. Opgesteld uit verscheiden' geloofwaardige Brieven en Berigten*. 1741.

"Verhaal van de overstroomingen heir te lande. En byzonderlyk van den Alblasserwaard in 1740 en 1741. Opgesteld uit verscheiden' geloofwaardige Brieven en Berigten." In *Verzameling van historische en politieke tractaaten*. Amsterdam: Yntema en Tieboel, 1779.

Verhaal van de wyze, op welke de penningen, ten behoeve der noodlydenden, in de Overstroomde Landen van Gelderland en Holland, in den aanvang des Jaars 1741, te Amsterdam en elders ingezameld, bestierd en besteed zijn. Amsterdam: Isaak Tirion, 1742.

Yarranton, Andrew. *England's Improvement by Sea and Land: To Out-Do the Dutch without Fighting, to Pay Debts without Moneys* ... London: R. Everingham, 1677.

Published Secondary Sources

Aalbers, J. "Het machtsverval van de Republiek der Verenigde Nederland 1713–1741." In *Machtverval in de internationale context*, edited by J. Aalbers and A. P. Goudoever, 7–37. Groningen: Wolters-Noordhoff/Forsten, 1986.

"Holland's Financial Problems (1713–1733) and the Wars against Louis XIV." In *Britain and the Netherlands*, Vol. 6, edited by A. C. Duke and C. A. Tamse, 79–93. The Hague: Martinus Nijhoff, 1978.

Abbing, Cornelis Alard. *Geschiedenis der stad Hoorn, hoofdstad van West-Vriesland gedurende het grootste gedeelte der 17e en 18e Eeuw, of vervolg op Velius Chronyk*, Vol. 2. Hoorn: Gebr. Vermande, 1842.

Adey, Peter, Ben Anderson, and Stephen Graham. "Introduction: Governing Emergencies: Beyond Exceptionality." *Theory, Culture, & Society* 32, no. 2 (2015): 3–17.

Adger, W. Neil, Jon Barnett, Katrina Brown, Nadine Marshall, and Karen O'Brien. "Cultural Dimensions of Climate Change Impacts and Adaptation." *Nature Climate Change* 3, no. 2 (2013): 112–17.

Adger, W. Neil, Suraje Dessai, Marisa Goulden et al. "Are There Social Limits to Adaptation to Climate Change?" *Climatic Change* 93, no. 3–4 (2009): 335–54.

Ahn, Doohwan. "The Anglo-French Treaty of Commerce of 1713: Tory Trade Politics and the Question of Dutch Decline." *History of European Ideas* 36, no. 2 (2010): 167–80.

Alfani, Guido and Cormac Ó Gráda, "Famines in Europe: An Overview." In *Famine in European History*, edited by Guido Alfani and Cormac Ó Gráda, 1–24. Cambridge: Cambridge University Press, 2017.

Alfani, Guido and Wouter Ryckbosch. "Growing Apart in Early Modern Europe? A Comparison of Inequality Trends in Italy and the Low Countries, 1500–1800." *Explorations in Economic History* 62 (2016): 143–53.

Allemeyer, Maria Louisa. '*Kein Land ohne Deich–!*': *Lebenswelten einer Küstengesellschaft in der Frühen Neuzeit*. Göttingen: Vandenhoeck & Ruprecht, 2006.

"Nature in Conflict. Disputes Surrounding the Dike in 17th Century Northern Frisia as a Window into an Early Modern Coastal Society." In *Historians and Nature. Comparative Approaches to Environmental History*, edited by Ursula Lehmkuhl and Hermann Wellenreuther, 92–109. New York: Oxford University Press, 2007.

"Profane Hazard or Divine Judgement? Coping with Urban Fire in the 17th Century." *Historical Social Research/Historische Sozialforschung* 32, no. 3 (2007): 145–68.

"The World According to Harro: Mentalities, Politics, and Social Relations in an Early Modern Coastal Society." *Bulletin of the German Historical Institute* 3 (2006): 53–76.

Anderson, Ben. "Emergency Futures: Exception, Urgency, Interval, Hope." *The Sociological Review* 65, no. 3 (2017): 463–77.

Appelqvist, Christin and Jonathan N. Havenhand. "A Phenological Shift in the Time of Recruitment of the Shipworm, *Teredo navalis* L., Mirrors Marine Climate Change." *Ecology and Evolution* 6, no. 12 (2016): 3862–70.

Appuhn, Karl. "Ecologies of Beef: Eighteenth-Century Epizootics and the Environmental History of Early Modern Europe." *Environmental History* 15, no. 2 (2010): 268–87.

Assmann, Jan. "Communicative and Cultural Memory." In *Cultural Memory Studies: An International and Interdisciplinary Handbook*, edited by Astrid Erll and Ansgar Nünning, 109–18. Berlin: Walter de Gruyter, 2008.

Ayalon, Yaron. *Natural Disasters in the Ottoman Empire*. Cambridge: Cambridge University Press, 2014.

Baan, Paul J. A. and Frans Klijn, "Flood Risk Perception and Implications for Flood Risk Management in the Netherlands." *International Journal of River Basin Management* 2, no. 2 (2004): 113–22.

Baars, C. "De paalwormfurie van 1731–1732 en de schade aan de West-Fries zeedijk." *Waterschapsbelangen* 73 (1988): 809–15.
"Het dijkherstel onder leiding van de Staten van Holland." *Waterschapsbelangen* 74 (1989): 196–204.
"Nabeschouwing over de paalwormplaag van 1731–32 en de gevolgen daarvan." *Waterschapsbelangen* 75 (1990): 504–9.
Bakker, J. A. *Megalithic Research in the Netherlands, 1547–1911: From "Giant's Beds" and "Pillars of Hercules" to Accurate Investigations*. Leiden: Sidestone Press, 2010.
Bakker, Piet. "Crisis? Welke crisis? Kanttekeningen bij het economisch verval van de schilderkunst in Leiden na 1660." *De Zeventiende Eeuw* 27, no. 2 (2012): 232–69.
Balakrishnan Nair, N. "Ecology of Marine Fouling and Wood-Boring Organisms of Western Norway." *Sarsia* 8, no. 1 (1962): 1–88.
Bankoff, Greg. "Bordering on Danger: An Introduction." In *Natural Hazards and Peoples in the Indian Ocean World: Bordering on Danger*, edited by Greg Bankoff and Joseph Christensen, 1–30. New York: Palgrave, 2016.
"Comparing Vulnerabilities: Toward Charting an Historical Trajectory of Disasters." *Historical Social Research* 3, no. 32 (2007): 103–14.
"Constructing Vulnerability: The Historical, Natural and Social Generation of Flooding in Metropolitan Manila." *Disasters* 27, no. 3 (2003): 224–38.
Cultures of Disaster: Society and Natural Hazard in the Philippines. New York: Taylor & Francis, 2003.
"The 'English Lowlands' and the North Sea Basin System: A History of Shared Risk." *Environment and History* 19, no. 1 (2013): 3–37.
"Time Is of the Essence: Disaster, Vulnerability, and History." *International Journal of Mass Emergencies and Disasters* 22, no. 3 (2004): 23–42.
Barendsen, W. "De zeedijk van zijn ontstaan tot het jaar 1730." *OTAR* 45, no. 9 (1961): 195–205.
Barrett, Tom, Paul-Pierre Pastoret, and William P. Taylor. *Rinderpest and Peste Des Petits Ruminants: Virus Plagues of Large and Small Ruminants*. London: Elsevier Science, 2005.
Bartels, Michiel. "Het bolwerk tegen de woede van de zee." In *Dwars door de dijk: Archeologie en geschiedenis van de Westfriese Omringdijk tussen Hoorn en Enkhuizen*, edited by Michiel Bartels, 122–75. Hoorn: Stichting Archeologie West Friesland, 2016.
Bartels, Michiel H., Peter Swart, and H. de Weerd. "Wormspijkers in het Medemblikker havenhoofd: Archeologisch en historisch onderzoek naar de maatregelen tegen de paalworm in het noordelijk havenhoofd van Medemblik, West-Friesland." *West-Friese Archaeologische Rapporten* 80 (2015).
Barten, Egbert. "... en rukte overal alles weg, waer zij trof ...": *De runderpest in Nederland, met nadruk op de Langedijk, in de 18de Eeuw (1713–1784)*. Noord-Scharwoude: De Schrijfcaemer, 1983.
Bazin, Hervé. *Vaccinations: A History: From Lady Montagu to Jenner and Genetic Engineering*. Montrouge: John Libbey Eurotext, 2011.

Beck, Ulrich. *Risk Society: Towards a New Modernity*. Translated by Mark Ritter. London: Sage, 1992.
Beekman, A. A. *Het dijk- en waterschapsrecht in Nederland vóór 1795*. Vol. 2. 's Gravenhage: M. Nijhoff, 1907.
Behringer, Wolfgang. "Climatic Change and Witch-Hunting: The Impact of the Little Ice Age on Mentalities." *Climatic Change* 43, no. 1 (1999): 335–51.
A Cultural History of Climate. Cambridge: Polity, 2010.
"Der Krise von 1570. Ein Beitrag zur Krisengeschichte der Neuzeit." In *Um Himmels Willen: Religion in Katastrophenzeiten*, edited by Manfred Jakubowski-Tiessen and Harmut Lehmann, 51–156. Göttingen: Vandenhoeck & Ruprecht, 2003.
Berendsen, H. J. A. "Birds-Eye View of the Rhine-Meuse Delta (The Netherlands)." *Journal of Coastal Research* 14, no. 3 (1998): 741–52.
Bieleman, Jan. *Boeren in Nederland: Geschiedenis van de landbouw 1500–2000*. Amsterdam: Boom, 2008.
Boeren op het Drentse zand, 1600–1910: een nieuwe visie op de "oude" landbouw. Wageningen: Afd. Agrarische Geschiedenis, Landbouwuniversiteit, 1987.
Five Centuries of Farming: A Short History of Dutch Agriculture, 1500–2000. Wageningen: Academic Publishers, 2010.
Geschiedenis van de landbouw in Nederland, 1500–1950: Veranderingen En Verscheidenheid. Meppel: Boom, 1992.
"Rural Change in the Dutch Province of Drenthe in the Seventeenth and Eighteenth Centuries." *The Agricultural History Review* 33, no. 2 (1985): 105–17.
Biesterveld, P. and H. H. Kuyper. *Kerkelijk Handboekje: bevattende de bepalingen der Nederlandsche synoden en andere stukken van beteekenis voor de regeering der Kerk*. Kampen: J. H. Bos, 1905.
Birkland, Thomas A. *After Disaster: Agenda Setting, Public Policy, and Focusing Events*. Washington, DC: Georgetown University Press, 1997.
Lessons of Disaster: Policy Change after Catastrophic Events. Washington, DC: Georgetown University Press, 2006.
Blaikie, Piers, Terry Cannon, Ian Davis, and Ben Wisner. *At Risk: Natural Hazards, People's Vulnerability and Disasters*. New York: Routledge, 1994.
Blancou, Jean. *History of the Surveillance and Control of Transmissible Animal Diseases*. Paris: Office International des Épizooties, 2003.
Blink, H. *Geschiedenis van den boerenstand en den landbouw in Nederland*. Groningen: J. B. Wolters, 1904.
Boersma, Erica. "Noodhulpbeleid bij stads- en dorpsrampen in de Republiek." In *Crisis en catastrofe. De Nederlandse omgang met rampen in de lange negentiende eeuw*, edited by Lotte Jensen, 187–206. Amsterdam: Amsterdam University Press, 2020.
Boon, J. G. M. "Historische aantekeningen van een pastoor." *Heemtydinghen: orgaan voor de streekgeschiedenis van het Stichts-Hollandse grensgebied* 8, no. 15 (1965): 6–9.
Boon, L. J. *Dien godlosen hoop van menschen: Vervolging van homoseksuelen in De Republiek in de jaren dertig van de achttiende eeuw*. Amsterdam: De Bataafsche Leeuw, 1997.

"Those Damned Sodomites: Public Images of Sodomy in the Eighteenth Century Netherlands." *Journal of Homosexuality* 16, no. 1–2 (1988): 237–48.

Boon, Piet. "Voorland en inlagen: de Westfriese strijd tegen het buitenwater." *West-Frieslands Oud en Nieuw* 58 (1991): 78–113.

Borger, Guus J. "De betekenis van de Kerstvloed van 1717 voor de gebieden rond het IJ." *Geografisch Tijdschrift* 1 (1967): 97–103.

Borger, Guus J. and Willem A. Ligtendag. "The Role of Water in the Development of the Netherlands – A Historical Perspective." *Journal of Coastal Conservation* 4, no. 2 (1998): 109–14.

Borger, Guus J., Philippus Breuker, and Hylkje de Jong, eds. *Van Groningen tot Zeeland: geschiedenis van het cultuurhistorisch onderzoek naar het kustlandschap*. Hilversum: Verloren, 2010.

Borges, Luísa M. S., Lucas M. Merckelbach, Íris Sampaio, and Simon M. Cragg. "Diversity, Environmental Requirements, and Biogeography of Bivalve Wood-Borers (Teredinidae) in European Coastal Waters." *Frontiers in Zoology* 11, no. 1 (2014): 1–13.

Bos, J. F., R. C Burghardt, J. K. H. van der Meer, and F. Timmerman, eds. *Handschrift Schoemaker: een achttiende-eeuwse kijk op de Drentse geschiedenis*. Assen: Van Gorcum, 2004.

Bosch, Toon. "Natuur en cultuur: modernising van hulpverlening na catastrofale overstromingen in de Nederlandse Delta, 1740–1861." *Tijdschrift voor Waterstaatsgeschiedenis* 21 (2012): 39–47.

Om de macht over het water: de nationale waterstaatdienst tussen staat en samenleving, 1798–1849. Zaltbommel: Europese Bibliotheek, 2000.

"Waar is Johanna van Beek? Over het herinneren en vergeten van calamiteiten in het rivierengebied." *Streven: Cultureel maatschappelijk maandblad* 75, no. 2 (2008): 111–122.

Boschma, C. ed. *Meesterlijk vee: Nederlandse veeschilders 1600–1900*. Zwolle: Waanders, 1988.

Bots, J. *Tussen Descartes en Darwin: Geloof en natuurwetenschap in de achttiende eeuw in Nederland*. Assen: Van Gorcum, 1972.

Boxer, C. R. "The Dutch Economic Decline." In *The Economic Decline of Empires*, edited by Carlo Cipolla, 253–63. London: Taylor & Francis, 1970.

Brázdil, Rudolf, Zbigniew W. Kundzewicz, and Gerardo Benito. "Historical Hydrology for Studying Flood Risk in Europe." *Hydrological Sciences Journal* 51, no. 5 (2006): 739–64.

Breuker, Ph. H. "De achttjinde-ieuske yninting tsjin de feepest yn West-Europa." In *Freonen om ds. J.J. Kalma hinne: Stúdzjes, Meast oer Fryslân, foar syn Fiifensantichste Jierdei*, edited by Ph. H. Breuker and M. Zeeman, 325–50. Leeuwarden: Tille, 1982.

Broad, John. "Cattle Plague in Eighteenth-Century England." *The Agricultural History Review* 31, no. 2 (1983): 104–15.

Brönnimann, Stefan, Sam White, and Victoria Slonosky. "Climate from 1800 to 1970 in North America and Europe." In *The Palgrave Handbook of Climate History*, edited by Sam White, Christian Pfister, and Franz Mauelshagen, 309–20. London: Palgrave Macmillan, 2018.

Brouwer, Judith. *Levenstekens: Gekaapte brieven uit het Rampjaar 1672*. Hilversum: Verloren, 2014.
Brusse, Paul. "Overleven door ondernemen: De agrarische geschiedenis van de Over-Betuwe 1650–1850." PhD diss., Landbouwuniversiteit Wageningen, 1999.
"Property, Power and Participation in Local Administration in the Dutch Delta in the Early Modern Period." *Continuity and Change* 33, no. 1 (2018): 59–86.
Buisman, Jan. *Duizend jaar weer, wind en water in de Lage Landen: 1300–1450*, Vol. 2. Franeker: Van Wijnen, 1996.
Duizend jaar weer, wind en water in de Lage Landen: 1575–1675, Vol. 4. Franeker: Van Wijnen, 2000.
Duizend jaar weer, wind en water in de Lage Landen: 1675–1750, Vol. 5. Franeker: Van Wijnen, 2006.
Duizend jaar weer, wind en water in de Lage Landen: 1751–1800, Vol. 6. Franeker: Van Wijnen, 2015.
Buisman, Jan Willem. *Tussen Vroomheid en Verlichting: een cultuurhistorische en- sociologisch onderzoek naar enkele aspecten van de Verlichting in Nederland (1755–1810)*. Zwolle: Waanders, 1992.
Burgers, Ton. *Nederlands grote rivieren. Drie eeuwen strijd tegen overstromingen*. Utrecht: Uitgeverij Matrijs, 2014.
Burton, Ian, Robert W. Kates, and Gilbert F. White, eds. *The Environment as Hazard*. New York: The Guilford Press, 1993.
Cabato, Regine and Jason Samenow. "Super Typhoon Goni, World's Most Powerful Storm in Four Years, Smashes into the Philippines." *Washington Post*, November 1, 2020.
Camenisch, Chantal. "Two Decades of Crisis: Famine and Dearth during the 1480s and 1490s in Western and Central Europe." In *Famines during the 'Little Ice Age' (1300–1800): Socionatural Entanglements in Premodern Societies*, edited by Dominik Collet and Maximilian Schuh, 69–90. Cham: Springer, 2018.
Camenisch, Chantal, Kathrin M. Keller, Melanie Salvisberg et al. "The 1430s: A Cold Period of Extraordinary Internal Climate Variability during the Early Spörer Minimum with Social and Economic Impacts in North-Western and Central Europe." *Climate of the Past* 12, no. 11 (2016): 2107–26.
Camenisch, Chantal and Christian Rohr. "When the Weather Turned Bad. The Research of Climate Impacts on Society and Economy during the Little Ice Age in Europe. An Overview." *Cuadernos de Investigación Geográfica* 44, no. 1 (2018): 99–114.
Campbell, Bruce M. S. "The European Mortality Crises of 1346–52 and Advent of the Little Ice Age." In *Famines During the 'Little Ice Age' (1300–1800): Socionatural Entanglements in Premodern Societies*, edited by Dominik Collet and Maximilian Schuh, 19–41. Cham: Springer, 2018.
The Great Transition: Climate, Disease and Society in the Late-Medieval World. Cambridge: Cambridge University Press, 2016.
Cannon, Terry. "Vulnerability Analysis and the Explanation of 'Natural Disasters'." In *Disasters, Development and Environment*, edited by Ann Varley, 13–30. Chichester: John Wiley and Sons, 1994.

Carey, Mark. "Climate and History: A Critical Review of Historical Climatology and Climate Change Historiography." *Wiley Interdisciplinary Reviews: Climate Change* 3, no. 3 (2012): 233–49.
Carlton, James T. "Biological Invasions and Cryptogenic Species." *Ecology* 77, no. 6 (1996): 1653–55.
"Molluscan Invasions in Marine and Estuarine Communities." *Malacologia* 41, no. 2 (1999): 439–54.
"The Scale and Ecological Consequences of Biological Invasions in the World's Oceans." In *Invasive Species and Biodiversity Management*, edited by Odd Terje Sandlund, Peter Johan Schei, and Åslaug Viken, 195–212. Dordrecht: Kluwer Academic Publishers, 1999.
Carlton, James T. and J. Hodder. "Biogeography and Dispersal of Coastal Marine Organisms: Experimental Studies on a Replica of a 16th-Century Sailing Vessel." *Marine Biology* 121, no. 4 (1995): 721–30.
Castonguay, Stéphane. "The Production of Flood as Natural Catastrophe: Extreme Events and the Construction of Vulnerability in the Drainage Basin of the St. Francis River (Quebec), Mid-Nineteenth to Mid-Twentieth Century." *Environmental History* 12, no. 4 (2007): 820–44.
Casty, Carlo, Heinz Wanner, Jürg Luterbacher, Jan Esper, and Reinhard Böhm. "Temperature and Precipitation Variability in the European Alps since 1500." *International Journal of Climatology* 25, no. 14 (2005): 1855–80.
Clark, G. N. "War Trade and Trade War, 1701–1713." *The Economic History Review* 1, no. 2 (1928): 278–9.
Clarke, Michele L. and Helen M. Rendell. "The Impact of North Atlantic Storminess on Western European Coasts: A Review." *Quaternary International* 195, no. 1–2 (2009): 31–41.
Collenteur, G. A. C., Pim Kooij, and P. Kooij, eds. *Stad en Regio*. Assen: Van Gorcum, 2010.
Collier, Kiah. "Can the 'Masters of the Flood' Help Texas Protect its Coast from Hurricanes?" *The Texas Tribune*, July 15, 2019. www.texastribune.org/2019/07/15/can-masters-flood-help-texas-protect-its-coast-hurricanes/
Cornes, R. C. "Historic Storms of the Northeast Atlantic since circa 1700: A Brief Review of Recent Research." *Weather* 69, no. 5 (2014): 121–25.
CRED. *Natural Disasters 2019*. Brussels: CRED, 2020.
CRED and UNDRR. *Human Cost of Disasters: An Overview of the Last 20 Years, 2000–2019*. https://cred.be/sites/default/files/CRED-Disaster-Report-Human-Cost2000-2019.pdf
Crooks, Jeffrey A. "Lag Times and Exotic Species: The Ecology and Management of Biological Invasions in Slow-Motion." *Ecoscience* 12, no. 3 (2005): 316–29.
Crosby, Alfred W. *The Columbian Exchange: Biological and Cultural Consequences of 1492*. Westport, CT: Greenwood Publishing Group, 1972.
Curth, Louise Hill. *The Care of Brute Beasts: A Social and Cultural Study of Veterinary Medicine in Early Modern England*. Leiden: Brill, 2010.
Curtis, Daniel R. "Danger and Displacement in the Dollard: The 1509 Flooding of the Dollard Sea (Groningen) and Its Impact on Long-term Inequality in the

Distribution of Property." *Environment and History* 22, no. 1 (2016): 103–35.
"Was Plague an Exclusively Urban Phenomenon? Plague Mortality in the Seventeenth-Century Low Countries." *Journal of Interdisciplinary History* 47, no. 2 (2016): 139–70.
Curtis, Daniel R. and Jessica Dijkman. "The Escape from Famine in the Northern Netherlands: A Reconsideration Using the 1690s Harvest Failures and a Broader Northwest European Perspective." *The Seventeenth Century* 34, no. 2 (2019): 229–58.
Curtis, Daniel R., Bas van Bavel, and Tim Soens. "History and the Social Sciences: Shock Therapy with Medieval Economic History as the Patient." *Social Science History* 40, no. 4 (2016): 751–74.
Curtis, Daniel R., Jessica Dijkman, Thijs Lambrecht, and Eric Vanhaute. "Low Countries." In *Famine in European History*, edited by Guido Alfani and Cormac O Grada, 119–40. Cambridge: Cambridge University Press, 2017.
Cutter, Susan L. "Compound, Cascading, or Complex Disasters: What's in a Name?" *Environment: Science and Policy for Sustainable Development* 60, no. 6 (2018): 16–25.
Dahlberg, Rasmus, Olivier Rubin, and Morten Thanning Vendelø, eds. *Disaster Research: Multidisciplinary and International Perspectives*. London: Routledge, 2016.
Dalby, Simon. "Anthropocene Formations: Environmental Security, Geopolitics and Disaster." *Theory, Culture & Society* 34, no. 2–3 (2017): 233–52.
Darling, John A. and James T. Carlton. "A Framework for Understanding Marine Cosmopolitanism in the Anthropocene." *Frontiers in Marine Science* 5 (2018): 293.
Davids, Karel. "River Control and the Evolution of Knowledge: A Comparison between Regions in China and Europe, c. 1400–1850." *Journal of Global History* 1, no. 1 (2006): 59–79.
The Rise and Decline of Dutch Technological Leadership: Technology, Economy and Culture in the Netherlands, 1350–1800. Leiden: Brill, 2008.
Davis, Stephen A., Herwig Leirs, Roger Pech, Zhibin Zhang, and Nils Chr. Stenseth. "On the Economic Benefit of Predicting Rodent Outbreaks in Agricultural Systems." *Crop Protection* 23, no. 4 (2004): 305–14.
De Bruijn, J. G. *Inventaris van de prijsvragen uitgeschreven door de Hollandsche Maatschappij der Wetenschappen 1753–1917*. Haarlem: H. D. Tjeenk Willink, 1977.
De Haan, J. A. B. *De Hollandsche Maatschappij der Wetenschappen, 1752–1952*. Haarlem: J. Enschedé en Zonen, 1952.
De Jong, J., L. Kooijmans, and H. F. de Wit. "Schuld en boete in de Nederlandse Verlichting." *Kleio* 19 (1978): 237–44.
De Jong, Rixt, Svante Björck, Leif Björkman, and Lars B. Clemmensen. "Storminess Variation during the Last 6500 Years as Reconstructed from an Ombrotrophic Peat Bog in Halland, Southwest Sweden." *Journal of Quaternary Science* 21, no. 8 (2006): 905–19.
De Jongste, J. A. F. *Onrust aan het Spaarne: Haarlem in de jaren 1747–1751*. Dieren: Bataafsche Leeuw, 1984.

De Kraker, Adriaan M. J. "Flood Events in the Southwestern Netherlands and Coastal Belgium, 1400–1953." *Hydrological Sciences Journal* 51, no. 5 (2006): 913–29.
"Flooding in River Mouths: Human Caused or Natural Events? Five Centuries of Flooding Events in the SW Netherlands, 1500–2000." *Hydrology & Earth System Sciences* 19, no. 6 (2015): 1–12.
"Ice and Water. The Removal of Ice on Waterways in the Low Countries, 1330–1800." *Water History* 9, no. 2 (2017): 109–28.
De Kraker, Adriaan M. J. "Reconstruction of Storm Frequency in the North Sea Area of the Pre-industrial Period, 1400–1625 and the Connection with Reconstructed Time Series of Temperatures." *History of Meteorology* 2 (2005): 51–69.
"Storminess in the Low Countries, 1390–1725." *Environment and History* 19, no. 2 (2013): 149–71.
"Two Floods Compared: Perception of and Response to the 1682 and 1715 Flooding Disasters in the Low Countries." In *Forces of Nature and Cultural Responses*, edited by Katrin Pfeifer and Niki Pfeifer, 287–302. Dordrecht: Springer, 2012.
De Navorscher: een middel tot gedachtenwisseling en letterkundig verkeer tuschen allen, die iets weten, iets te vragen hebben of iets kunnen oplossen. Amsterdam: Frederick Muller, 1862.
De Nijs, Thimo and Eelco Beukers, eds. *Geschiedenis van Holland, Deel I.* Hilversum: Verloren, 2002.
De Voys, C. G. N. "Expected Biological Effects of Long-Term Changes in Temperatures on Benthic Ecosystems in Coastal Waters around the Netherlands." In *Expected Effects of Climatic Change on Marine Coastal Ecosystems*, edited by J. J. Beukema, W. J. Wolff, and J. J. W. M. Brouns, 72–82. Dordrecht: Kluwer Academic Publishers, 1990.
De Vriend, Huib J., Mark van Koningsveld, Stefan G. J. Aarninkhof et al. "Sustainable Hydraulic Engineering through Building with Nature." *Journal of Hydro-Environment Research* 9, no. 2 (2015): 159–71.
De Vries, Jan. "The Crisis of the Seventeenth Century: The Little Ice Age and the Mystery of the 'Great Divergence'." *Journal of Interdisciplinary History* 44, no. 3 (2013): 374–5.
"The Decline and Rise of the Dutch Economy, 1675–1900." In *Technique, Spirit, and Form in the Making of the Modern Economies: Essays in Honor of William N. Parker*, edited by Gary R. Saxonhouse and Gavin Wright, 149–89. London: JAI Press, 1984.
The Dutch Rural Economy in the Golden Age, 1500–1700. New Haven, CT: Yale University Press, 1974.
The Economy of Europe in an Age of Crisis, 1600–1750. Cambridge: Cambridge University Press, 1976.
"Measuring the Impact of Climate on History: The Search for Appropriate Methodologies." *The Journal of Interdisciplinary History* 10, no. 4 (1980): 599–630.

De Vries, Jan and Ad van der Woude. *The First Modern Economy: Success, Failure, and Perseverance of the Dutch Economy, 1500–1815*. Cambridge: Cambridge University Press, 1997.

De Vries, Johannes. *De economische achteruitgang der Republiek in de achttiende eeuw*. Leiden: Stenfert Kroese, 1968.

Deason, Gary B. "Reformation Theology and the Mechanistic Conception of Nature." In *God and Nature: Historical Essays on the Encounter between Christianity and Science*, edited by David C. Lindberg and Ronald L. Numbers, 167–91. Berkeley: University of California Press, 1986.

Deen, Femke, David Onnekink, and Michel Reinders. "Pamphlets and Politics: Introduction." In *Pamphlets and Politics in the Dutch Republic*, edited by Femke Deen, David Onnekink, and Michel Reinders, 3–30. Leiden: Brill, 2010.

Degroot, Dagomar. "Climate Change and Society in the 15th to 18th Centuries." *Wiley Interdisciplinary Reviews: Climate Change* 9, no. 3 (2018): e518.

The Frigid Golden Age: Climate Change, the Little Ice Age, and the Dutch Republic, 1560–1720. Cambridge: Cambridge University Press, 2018.

"'Never such Weather Known in These Seas': Climatic Fluctuations and the Anglo-Dutch Wars of the Seventeenth Century, 1652–1674." *Environment and History* 20, no. 2 (2014): 239–73.

Dekker, Rudolf. *Holland in beroering: oproeren in de 17de en 18de eeuw*. Baarn: Ambo, 1982.

DeLacy, Margaret. *The Germ of an Idea: Contagionism, Religion, and Society in Britain, 1660–1730*. New York: Palgrave Macmillan, 2016.

Diedriks, Herman. "Economic Decline and the Urban Elite in Eighteenth-Century Dutch Towns: A Review Essay." *Urban History Yearbook* 16 (1989): 78–81.

Drees, Marijke Meijer. "'Providential Discourse' Reconsidered: The Case of the Delft Thunderclap (1654)." *Dutch Crossing* 40, no. 2 (2016): 108–21.

"'Vechten voor het vaderland' in de literatuur, 1650–1750." In *Vaderland: Een geschiedenis van de vijftiende eeuw tot 1940*, edited by N. C. F. van Sas, 109–42. Amsterdam: Amsterdam University Press, 1999.

Dingeldein, W. H. "Iets over mond en klauwzeer in 1732 en vroeger." *Tijdschrift voor Diergeneeskunde* 36 (1933): 389–92.

Doedens, Anne. *Witte de With 1599–1658: Wereldwijde strijd op zee in de Gouden Eeuw*. Hilversum: Uitgeverij Verloren, 2008.

Draaisma, Kees. "Een nieuwe kijk op Caspars dijk." *It Beaken* 79, no. 1–2 (2017): 27–105.

Dreiskämper, Petra. *Redeloos, radeloos, reddeloos: de geschiedenis van het Rampjaar 1672*. Hilversum: Verloren, 1998.

Driessen, Anneke. "Hulpverlening na overstromingsrampen in het Nederlands rivierengebied." *Groniek* 33 (2000): 185–98.

Watersnood tussen Maas en Waal: Overstromingsrampen in het rivierengebied tussen 1780 en 1810. Zutphen: Walburg Pers, 1994.

Duiveman, Adriaan. "Praying for (the) Community: Disasters, Ritual and Solidarity in the Eighteenth-Century Dutch Republic." *Cultural and Social History* 16, no. 5 (2019): 543–60.

Edmondson, Charles H. "Teredinidae, Ocean Travelers." *Occasional Papers of Bernice P. Bishop Museum* 23, no. 3 (1962): 45.
Edwards, O. C. Jr. "Varieties of the Sermon: A Survey of Preaching in the Long Eighteenth Century." In *Preaching, Sermon and Cultural Change in the Long Eighteenth Century*, edited by Joris van Eijnatten, 3–53. Leiden: Brill, 2009.
Egner, H., M. Schorch, and M. Voss. "Can Societies Learn from Calamities?." In *Learning and Calamities: Practices, Interpretations, Patterns*, edited by H. Egner, M. Schorch, and M. Voss, 1–26. New York: Taylor & Francis, 2014.
Endfield, Georgina H. and Sarah L. O'Hara. "Conflicts over Water in 'The Little Drought Age' in Central Mexico." *Environment and History* 3, no. 3 (1997): 255–72.
Endfield, Georgina H., Sarah J. Davies, Isabel Fernández Tejedo, Sarah E. Metcalfe, and Sarah L. O'Hara. "Documenting Disaster: Archival Investigations of Climate, Crisis, and Catastrophe in Colonial Mexico." In *Natural Disasters, Cultural Responses: Case Studies toward a Global Environmental History*, edited by Christof Mauch and Christian Pfister, 305–25. New York: Lexington Books, 2009.
Engbers, K. A. M. and A. L. Hempenius. *Verzamelinventaris van de archieven van de "Kust"waterschappen inliggend in het waterschap Hunsingo en het waterschap Reitdiep (1805) 1856–1990 (1994)*. Groningen: Laurentius Archief & Geschiedenis, 2011.
Enthoven, Victor. "An Assessment of Dutch Transatlantic Commerce, 1585–1817." In *Riches from Atlantic Commerce: Dutch Transatlantic Trade and Shipping, 1585–1817*, edited by Johannes Postma and Victor Enthoven, 385–445. Leiden: Brill, 2003.
Ermus, Cindy. "The Spanish Plague That Never Was: Crisis and Exploitation in Cádiz during the Peste of Provence." *Eighteenth-Century Studies* 49, no. 2 (2016): 167–93.
Esser, Raingard. "Fear of Water and Floods in the Low Countries." In *Fear in Early Modern Society*, edited by William G. Naphy and Penny Roberts, 62–77. New York: Manchester University Press, 1997.
  "'Ofter gheen water op en hadde gheweest' – Narratives of Resilience on the Dutch Coast in the Seventeenth Century." *Dutch Crossing* 40, no. 2 (2016): 97–107.
Faber, J. A. "Cattle Plague in the Netherlands during the Eighteenth Century." *Mededelingen van de Landbouwhogeschool te Wageningen* 62, no. 11 (1962): 1–7.
  Drie eeuwen Friesland: Economische en sociale ontwikkelingen van 1500 tot 1800. Leeuwarden: De Tille, 1973.
Ferrières, Madeleine. *Sacred Cow, Mad Cow: A History of Food Fears*. New York: Columbia University Press, 2006.
Filkov, Alexander I., Tuan Ngo, Stuart Matthews et al. "Impact of Australia's Catastrophic 2019/20 Bushfire Season on Communities and Environment. Retrospective Analysis and Current Trends." *Journal of Safety Science and Resilience* 1, no. 1 (2020): 44–56.
Fischer, Norbert. *Wassersnot und Marschengesellschaft: Zur Geschichte der Deiche in Kehdingen*. Stade: Schriftenreihe des Landschaftsverbandes der ehemaligen HerzogtümerBremen und Verden, 2003.

Fix, Andrew. "Comets in the Early Dutch Enlightenment." In *The Early Enlightenment in the Dutch Republic, 1650–1750*, edited by Wiep van Bunge, 157–72. Leiden: Brill, 2003.
Fleischer, Alette. "The Garden behind the Dyke: Land Reclamation and Dutch Culture in the 17th Century." *Icon* 11 (2005): 16–32.
Folin, Marco and Monica Preti. *Wounded Cities: The Representation of Urban Disasters in European Art (14th–20th Centuries)*. Leiden: Brill, 2015.
Food and Agriculture Administration of the United Nations. "FAO makes gains in the fight against desert locusts in East Africa and Yemen but threat of a food security crisisremains." *FAO News*, November 5, 2020, www.fao.org/emergencies/fao-in-action/stories/stories-detail/en/c/1275091/
Fransen, Alfons. *Dijk onder spanning: De ecologische, politieke en fianciële geschiedenis vande Diemerdijk bij Amsterdam, 1591–1864*. Hilversum: Verloren, 2011.
Frijhoff, W. "Fiery Metaphors in the Public Space: Celebratory Culture and Political Consciousness around the Peace of Utrecht." In *Performances of Peace: Utrecht 1713*, edited by Renger Evert Bruin, Lotte Jensen, and David Onnekink, 223–48. Leiden: Brill, 2015.
Frijhoff, W. "Utrechts vreugdevuur, Masker voor 's lands neergang?." *De Achttiende Eeuw* 40 (2008): 5–20.
Frijhoff, W. and M. Spies. *Dutch Culture in a European Perspective: 1650, Hard-Won Unity*. Assen: Uitgeverij Van Gorcum, 2004.
Frijhoff, W., J. Kloek, and M. Spies. *Dutch Culture in a European Perspective: 1800, Blueprints for a National Community*. New York: Palgrave, 2004.
Fritschy, Wantje. "The Poor, the Rich, and the Taxes in Heinsius' Times." In *Anthonie Heinsius and the Dutch Republic, 1688–1720. Politics, War, and Finance*, edited by J. de Jongste and A. J. Veenendaal, 242–58. The Hague: Institute of Netherlands History, 2002.
Fruin, Robert. *De oorlog van 1672*. Groningen: Wolters-Noordhoff, 1972.
Galil, B. S., Agnese Marchini, Anna Occhipinti-Ambrogi et al. "International Arrivals: Widespread Bioinvasions in European Seas." *Ethology Ecology & Evolution* 26, no. 2–3 (2014): 152–71.
Galloway, James A. "Coastal Flooding and Socioeconomic Change in Eastern England in the Later Middle Ages." *Environment and History* 19 (2013): 173–207.
García-Herrera, Ricardo, David Barriopedro, David Gallego et al. "Understanding Weather and Climate of the Last 300 Years from Ships' Logbooks." *Wiley Interdisciplinary Reviews: Climate Change* 9, no. 6 (2018): e544.
Garnier, Emmanuel. "Historic Drought from Archives: Beyond the Instrumental Record." In *Drought: Science and Policy*, edited by Ana Iglesias, Dionysis Assimacopoulos, and Henny A. J. van Lanen, 45–67. Hoboken, NJ: Wiley, 2019.
Gawroński, J. De *"Equipagie" van de Hollandia en de Amsterdam: VOC-bedrijvigheid in 18de-eeuws Amsterdam*. Amsterdam: De Bataafsche Leeuw, 1996.
Gay, Peter. *Enlightenment*, Vol. 2. New York: Knopf Doubleday Publishing Group, 2013.

Gelderblom, Oscar, ed. *The Political Economy of the Dutch Republic.* Surrey: Ashgate, 2016.
Gelderblom, Oscar and Joost Jonker. "Public Finance and Economic Growth: The Case of Holland in the Seventeenth Century." *The Journal of Economic History* 71, no. 1 (2011): 1–39.
Gelderblom, Oscar, Abe De Jong, and Joost Jonker. "The Formative Years of the Modern Corporation: The Dutch East India Company VOC, 1602–1623." *The Journal of Economic History* 73, no. 4 (2013): 1050–76.
Gijsbers, Wilhelmina Maria. "Danish Oxen in Dutch Meadows: Beef Cattle Trading and Graziery in the Netherlands between 1580 and 1750." In *Facing the North Sea: West Jutland and the World; Proceeding of the Ribe Conference*, edited by Mette Guldberg, Poul Holm, and Per Kristian Madsen, 129–48. Esbjerg: Fiskeri- og Søfartsmuseet, 1993.
Gijsbers, Wilhelmina Maria and P. A. Koolmees. "Food on Foot: Long Distance Trade in Slaughter Oxen between Denmark and the Netherlands (14th–18th Century)." *Historia Medicinae Veterinariae* 26 (2001): 115–27.
Kapitale ossen: De internationale handel in slachtvee in Noordwest-Europa (1300–1750). Hilversum: Verloren, 1999.
Gijsbers, Wilhelmina and Bert Lambooij. "Oxen for the Axe. A Contemporary View on Historical Long-Distance Live Stock Transport." In *By, Marsk og Geest 17*, edited by Morten Søvsø and Jakob Kieffer-Olsen, 58–79. Ribe: Ribe Byhistoriske Arkiv & Denantikvariske Samling I Ribe Forlaget Liljeberget, 2005.
Glacken, Clarence J. *Traces on the Rhodian Shore: Nature and Culture in Western Thought from Ancient Times to the End of the Eighteenth Century.* Berkeley: University of California Press, 1967.
Glaser, R. and H. Stangl. "Historical Floods in the Dutch Rhine Delta." *Natural Hazards and Earth System Sciences* 3, no. 6 (2003): 605–13.
Gollasch, Stephan, Deniz Haydar, Dan Minchin, Wim J. Wolff, and Karsten Reise. "Introduced Aquatic Species of the North Sea Coasts and Adjacent Brackish Waters." In *Biological Invasions in Marine Ecosystems: Ecological, Management, and Geographic Perspectives*, edited by Gil Rilov and Jeffrey A. Crooks, 507–28. Berlin, Heidelberg: Springer Berlin Heidelberg, 2009.
Gottschalk, M. K. E. *Stormvloeden en rivieroverstromingen in Nederland.* Vol. 3. Amsterdam: Van Gorcum, 1977.
Grace, Delia and John McDermott. "Livestock Epidemic." In *Routledge Handbook of Hazards and Disaster Risk Reduction*, edited by Ben Wisner, J. C. Gaillard, and Ilan Kelman, 372–83. London: Routledge, 2011.
Gray, Stephen T., Lisa J. Graumlich, Julio L. Betancourt et al., "A Tree-Ring Based Reconstruction of the Atlantic Multidecadal Oscillation since 1567 AD." *Geophysical Research Letters* 31, no. 12 (2004): 1–4.
Grijzenhout, Frans. "A Myth of Decline." In *The Dutch Republic in the Eighteenth Century: Decline, Enlightenment, and Revolution*, edited by Margaret C. Jacob and Wijnand W. Mijnhardt, 324–37. Ithaca, NY: Cornell University Press, 1992.
Groenhuis, G. "Calvinism and National Consciousness: The Dutch Republic as the New Israel." In *Britain and the Netherlands: Vol VII, Church and State*

Since the Reformation, edited by A. C. Duke and C. A. Tamse, 118–33. The Hague: Martinus Nijhoff, 1981.

Hacquebord, L. and A. L. Hempenius. *Groninger dijken op deltahoogte*. Groningen: Wolters-Noordhoff, 1990.

Hagoort, Wim. *Het hoofd boven water. De geschiedenis van de Gelderse zeepolder Arkemheen, gemeenten Nijkerk en Putten 1356 (806)-1916*. Nijkerk: Nabij, 2018.

Haitsma Mulier, E. O. G. 'Het begrip 'vaderland' in de Nederlandse geschiedschrijving van de late zestiende tot de eerste helft van de achttiende." In *Vaderland: Een geschiedenis van de vijftiende eeuw tot 1940*, edited by N. C. F. van Sas, 163–80. Amsterdam: Amsterdam University Press, 1999.

Haks, Donald. *Journalistiek in crisistijd: De (Nieuwe) Nederlandsche Jaarboeken 1747–1822*. Hilversum: Verloren, 2017.

Vaderland en vrede, 1672–1713: Publiciteit over de Nederlandse Republiek in oorlog. Hilversum: Verloren, 2013.

Hameleers, Marc. *West-Friesland in oude kaarten*. Wormerveer: Stichting Uitgeverij Noord-Holland, 1987.

Hanska, Jussi. *Strategies of Sanity and Survival: Religious Responses to Natural Disasters in the Middle Ages*. Helsinki: Finnish Literature Society, 2002.

Hardwick, Joseph and Randall J. Stephens. "Acts of God: Continuities and Change in Christian Responses to Extreme Weather Events from Early Modernity to the Present." *Wiley Interdisciplinary Reviews: Climate Change* 11, no. 2 (2020): 1–16.

Hartsema, David. "De Kerstvloed van 1717." *Vroeger en later* 1 (1989): 29–35.

Hauer, Katrin. "Wahrnehmung, Deutung und Bewältigung von Starkwinden. Der Ostalpenraumund Holland im Vergleich (1600–1750)." PhD diss., Universität Salzburg, 2008.

Hekmeijer, Franciscus Cornelius. *Korte geschiedenis der runderpest: benevens eene opgave van al de over deze ziekte handelende geschriften, die van de vroegste tijden tot op heden zijn uitgekomen*. Amersfoort: Jacobs en Meijers, 1845.

Helmers, Helmer J. and Geert H. Janssen. "Introduction: Understanding the Dutch Golden Age." In *The Cambridge Companion to the Dutch Golden Age*, edited by Helmer J. Helmers and Geert H. Janssen, 1–12. Cambridge: Cambridge University Press, 2018.

Hesselink, Annika. "History Makes a River: Morphological Changes and Human Interference inthe River Rhine, The Netherlands." PhD diss., Utrecht University. Netherlands Geographical Studies, 2002.

Hewitt, Kenneth. "The Idea of Calamity in a Technocratic Age." In *Interpretations of Calamity*, edited by Kenneth Hewitt, 3–32. London: Unwin-Hyman, 1983.

Regions of Risk: A Geographical Introduction to Disasters. Harlow, Essex: Addison Wesley Longman, 1997.

Higuera, P. E. and J. T. Abatzoglou. "Record-Setting Climate Enabled the Extraordinary 2020 Fire Season in the Western United States." *Global Change Biology* 27 (2021): 1–2.

Hollestelle, Leo. "De zorg voor de zeewering van Walcheren ten tijde van de Republiek, 1574–1795." In *Duizend jaar Walcheren: Over gelanden, heren*

en geschot, over binnen- en buiten beheer, edited by A. Beenhakker, 103–21. Middelburg: Koninklijk ZeeuwschGenoots, 1996.
Hoppe, K. N. "Teredo Navalis – The Cryptogenic Shipworm." In *Invasive Aquatic Species of Europe; Distribution, Impacts, and Management*, edited by Erkki Leppäkoski, Stephan Gollasch, and Sergej Olenin, 116–19. Dordrecht: Springer, 2002.
Hoppenbrouwers, P. C. M. "Geen heer, geen boer: de bemoeienissen van Johannes le Francq van Berkhey met de Hollandse landbouw." *Holland: Regionaal-historisch Tijdschrift* 18, no. 3 (1986): 130–47.
Hudson, Paul, Hans Middelkoop, and Esther Stouthamer. "Flood Management along the Lower Mississippi and Rhine Rivers (The Netherlands) and the Continuum of Geomorphic Adjustment." *Geomorphology* 101, no. 1 (2008): 209–36.
Huisman, Cornelis. *Neerlands Israël: Het natiebesef der traditioneel-gereformeerden in de achttiende eeuw*. Dordrecht: J. P. van den Tol, 1983.
Huisman, Frank. "Medicine and Health Care in the Netherlands, 1500–1800." In *A History of Science in the Netherlands: Survey, Themes and Reference*, edited by Klaas van Berkel, Albert van Helden, and Lodewijk Palm, 239–70. Leiden: Brill, 1999.
Huizinga, Johan H. *Dutch Civilisation in the Seventeenth Century: And Other Essays*. New York: Harper & Row, 1969.
Hünemörder, Kai F. "Zwischen 'abergläubischem Abwehrzauber' und der 'Inokulation der Hornviehseuche'. Entwicklungslinien der Rinderpestbekämpfung im 18. Jahrhundert." In *Beten, Impfen, Sammeln: Zur Viehseuchen- und Schädlingsbekämpfung in der Frühen Neuzeit. Graduiertenkolleg Interdisziplinäre Umweltgeschichte*, edited by Dominik Hünniger, Katharina Engelken, and Steffi Windelen, 21–56. Göttingen: Universitätsverlag Göttingen, 2007.
Hünniger, Dominik. *Die Viehseuche von 1744–52: Deutungen und Herrschaftspraxis inKrisenzeiten*. Neumünster: Wacholtz Verlag, 2011.
"Policing Epizootics Legislation and Administration during Outbreaks of Cattle Plague in Eighteenth-Century Northern Germany as Continuous Crisis Management." In *Healing the Herds: Disease, Livestock Economies, and the Globalization of Veterinary Medicine*, edited by K. Brown and D. Gilfoyle, 76–91. Athens: Ohio University Press, 2010.
"Umweltgeschichte kulturhistorisch: Tierseuchen in den Lebenswelten des 18. Jahrhunderts." In *Natur und Gesellschaft: Perspektiven der interdisziplinären Umweltgeschichte*, edited by Manfred Jakubowski-Tiessen and Jana Sprenger, 173–90. Göttingen: Universitätsverlag Göttingen, 2014.
Huussen, A. H. Jr. "Prosecution of Sodomy in Eighteenth Century Frisia, Netherlands." *Journal of Homosexuality* 16, no. 1–2 (1988): 249–62.
"Sodomy in the Dutch Republic during the Eighteenth Century." In *Hidden from History: Reclaiming the Gay and Lesbian Past*, edited by Martin Duberman, Martha Chauncey, and George Vicinus, 141–9. Ann Arbor: University of Michigan Press, 1990.
Huygelen, C. "The Immunization of Cattle against Rinderpest in Eighteenth-Century Europe." *Medical History* 41, no. 2 (1997): 182–96.

Ihalainen, Pasi. *Protestant Nations Redefined: Changing Perceptions of National Identity in the Rhetoric of the English, Dutch and Swedish Public Churches, 1685–1772.* Leiden: Brill, 2005.
Israel, Jonathan. *Conflicts of Empires: Spain, the Low Countries and the Struggle for World Supremacy, 1585–1713.* London: The Hambledon Press, 1997.
Dutch Primacy in World Trade, 1585–1740. New York: Oxford University Press, 1989.
The Dutch Republic: Its Rise, Greatness, and Fall, 1477–1806. New York: Oxford University Press, 1995.
Enlightenment Contested: Philosophy, Modernity, and the Emancipation of Man 1670–1752. Oxford: Oxford University Press, 2006.
Radical Enlightenment: Philosophy and the Making of Modernity, 1650–1750. Oxford: Oxford University Press, 2001.
Jacob, Margaret C. *Scientific Culture and the Making of the Industrial West.* Oxford: Oxford University Press, 1997.
Jacob, Margaret C. and Wijnand W. Mijnhardt, eds. *The Dutch Republic in the Eighteenth Century: Decline, Enlightenment, and Revolution.* Ithaca, NY: Cornell University Press, 1992.
Jakubowski-Tiessen, Manfred. "'Harte Exempel göttlicher Strafgerichte'. Kirche und Religion in Katastrophenzeiten: Die Weihnachtsflut von 1717." *Niedersächsisches Jahrbuch für Landesgeschichte* 73 (2001): 119–32.
Sturmflut 1717: Die Bewältigung einer Naturkatastrophe in der Frühen Neuzeit. Oldenbourg: Oldenbourg Wissenschaftsverlag, 1992.
Janković, Vladimir. *Reading the Skies: A Cultural History of English Weather, 1650–1820.* Chicago: University of Chicago Press, 2000.
Janse, Maartje. *De geest van Jan Salie: Nederland in verval?* Hilversum: Verloren, 2002.
Janssens, Uta. "Matthieu Maty and the Adoption of Inoculation for Smallpox in Holland." *Bulletin of the History of Medicine* 55, no. 2 (1981): 246–56.
Jelgersma, S. M. J. F., M. J. F. Stive, and L. Van der Valk. "Holocene Storm-Surge Signatures in the Coastal Dunes of the Western Netherlands." *Marine Geology* 125, no. 1–2 (1995): 95–110.
Jensen, Lotte. *Celebrating Peace: The Emergence of Dutch Identity, 1648–1815.* Nijmegen: Vantilt, 2017.
Vieren van vrede: Het ontstaan van de Nederlandse identiteit, 1648–1815. Nijmegen: Uitgeverij Vantilt, 2016.
Wij tegen het water: Een eeuwenoude strijd. Nijmegen: Uitgeverij Vantilt, 2018.
Johns, Alessa, ed. *Dreadful Visitations: Confronting Natural Catastrophe in the Age of Enlightenment.* New York: Routledge, 1999.
Jones, P. D. and K. R. Briffa. "Unusual Climate in Northwest Europe during the Period 1730 to 1745 Based on Instrumental and Documentary Data." *Climatic Change* 79, no. 3/4 (2006): 361–79.
Jorink, Eric. *Reading the Book of Nature in the Dutch Golden Age, 1575–1715*, Translated by Peter Mason. Leiden: Brill, 2010.
Juneja, Monica and Gerrit Jasper Schenk, eds. *Disaster as Image: Iconographies and Media Strategies Across Europe and Asia.* Regensburg: Verlag Schnell & Steiner, 2014.

Juneja, Monica and Franz Mauelshagen. "Disasters and Pre-Industrial Societies: Historiographic Trends and Comparative Perspectives." *The Medieval History Journal* 10, no. 1/2 (2007): 1–31.

Karskens, Grace. "Floods and Flood-Mindedness in Early Colonial Australia." *Environmental History* 21, no. 2 (2016): 315–42.

Kempe, Michael. "'Mind the Next Flood!' Memories of Natural Disasters in Northern Germany from the Sixteenth Century to the Present." *The Medieval History Journal* 10, no. 1/2 (2007): 327–54.

"Noah's Flood: The Genesis Story and Natural Disasters in Early Modern Times." *Environment and History* 9, no. 2 (2003): 151–71.

Kettering, Alison M. "After Life: Rembrandt's Slaughtered Ox." *Artibus et Historiae* 79 (2019): 267–286.

Khanal, Sonu, Arthur F. Lutz, Walter W. Immerzeel et al., "The Impact of Meteorological and Hydrological Memory on Compound Peak Flows in the Rhine River Basin." *Atmosphere* 10, no. 4 (2019): 1–19.

Kind, J. M. "Economically Efficient Flood Protection Standards for the Netherlands." *Journal of Flood Risk Management* 7, no. 2 (2014): 103–17.

Kist, Nicholaas Christiaan. *Neêrland's bededagen en biddagsbrieven: eene bijdrage ter opbouwing der geschiedenis van staat en kerk in Nederland. Deel 2.* Leiden: Luchtmans, 1849.

Kjærgaard, Thorkild. *The Danish Revolution, 1500–1800: An Ecohistorical Interpretation.* New York: Cambridge University Press, 2006.

Kleinhans, Maarten G., Henk J. T. Weerts, and Kim M. Cohen. "Avulsion in Action: Reconstruction and Modelling Sedimentation Pace and Upstream Flood Water Levels Following a Medieval Tidal-River Diversion Catastrophe (Biesbosch, The Netherlands, 1421–1750 AD)." *Geomorphology* 118, no. 1–2 (2010): 65–79.

Klooster, Wim. *The Dutch Moment: War, Trade, and Settlement in the Seventeenth-Century Atlantic World.* Ithaca, NY: Cornell University Press, 2016.

Knottnerus, Otto S. "Angst voor de zee: Veranderende culturele patronen langs de Nederlandse en Duitse waddenkust (1500–1800)." In *De Republiek tussen zee en vasteland: Buitenlandse invloeden op cultuur, economie en politiek in Nederland 1580–1800*, edited by Karel Davids, Marjolein 't Hart, Henk Kleijer, and Jan Luccassen, 57–81. Leuven: Garant, 1995.

"History of Human Settlement, Cultural Change and Interference with the Marine Environment." *Helgoland Marine Research* 59, no. 1 (2005): 2–8.

"Malaria around the North Sea: A Survey." In *Climatic Development and History of the North Atlantic Realm: Hanse Conference Report*, edited by Gerold Wefer, Wolfgang H. Berger, Karl-Ernst Behre, and Eynstein Jansen, 339–53. Berlin: Springer-Verlag, 2002.

"Vroegmoderne cultuurgebieden in Nederland en Noordwest-Duitsland gedachten over behoudzucht en dynamiek." *De Zeventiende Eeuw* 16 (2000): 14–28.

"Yeomen and Farmers in the Wadden Sea Coastal Marshes, c. 1500–c. 1900." In *Landholding and Land Transfer in the North Sea Area (Late Middle Ages–19th Century)*, edited by P. Hoppenbrouwers and Bas van Bavel, 149–86. Turnhout: Brepols, 2004.

Knowles, Scott G. "Learning from Disaster? The History of Technology and the Future of Disaster Research." *Technology and Culture* 55, no. 4 (2014): 780–1.

Koene, Bert. *Goede luiden en gemene onderzaten: Assendelft vanaf zijn ontstaan tot de nadagen van de Gouden Eeuw*. Hilversum: Verloren, 2010.

Kole, Heleen. *Polderen of niet? Participatie in het bestuur van de waterschappen Bunschoten en Mastenbroek vóór 1800*. Hilversum: Verloren, 2017.

Koolmees, P. A. "Epizootic Diseases in the Netherlands, 1713–2002: Veterinary Science, Agricultural Policy, and Public Response." In *Healing the Herds: Disease, Livestock Economies, and the Globalization of Veterinary Medicine*, edited by Karen Brown and Daniel Gilfoyle, 19–41. Athens: Ohio University Press, 2010.

Kooper, Johan. *Het waterstaatsverleden van de provincie Groningen*. Groningen: J. B. Wolters, 1939, 99–102.

Koopmans, Joop. "The Early 1730s Shipworm Disaster in Dutch News Media." *Dutch Crossing* 40, no. 2 (2016): 139–50.

Koppenol, Johan. "Noah's Ark Disembarked in Holland: Animals in Dutch Poetry, 1500–1700." In *Early Modern Zoology: The Construction of Animals in Science, Literature and the Visual Art*, 451–528. Leiden: Brill, 2007.

Kossmann, E. H. "The Dutch Republic in the Eighteenth Century." In *The Dutch Republic in the Eighteenth Century: Decline, Enlightenment, and Revolution*, edited by Margaret C. Jacob and Wijnand W. Mijnhardt, 19–31. Ithaca, NY: Cornell University Press, 1992.

Krüger, Fred, Greg Bankoff, Benedikt Orlowski, and Terry Cannon, eds. *Cultures and Disasters: Understanding Cultural Framings in Disaster Risk Reduction*. New York: Routledge, 2014.

Labrijn, A. "Het klimaat van Nederland gedurende de laatste twee en een halve eeuw." *Mededeelingen en Verhandelingen* 102 (1945): 11–114.

Laidlaw, Frederick B. "The History of the Prevention of Fouling." In *Marine Fouling and Its Prevention*, edited by Woods Hole Oceanographic Institution, 211–23. Annapolis, MD: United States Naval Institute, 1952.

Lamb, H. H. and Knud Frydendahl. *Historic Storms of the North Sea, British Isles and Northwest Europe*. Cambridge: Cambridge University Press, 1991.

Leemans, Inger and Gert-Jan Johannes. "Gnawing Worms and Rolling Thunder: The Unstable Harmony of Dutch Eighteenth-Century Literature." In *Discord and Consensus in the Low Countries, 1700–2000*, edited by Jane Fenoulhet, Gerdi Quist, and Ulrich Tiedau, 20–37. London: UCL Press, 2016.

Worm en donder: Geschiedenis van de Nederlandse literatuur, 1700–1800: De Republiek. Amsterdam: Bert Bakker, 2013.

Lesger, Clé. *Hoorn als stedelijk knooppunt: Stedensystemen tijdens de late middeleeuwen en vroegmoderne tijd*. Hilversum: Verloren, 1990.

The Rise of the Amsterdam Market and Information Exchange: Merchants, Commercial Expansion and Change in the Spatial economy of the Low Countries, c. 1550–1630. Surrey: Ashgate, 2006.

Lindeboom, Hans J. "Changes in Coastal Zone Ecosytems." In *Climate Development and History of the North Atlantic Realm*, edited by Gerold

Wefer, Wolfgang H. Berger, Karl-Ernst Behre, and Eystein Jansen, 447–55. Berlin: Springer, 2002.
Lindemann, Mary. "Medical Practice and Public Health." In *A Companion to Eighteenth Century Europe*, edited by Peter H. Wilson, 158–75. Oxford: Blackwell, 2008.
Lindemann, Ton. "De stormvloed van 25 December 1717: Astronomische, hydrologische en meteorologische achtergronden." Meteo Maarssen, Report MM-17.3 (2017): 1–69.
Ljungqvist, Fredrik Charpentier, Andrea Seim, and Heli Huhtamaa. "Climate and Society in European History." *Wiley Interdisciplinary Reviews: Climate Change* 12, no. 2 (2021): e691.
Lopez-Anido, Roberto, Antonis P. Michael, Barry Goodell, and Thomas C. Sandford. "Assessment of Wood Pile Deterioration due to Marine Organisms." *Journal of Waterway, Port, Coastal, and Ocean Engineering* 130, no. 2 (2004): 70–6.
Lukasiewicz, Anna. "The Emerging Imperative of Disaster Justice." In *Natural Hazards and Disaster Justice*, edited by A. Lukasiewicz and C. Baldwin, 3–23. Singapore: Palgrave Macmillan, 2020.
Luterbacher, Jürg, Daniel Dietrich, Elena Xoplaki, Martin Grosjean, and Heinz Wanner. "European Seasonal and Annual Temperature Variability, Trends, and Extremes since 1500." *Science* 303, no. 5663 (2004): 1499–503.
Luterbacher, Jürg, Ralph Rickli, Elena Xoplaki et al. "The Late Maunder Minimum (1675–1715) – A Key Period for Studying Decadal Scale Climatic Change in Europe." *Climatic Change* 49, no. 4 (2001): 441–62.
Malechek, John C. and Benton M. Smith. "Behavior of Range Cows in Response to Winter Weather." *Rangeland Ecology & Management/Journal of Range Management Archives* 29, no. 1 (1976): 9–12.
Masters, Jeff. "A Look Back at the Horrific 2020 Atlantic Hurricane Season." *Yale Climate Connections*, December 1, 2020, https://yaleclimateconnections.org/2020/12/a-look-back-at-the-horrific-2020-atlantic-hurricane-center/
Mauch, Christof. "Introduction." In *Natural Disasters, Cultural Responses: Case Studies toward a Global Environmental History*, edited by Christof Mauch and Christian Pfister, 1–16. New York: Lexington Books, 2009.
Mauelshagen, Franz. "Defining Catastrophes." In *Catastrophe and Catharsis: Perspectives on Disaster and Redemption in German Culture and Beyond*, edited by Katharina Gerstenberger and Tanja Nusser, 172–90. Rochester, NY: Camden House, 2015.
"Disaster and Political Culture in Germany since 1500." In *Natural Disaster, Cultural Responses: Case Studies toward a Global Environmental History*, edited by Christof Mauch and Christian Pfister, 41–75. New York: Lexington Books, 2009.
"Flood Disasters and Political Cultura at the German North Sea Coast: A Long-Term Historical Perspective." *Historical Social Research / Historische Sozialforschung* 32, no. 3 (2007): 133–44.
Maurya, Vijai P., Veerasamy Sejian, Kamal Kumar et al. "Walking Stress Influence on Livestock Production." In *Environmental Stress and*

Amelioration in Livestock Production, edited by Veerasamy Sejian, S. M. K. Naqvi, Thaddeus Ezeji, Jeffrey Lakritz, and Rattan Lal, 75–95. Berlin, Heidelberg: Springer Berlin Heidelberg, 2012.

McCants, Anne E. C. "Inequality Among the Poor of Eighteenth Century Amsterdam." *Explorations in Economic History* 44, no. 1 (2007): 1–21.

McQuaid, John. "What the Dutch Can Teach Us about Weathering the Next Katrina?." *Mother Jones*, August 28, 2007. www.motherjones.com/environment/2007/08/what-dutch-can-teach-us-about-weathering-next-katrina/

Meihuizen, L. S. "Sociaal-economische geschiedenis van Groningerland." In *Historie van Groningen Stad en Land*, edited by Wiebe Jannes Formsma and M. G. Buist, 293–330. Groningen, 1981.

Mellado-Cano, Javier, David Barriopedro, Ricardo García-Herrera et al. "Euro-Atlantic Atmospheric Circulation during the Late Maunder Minimum." *Journal of Climate* 31, no. 10 (2018): 3849–63.

Mijnhardt, Wijnand. "The Dutch Enlightenment: Humanism, Nationalism, and Decline." In *The Dutch Republic in the Eighteenth Century: Decline, Enlightenment, and Revolution*, edited by Margaret Jacob and Wijnand W. Mijnhardt, 197–223. Ithaca, NY: Cornell University Press, 1992.

Tot Heil van 't menschdom: Culturele genootschappen in Nederland, 1750–1815. Utrecht: Rodopi, 1987.

Miller, Genevieve. *The Adoption of Inoculation for Smallpox in England and France*. Philadelphia: University of Pennsylvania Press, 1957.

Moelker, H. P. "De Diemerdijk: de gevolgen van paalwormvraat in de 18e Eeuw." *Tijdschrift voor Waterstaatsgeschiedenis* 6 (1997): 46–51.

Moreno-Chamarro, Eduardo, Davide Zanchettin, Katja Lohmann et al. "Winter Amplification of the European Little Ice Age Cooling by the Subpolar gyre." *Scientific Reports* 7, no. 1 (2017): 1–8.

Morgan, John Emrys. "Understanding Flooding in Early Modern England." *Journal of Historical Geography* 50 (2015): 37–50.

Mouthaan, José. "The Appearance of a Strange Kind of Sea Worm at the Dutch Coast, 1731." *Dutch Crossing* 27, no. 1 (2003): 3–22.

Muis, Sanne, Martin Verlaan, Hessel C. Winsemius, Jeroen C. J. H. Aerts, and Philip J. Ward. "A Global Reanalysis of Storm Surges and Extreme Sea Levels." *Nature Communications* 7 (2016): 11969.

Munt, Annette. "The Impact of the Rampjaar on Dutch Golden Age Culture." *Dutch Crossing* 21, no. 1 (2016): 3–51.

Nair, N. B. and M. Saraswathy. "The Biology of Wood-Boring Teredinid Molluscs." *Advances in Marine Biology* 9 (1971): 335–509.

Nelson, Derek Lee and Adam Sundberg. "Shipworm Ecology in the Age of Sail." In *Maritime Animals: Ships, Species, Stories* (forthcoming).

Nesje, Atle and Svein Olaf Dahl. "The 'Little Ice Age'– Only Temperature?." *The Holocene* 13, no. 1 (2003): 139–45.

Neukom, Raphael, Joëlle Gergis, David J. Karoly, et al. "Inter-Hemispheric Temperature Variability over the Past Millennium." *Nature Climate Change* 4, no. 5 (2014): 362–7.

Newfield, Timothy P. "A Cattle Panzootic in Early Fourteenth-Century Europe." *Agricultural History Review* 57, no. 2 (2009): 155–90.

"Domesticates, Disease and Climate in Early Post-classical Europe: The Cattle Plague of c. 940 and Its Environmental Context." *Post-Classical Archaeologies* 5 (2015): 95–126.

"Early Medieval Epizootics and Landscapes of Disease: The Origins and Triggers of European Livestock Pestilences." In *Landscapes and Societies in Medieval Europe East of the Elbe: Interactions between Environmental Settings and Cultural Transformations*, edited by Sunhild Kleingärtner, Timothy P. Newfield, Sébastien Rossignol, and Donat Wehner, 73–113. Toronto: Pontifical Institute of Mediaeval Studies, 2013.

Nienhuis, Piet H. *Environmental History of the Rhine-Meuse Delta: An Ecological Story on Evolving Human-environmental Relations Coping with Climate Change and Sea-Level Rise*. Dordrecht: Springer, 2008.

Nijenhuis, Ida J. A. "Captured by the Commercial Paradigm: Physiocracy Going Dutch." In *The Economic Turn: Recasting Political Economy in Enlightenment Europe*, edited by Sophus A. Reinert and Steven L. Kaplan, 635–56. New York: Anthem Press, 2019.

"Shining Comet, Falling Meteor: Contemporary Reflections on the Dutch Republic as a Commercial Power during the Second Stadholderless Era." In *Anthonie Heinsius and the Dutch Republic 1688–1720. Politics, War, and Finance*, edited by J. A. F. de Jongste and Augustus J. Veenendaal Jr., 115–31. The Hague: Institute of Netherlands History, 2019.

Nixon, Rob. *Slow Violence and the Environmentalism of the Poor*. Cambridge: Harvard University Press, 2011.

Njeumi, F., W. Taylor, A. Diallo et al. "The Long Journey: A Brief Review of the Eradication of Rinderpest." *Revue Scientifique et Technique* 31, no. 3 (2012): 729–46.

Noordam, D. J. "Homosocial Relations in Leiden (1533–1811)." In *Among Men, Among Women: Sociological and Historical Recognition of Homosocial Arrangements*, edited by Mattias Duyves, Gert Hekma, and Paula Koelemij, 218–23. Amsterdam: University of Amsterdam, 1983.

Noordegraaf, Leo. "Of bidden helpt? Bededagen als reactie op rampen in de Republiek." In *Of bidden helpt? Tegenslag en cultuur in Europa circa 1500–2000*, edited by Marijke Gijswijt-Hofstra and Florike Egmond, 29–42. Amsterdam: Amsterdam University Press, 1997.

Noordhoff, L. J. "Thomas van Seeratt, zijn levensloop en zijn betekenis voor de provincie Groningen." *Groningse Volksalmanak* (1961): 49–70.

November, Valérie, Marion Penelas, and Pascal Viot. "When Flood Risk Transforms a Territory: The Lully Effect." *Geography* 94, no. 3 (2009): 189–97.

O'Brien, Patrick. "Mercantilism and Imperialism in the Rise and Decline of the Dutch and British Economies, 1585–1815." *De Economist* 148, no. 4 (2000): 469–501.

O'Neill, Saffron and Sophie Nicholson-Cole. "'Fear Won't Do It': Promoting Positive Engagement with Climate Change through Visual and Iconic Representations." *Science Communication* 30, no. 3 (2009): 355–79.

Offringa, Cornelis. *Van Gildestein naar Uithof: 150 jaar diergeneeskundig onderwijs in Utrecht. 1. 's Rijksveeartsenischool (1821–1918): Veeartsenijkundige Hoogeschool (1918–1925)*. Utrecht: Rijksuniversiteit te Utrecht, 1971.

Old World Drought Atlas. Accessed 23 July 2020. http://drought.memphis.edu/OWDA/
Oliver-Smith, Anthony. "Theorizing Disasters: Nature, Power, and Culture." In *Catastrophe & Culture: The Anthropology of Disaster*, edited by Susanna M. Hoffman and Anthony Oliver-Smith, 23–48. Oxford: James Currey, 2002.
"'What Is a Disaster?': Anthropological Perspectives on a Persistent Question." In *The Angry Earth: Disaster in Anthropological Perspective*, edited by Anthony Oliver-Smith and Susanna Hoffman, 32–48. London: Routledge, 1999.
Oliver-Smith, Anthony and Susanna Hoffman, eds. *The Angry Earth: Disaster in Anthropological Perspective*. London: Routledge, 1999.
Onnekink, David. *Reinterpreting the Dutch Forty Years War, 1672–1713*. London: Springer, 2017.
Onnekink, David and Gijs Rommelse. *The Dutch in the Early Modern World: A History of a Global Power*. Cambridge: Cambridge University Press, 2019.
Oostindie, Gert and Jessica Vance Roitman. "Repositioning the Dutch in the Atlantic, 1680–1800." *Itinerario* 36, no. 2 (2012): 129–60.
Op 't Hof, Willem Jan. "Lusthof des Gemoets in Comparison and Competition with De Practycke ofte oeffeninghe der godtzaligheydt: Vredestad and Reformed Piety in Seventeenth-Century Dutch Culture." In *Religious Minorities and Cultural Diversity in the Dutch Republic*, edited by August den Hollander, Mirjam van Veen, Anna Voolstra, and Alex Noord, 133–49. Leiden: Brill, 2014.
Oram, Richard. "'The Worst Disaster Suffered by the People of Scotland in Recorded History': Climate Change, Dearth and Pathogens in the Long 14th Century." *Proceedings of the Society of Antiquaries of Scotland* 144 (2014): 223–44.
Orme, Lisa C., Liam Reinhardt, Richard T. Jones, Dan J. Charman, Andrew Barkwith, and Michael A. Ellis. "Aeolian Sediment Reconstructions from the Scottish Outer Hebrides: Late Holocene Storminess and the Role of the North Atlantic Oscillation." *Quaternary Science Reviews* 132 (2016): 15–25.
Ormrod, David and Gijs Rommelse. "Introduction: Anglo-Dutch Conflict in the North Sea and Beyond." In *War, Trade and the State: Anglo-Dutch Conflict, 1652–89*, edited by David Ormrod and Gijs Rommelse, 3–33. Rochester, NY: Boydell and Brewer, 2020.
Ortega, Pablo, Flavio Lehner, Didier Swingedouw et al. "A Model-Tested North Atlantic Oscillation Reconstruction for the Past Millennium." *Nature* 523, no. 7558 (2015): 71–4.
Paalvast, Peter. "Ecological Studies in a Man-Made Estuarine Environment, the Port of Rotterdam." PhD diss., Radboud Universiteit Nijmegen, 2014.
Paalvast, Peter and Gerard van der Velde. "New Threats of an Old Enemy: The Distribution of the Shipworm *Teredo navalis* L. (Bivalvia: Teredinidae) Related to Climate Change in the Port of Rotterdam Area, the Netherlands." *Marine Pollution Bulletin* 62, no. 8 (2011): 1822–29.
Paimans, W. J. "De Veeartsenijkunde in Nederland vóór de stichting Der Veeartsenijschool de Utrecht." In *Een eeuw veeartsenijkundig onderwijs. 's Rijks-Veeeartsenijschool, Veeartsenijkundige Hoogeschool, 1821–1921*,

edited by H. M. Kroon, W. J. Paimans, and J. E. W. Ihle, 1–24. Utrecht: Senaat der Veeartsenijkundig Hoogeschool te Utrecht, 1921.

Paping, R. F. J. "De Kerstvloed van 1717 in Vliedorp en Wierhuizen." *Stad en Lande: Cultuurhistorisch Tijdschrift voor Groningen* 26, no. 4 (2017): 10–13.

"De ontwikkeling van de veehouderij in Groningen in de achttiende en negentiende eeuw; een grove schets." *Argos* 24 (2001): 175–85.

"De zijlschotregisters, dijkrollen en registers van schouwbare objecten in Groningen tot circa 1800." In *Broncommentaren 4: Bronnen betreffende de registratie van onroerend goed in middeleeuwen en Ancien Régime*, edited by G. A. M. van Synghel, 277–310. The Hague: Instituut voor Nederlandse Geschiedenis, 2001.

Parker, Geoffrey. *Global Crisis: War, Climate Change and Catastrophe in the Seventeenth Century*. New Haven, CT: Yale University Press, 2013.

Parthesius, Robert. *Dutch Ships in Tropical Waters: The Development of the Dutch East India Company (VOC) Shipping Network in Asia 1595–1660*. Amsterdam: Amsterdam University Press, 2010.

Pauling, Andreas, Jürg Luterbacher, Carlo Casty, and Heinz Wanner. "Five Hundred Years of Gridded High-Resolution Precipitation Reconstructions over Europe and the Connection to Large-Scale Circulation." *Climate Dynamics* 26, no. 4 (2006): 387–405.

Pedersen, Karl Peder. "Als Gott sein Strafendes Schwert über dem Dänischen Sahnestück Fünen Schwang. Über Verlauf und Bekämpfung der Viehseuche auf Fünen 1745–1770 unter besonderer Berücksichtigung des Bauernschreibebuchs von Peder Madsen auf Munkgaarde." In *Beten, Impfen, Sammeln. Zur Viehseuchen-und Schädlingsbekämpfung in der Frühen Neuzeit. Graduiertenkolleg Interdisziplinäre Umweltgeschichte*, edited by Katharina Engelken, Dominik Hünniger, and Steffi Windelen, 57–78. Göttingen: Universitätsverlag Göttingen, 2007.

Peeters, Thérèse. "'Sweet Milk-Cows' in Huizen and 'Memorable Incidents' in Oost Zaandam: Identity and Responsibility in Two Eighteenth-Century Rural Chronicles." *Volkskunde* 2 (2014): 163–79.

Pfister, Christian. "Climatic Extremes, Recurrent Crises and Witch Hunts: Strategies of European Societies in Coping with Exogenous Shocks in the Late Sixteenth and Early Seventeenth Centuries." *The Medieval History Journal* 10, no. 1–2 (2006): 33–73.

"Learning from Nature-Induced Disasters: Theoretical Considerations and Case Studies from Western Europe." In *Natural Disasters, Cultural Responses: Case Studies toward a Global Environmental History*, edited by Christof Mauch and Christian Pfister, 17–40. New York: Lexington Books, 2009.

"Little Ice Age-Type Impacts and the Mitigation of Social Vulnerability to Climate in the Swiss Canton of Bern prior to 1800." In *Sustainability of Collapse? An Integrated History and Future of People on Earth*, edited by Robert Costanza, Lisa J. Graumilch, and Will Steffen, 197–212. Cambridge: MIT Press, 2007.

"The Monster Swallows You: Disaster Memory and Risk Culture in Western Europe, 1500–2000." *Rachel Carson Center Perspectives* 1 (2011): 1–23.

"Weeping in the Snow: The Second Period of Little Ice Age-Type Impacts, 1570–1630." In *Kulturelle Konsequenzen der Kleinen Eiszeit*, edited by Wolfgang Behringer, Hartmut Lehmann, and Christian Pfister, 31–85. Göttingen: Vandenhoeck & Ruprecht, 2005.

Pfister, Christian and Rudolf Brázdil. "Social Vulnerability to Climate in the 'Little Ice Age': An Example from Central Europe in the Early 1770s." *Climate of the Past* 2, no. 2 (2006): 115–29.

Phillips, Carly A., Astrid Caldas, Rachel Cleetus et al. "Compound Climate Risks in the COVID-19 Pandemic." *Nature Climate Change* 10 (2020): 586–88.

Pieters, Harm. "Herinneringscultuur van overstromingsrampen, gedenkboeken van overstromingen van 1775, 1776 en 1825 in het Zuiderzeegebied." *Tijdschrift voor Waterstaatsgeschiedenis* 21 (2012): 48–57.

Pinto, Joaquim G. and Christoph C. Raible. "Past and Recent Changes in the North Atlantic Oscillation." *Wiley Interdisciplinary Reviews: Climate Change* 3, no. 1 (2012): 79–90.

Porter, Roy. "Lay Medical Knowledge in the Eighteenth Century: The Evidence of the Gentleman's Magazine." *Medical History* 29, no. 2 (1985): 138–68.

"The Eighteenth Century." In *The Western Medical Tradition: 800 BC–AD 1800*, Vol.1, edited by Lawrence I. Conrad, Michael Neve, Vivian Nutton, Roy Porter, and Andrew Wear, 371–472. Cambridge: Cambridge University Press, 1995.

Post, John D. "Climatic Variability and the European Mortality Wave of the Early 1740s." *Journal of Interdisciplinary History* 15, no. 1 (1984): 1–30.

Food Shortage, Climatic Variability, and Epidemic Disease in Preindustrial Europe: The Mortality Peak in the Early 1740s. Ithaca, NY: Cornell University Press, 1985.

Poulsen, Bo. *Dutch Herring: An Environmental History, c.1600–1860*. Amsterdam: Aksant, 2008.

Prak, Maarten. *The Dutch Republic in the Seventeenth Century: The Golden Age*. Cambridge: Cambridge University Press, 2005.

Prak, Maarten and J. L. van Zanden. *Nederland en het poldermodel: Sociaal-economische geschiedenis van Nederland, 1000–2000*. Amsterdam: Bert Bakker, 2013.

Price, J. L. *Dutch Culture in the Golden Age*. London: Reaktion Books, 2012.

"The Dutch Republic." In *A Companion to Eighteenth-Century Europe*, edited by Peter H. Wilson, 289–303. Oxford: Wiley, 2009.

Priester, Peter R. *Geschiedenis van de Zeeuwse landbouw circa 1600–1910*. Utrecht: HES, 1998.

Pyšek, Petr, Vojtěch Jarošík, Philip E. Hulme et al. "Disentangling the Role of Environmental and Human Pressures on Biological Invasions across Europe." *Proceedings of the National Academy of Sciences* 107, no. 27 (2010): 12157–62.

Raible, C. C., Masakazu Yoshimori, T. F. Stocker, and Carlo Casty. "Extreme Midlatitude Cyclones and Their Implications for Precipitation and Wind

Speed Extremes in Simulations of the Maunder Minimum versus Present Day Conditions." *Climate Dynamics* 28, no. 4 (2007): 409–23.
Rayes, Courtney A., James Beattie, and Ian C. Duggan. "Boring through History: An Environmental History of the Extent, Impact and Management of Marine Woodborers in a Global and Local Context, 500 BCE to 1930s CE." *Environment and History* 21, no. 4 (2015): 477–512.
Reidmiller, D. R., C. W. Avery, D. R. Easterling, et al., *Impacts, Risks, and Adaptation in the United States: Fourth National Climate Assessment*, Vol. II. Washington, DC: U.S. Global Change Research Program, 2018.
Reinders, Michel. "Burghers, Orangists and 'Good Government': Popular Political Opposition during the 'Year of Disaster' 1672 in Dutch Pamphlets." *Seventeenth Century* 23, no. 2 (2008): 315–46.
  Gedrukte chaos: populisme en moord in het Rampjaar 1672. Amsterdam: Balans, 2010.
Reinert, Erik S. "Emulating Success: Contemporary Views of the Dutch Economy before 1800." In *The Political Economy of the Dutch Republic*, edited by Oscar Gelderblom, 35–7. Surrey: Ashgate, 2013.
Reinert, Sophus and Steven Kaplan, eds. *The Economic Turn: Recasting Political Economy in Enlightenment Europe*. London: Anthem Press, 2019.
Reuss, Martin. "Learning from the Dutch: Technology, Management, and Water Resources Development." *Technology and Culture* 43, no. 3 (2002): 465–72.
Reynolds, David J., J. D. Scourse, P. R. Halloran et al., "Annually Resolved North Atlantic Marine Climate over the Last Millennium." *Nature Communications* 7, no. 1 (2016): 1–11.
Rieken, Bernd. "Learning from Disasters in an Unsafe World: Considerations from a Psychoanalytical Ethnological Perspective." In *Learning and Calamities: Practices, Interpretations, Patterns*, edited by Heike Egner, Marén Schorch, and Martin Voss, 27–41. New York: Routledge, 2014.
  'Nordsee ist Mordsee': Sturmfluten und ihre Bedeutung für die Mentalitätsgeschichte der Friesen. Münster: Waxmann Verlag, 2005.
Rijkswaterstaat. "Ruimte voor de rivieren," Room for the River Programme. Accessed December 31, 2020. www.rijkswaterstaat.nl/water/waterbeheer/bescherming-tegen-het-water/maatregelen-om-overstromingen-te-voorkomen/ruimte-voor-de-rivieren/index.aspx
Riley, J. C. "The Dutch Economy after 1650: Decline or Growth?" *Journal of European Economic History* 13 (1984): 521–69.
  The Eighteenth-Century Campaign to Avoid Disease. London: Macmillan, 1987.
Ritzema, H. P. and J. M. van Loon-Steensma. "Coping with Climate Change in a Densely Populated Delta: A Paradigm Shift in Flood and Water Management in the Netherlands." *Irrigation and Drainage* 67 (2018): 52–65.
Rivera, Jason D. "Resistance to Change: Understanding Why Disaster Response and Recovery Institutions are Set in Their Ways." *Journal of Critical Incident Analysis* 4, no. 1 (2014): 44–65.
Roberts, Benjamin B. and Leendert F. Groenendijk. "Moral Panic and Holland's Libertine Youth of the 1650s and 1660s." *Journal of Family History* 30, no. 4 (2005): 327–46.

Roessingh, H. K. "Landbouw in de Noordelijke Nederlanden, 1650–1815." *Algemene geschiedenis der Nederlanden* 8 (1979): 16–72, 451–3.
Rohr, Christian. "Ice Jams and Their Impact on Urban Communities from a Long-term Perspective (Middle Ages to 19th Century)." In *The Power of Urban Water: Studies in Premodern Urbanism*, edited by Nicola Chiarenza, Annette Haug, and Ulrich Müller, 197–212. Berlin: Walter de Gruyter, 2020.
"Writing a Catastrophe. Describing and Constructing Disaster Perception in Narrative Sources from the Late Middle Ages." *Historical Social Research/ Historische Sozialforschung* 32, no. 3 (2007): 88–102.
Rommelse, Gijs. "Prizes and Profits: Dutch Maritime Trade during the Second Anglo-Dutch War." *International Journal of Maritime History* 19 (2007): 139–60.
Rommes, Ronald. "'Geen vrolyk geloei der melkzwaare koeijen': Runderpest in Utrecht inde achttien de eeuw." *Jaarboek Oud Utrecht* (2001): 87–136.
"Twee eeuwen runderpest in Nederland (1700–1900)." *Argos* 31 (2004): 33–40.
Rossiter, P. B. and A. D. James. "An Epidemiological Model of Rinderpest. II. Simulations of the Behavior of Rinderpest Virus in Populations." *Tropical Animal Health and Production* 21 (1989): 69–84.
Rowen, H. H. *Low Countries in Early Modern Times*. London: Palgrave Macmillan, 1972.
Royal Netherlands Meteorological Institute (KNMI). "Monthly, Seasonal, and Annual Means of the Air Temperature in Tenths of Centigrade in De Bilt, Netherlands, 1706." De Bilt, KNMI, 2007. http://projects.knmi.nl/klimato-logie/daggegevens/antieke_wrn/index.html
Scarborough, John. *Medical and Biological Terminologies: Classical Origins*. Norman: University of Oklahoma Press, 1992.
Schama, Simon. *The Embarrassment of Riches: An Interpretation of Dutch Culture in the Golden Age*. New York: Knopf, 1987.
Schenk, Gerrit Jasper. *Historical Disaster Experiences: Towards a Comparative and Transcultural History of Disasters Across Asia and Europe*. Heidelberg: Springer International Publishing, 2017.
"Historical Disaster Research. State of Research, Concepts, Methods and Case Studies." *Historical Social Research/Historische Sozialforschung* 32, no. 3 (2007): 9–31.
Schilling, H. *Civic Calvinism in Northwestern Germany and the Netherlands: Sixteenth to Nineteenth Centuries*. Ann Arbor: University of Michigan Press, 1991.
Schilstra, Johannes Jouke. *In de ban van de Dijk: De Westfriese omringdijk*. Hoorn: West-Friesland, 1982.
Wie water deert: Het Hoogheemraadschap van de Uitwaterende Sluizen in Kennemerland en West-Friesland, 1544–1969. Wormerveer: Meijer, 1969.
Schlimme, J. F. C. ed. *Het dagboek van Jacob Pos*. Hilversum, 1992.
Schöffer, Ivo. "Did Holland's Golden Age Coincide with a Period of Crisis?" In *The General Crisis of the Seventeenth Century*, edited by Parker Geoffrey and Lesley M. Smith, 83–109. London: Routledge, 1985.

Schoorl, H. *Zeshonderd jaar water en land: bijdrage tot de historische geo- en hydrografie van de Kop van Noord-Holland in de periode ± 1150–1750*. Groningen: Wolters-Noordhoff, 1973.

Schrickx, C. P. "Een kommerlijkste toestand en groot gevaar: Archeologie en historie van de Westfriese Omringdijk tussen Hoorn en Schellinkhout." *West-Friese Archeologische Rapporten* 16 (2010): 1–73.

"Tot behoud van stad en vaderland. Hout en houtgebruik in de Drechterlandse dijk." In *Dwars door de dijk: Archeologie en geschiedenis van de Westfriese Omringdijk tussen Hoorn en Enkhuizen*, edited by Michiel Bartels, 85–119. Hoorn: Stichting Archeologie West Friesland, 2016.

Shorto, Russell. "How to Think Like the Dutch in a Post-Sandy World." *The New York Times*, April 9, 2014. www.nytimes.com/2014/04/13/magazine/how-to-think-like-the-dutch-in-a-post-sandy-world.html

Siebert, Ernst. "Entwicklung des Deichwesens von Mittelalter bis zur Gegenwart." In *Ostfriesland im Schutze des Deiches: Beiträge zur Kultur-und Wirtschaftsgeschichte des Ostfriesischen Küstenlandes*, Vol. IV, edited by Jannes Ohling, 79–246. Leer: Gerhard Rautenberg, 1969.

Siemens, B. W. *Dijkrechten en zijlvesten*. Groningen: Tjeenk Willink, 1974.

Sigl, Michael, Mai Winstrup, Joseph R. McConnell et al. "Timing and Climate Forcing of Volcanic Eruptions for the Past 2,500 Years." *Nature* 523, no. 7562 (2015): 543–49.

Slavin, P. "The Great Bovine Pestilence and Its Economic and Environmental Consequences in England and Wales, 1318–50." *The Economic History Review* 65 (2012): 1239–66.

Slicher van Bath, B. H. "Agriculture in the Vital Revolution." In *The Cambridge Economic History of Europe*, edited by E. Rich and C. Wilson, 42–132. Cambridge: Cambridge University Press, 1977.

"Die europaischen Agrarverhältnisse im 17. und der ersten Hälfte des 18. Jahrhunderts." *A.A.G. Bijdragen* 13 (1965): 134–48.

Een samenleving onder spanning: geschiedenis van het platteland in Overijssel. Utrecht: HES Publishers, 1957.

Soens, Tim. *De spade in de dijk?: waterbeheer en rurale samenleving in de Vlaamse kustvlakte (1280–1580)*. Ghent: Academia Press, 2009.

"Flood Security in the Medieval and Early Modern North Sea Area: A Question of Entitlement?" *Environment and History* 19, no. 2 (2013): 209–32.

"Resilience in Historical Disaster Studies: Pitfalls and Opportunities." In *Strategies, Dispositions and Resources of Social Resilience: A Dialogue Between Medieval Studies and Sociology*, edited by Martin Endreß, Lukas Clemens, and Benjamin Rampp, 253–74. Wiesbaden: Springer VS, 2020.

"Resilient Societies, Vulnerable People: Coping with North Sea Floods before 1800." *Past & Present* 241, no. 1 (2018): 143–77.

"Waddenzee wordt moordzee: De Kerstvloed van 1717 en de kwetsbaarheid voor stormvloeden." *Tijdschrift voor Geschiedenis* 131, no. 4 (2018): 605–30.

Sørensen, Birgine Refslund and Kristoffer Albris. "The Social Life of Disasters: An Anthropological Approach." In *Disaster Research: Multidisciplinary and*

International Perspectives, edited by Rasmus Dahlberg, Olivier Rubin, and Morten Thanning Vendelø, 66–81. London: Routledge, 2016.
Speeleveldt, Theodorus. *Brieven over het eiland Walcheren*. The Hague: Immerzeel en comp., 1808.
Spinage, Clive A. *Cattle Plague: A History*. New York: Kluwer Academic/Plenum Publishers, 2003.
Spinks J. and C. Zika. *Disaster, Death and the Emotions in the Shadow of the Apocalypse, 1400–1700*. London: Palgrave Macmillan UK, 2016.
Spufford, Margaret. "Literacy, Trade and Religion in the Commercial Centres of Europe." In *A Miracle Mirrored: The Dutch Republic in European Perspective*, edited by Karel Davids and Jan Lucassen, 229–84. Cambridge: Cambridge University Press, 1995.
Stapelbroek, Koen. "Dutch Decline as a European Phenomenon." *History of European Ideas* 36, no. 2 (2010): 139–52.
Stapelbroek, Koen and Jani Marjanen, eds. *The Rise of Economic Societies in the Eighteenth Century: Patriotic Reform in Europe and North America*. New York: Palgrave Macmillan, 2012.
Steinberg, Ted. *Acts of God: The Unnatural History of Natural Disaster in America*. Oxford: Oxford University Press, 2000.
Stenvert, Ronald, Chris Kolman, Sabine Broekhoven, Saskia van Ginkel-Meester, and Yme Kuiper. *Monumenten in Nederland. Fryslân. Rijksdienst voor de Monumentenzorg*. Zwolle: Zeist/Waanders Uitgevers, 2000.
Stone, Deborah A. "Causal Stories and the Formation of Policy Agendas." *Political Science Quarterly* 104, no. 2 (1989): 281–300.
Stühring, Carsten. "Managing Epizootic Diseases in 18th Century Bavaria." In *Economic and Biological Interactions in Pre-Industrial Europe from the 13th to the 18th Century, 1000–1008*, edited by Simonetta Cavaciocchi, 473–80. Florence: Firenze University Press, 2010.
Sundberg, Adam. "Claiming the Past: History, Memory, and Innovation Following the Christmas Flood of 1717." *Environmental History* 20, no. 2 (2015): 238–61.
———. "Dikes, Ships, and Worms: Testing the Limits of Envirotechnical Transfer during the Dutch Shipworm Epidemic of the 1730s." In *Disaster in the Early Modern World: Examinations, Representations, Interventions* (forthcoming).
———. "Gemeenschappelijke verantwoordelijkheid en weerstand: De Kerstvloed van 1717 in Groningen." *Historisch Jaarboek Groningen* (2018): 32–49.
———. "Molluscan Explosion: The Dutch Shipworm Epidemic of the 1730s." Environment & Society Portal, *Arcadia* 14 (2015). Rachel Carson Center for Environment and Society. https://doi.org/10.5282/rcc/7307
———. "An Uncommon Threat: Shipworms as a Novel Disaster." *Dutch Crossing* 40, no. 2 (2016): 122–38.
Sutter, Paul S. "The World with Us: The State of American Environmental History." *Journal of American History* 100, no. 1 (2013): 94–119.
Sutton, Peter. "The Noblest of Livestock." *J. Paul Getty Museum Journal* 15 (1987): 97–110.

Swabe, Joana. *Animals, Disease and Human Society: Human–Animal Relations and the Rise of Veterinary Medicine*. London: Routledge, 2002.
Swart, Koenraad Walter. *The Miracle of the Dutch Republic as Seen in the Seventeenth Century: An Inaugural Lecture Delivered at Univ. Coll. London, 6 Nov. 1967*. London: H. K. Lewis, 1969.
't Hart, Marjolein. "Town and Country in the Netherlands, 1550–1750." In *Town and Country in Europe, 1300–1800*, edited by S. R. Epstein, 80–105. Cambridge: Cambridge University Press, 2001.
TeBrake, William H. *Medieval Frontier: Culture and Ecology in Rijnland*. College Station: Texas A&M University Press, 1985.
——— "Taming the Waterwolf: Hydraulic Engineering and Water Management in the Netherlands during the Middle Ages." *Technology and Culture* 43, no. 3 (2002): 475–99.
Ten Brinke, Wildred. *The Dutch Rhine: A Restrained River*. Diemen: Veen Magazines, 2005.
Thiers, Ottie. *'T Putje van Heiloo: Bedevaarten naar O.L. Vrouw ter Nood*. Hilversum: Verloren, 2005.
Thoen, Erik. "Clio Defeating Neptune: A Pyrrhic Victory? Men and Their Influence on the Evolution of Coastal Landscapes in the North Sea Area." In *Landscapes or Seascapes? The History of the Coastal Environment in the North Sea Area Reconsidered*, edited by Erik Thoen, Guus J. Borger, Adriaan M. J. de Kraker, Tim Soens, Dries Tys, Lies Vervaet, and Henk J. T. Weerts, 397–428. Turnhout: Brepols, 2013.
Tiegs, Robert. "Hidden Beneath the Waves: Commemorating and Forgetting the Military Inundations during the Siege of Leiden." *Canadian Journal of Netherlandic Studies* 35 (2014): 1–27.
——— "Wrestling with Neptune: The Political Consequences of the Military Inundations during the Dutch Revolt." PhD diss., Louisiana State University, 2016.
Tjalsma, Erik Jan and Ronald Rommes. "Dominees en de runderpest in de tweede helft van de 18e eeuw." *Argos* 59 (2018): 356–63.
Toebast, Judith. "Voor als de dijken doorgingen: maatregelen tegen rivieroverstromingen bij boerderijen, zeventiende-negentiende eeuw." *Tijdschrift voor Waterstaatsgeschiedenis* 21 (2012): 11–22.
Tol, Richard and Andreas Langen, "A Concise History of Dutch River Floods." *Climatic Change* 46, no. 3 (2000): 357–69.
Tolle, C. "Tweedracht maakt zacht. Conflicten binnen een Fries waterschap (1533–1573)." *Leidschrift* 28, no. 2 (2010): 41–57.
Toonen, Willem H. J. "Flood Frequency Analysis and Discussion of Non-Stationarity of the Lower Rhine Flooding Regime (AD 1350–2011): Using Discharge Data, Water Level Measurements, and Historical Records." *Journal of Hydrology* 528 (2015): 490–502.
Toonen, Willem H. J., Hans Middelkoop, Tiuri Y. M. Konijndijk, Mark G. Macklin, and Kim M. Cohen. "The Influence of Hydroclimatic Variability on Flood Frequency in the Lower Rhine." *Earth Surface Processes and Landforms* 41 (2016): 1266–75.

Trouet, Valérie, J. D. Scourse, and C. C. Raible. "North Atlantic Storminess and Atlantic Meridional Overturning Circulation during the Last Millennium: Reconciling Contradictory Proxy Records of NAO Variability." *Global and Planetary Change* 84 (2012): 48–55.

Trouet, Valérie, Jan Esper, Nicholas E. Graham, J. D. Scourse, and D. C. Frank. "Persistent Positive North Atlantic Oscillation Mode Dominated the Medieval Climate Anomaly." *Science* 324, no. 5923 (2009): 78–80.

Ufkes, Tonko. "De Kerstvloed van 1717: Oorzaken en gevolgen van een natuurramp." PhD diss., Rijksuniversiteit Groningen, 1984.

"De opstand in Aduard in 1718." *Groningse Volksalmanak* (1985): 91–101.

Unger, Richard W. *A History of Brewing in Holland, 900–1900: Economy, Technology, and the State*. Leiden: Brill, 2001.

"Dutch Herring, Technology, and International Trade in the Seventeenth Century." *The Journal of Economic History* 40, no. 2 (1980): 253–80.

Van Adrichem, D. "Een nieuwe devotie door de Vee-Ziekte van 1744." *Archief voor de geschiedenis van het Aartsbisdom Utrecht* 42 (1916): 373–83.

Van Alphen, Sander. "Room for the River: Innovation, or Tradition? The Case of the Noordwaard." In *Adaptive Strategies for Water Heritage*, edited by Carola Hein, 308–23. Cham: Springer, 2020.

Van Asperen, Hanneke. "Charity after the Flood: The Rijksmuseum's St Elizabeth and St Elizabeth's Flood Altar Wings." *The Rijksmuseum Bulletin* 67, no. 1 (2019): 30–53.

"Disaster and Discord: Romeyn de Hooghe and the Dutch State of Ruination in 1675." *Dutch Crossing* (2020): 1–22.

Van Bavel, Bas. "The Economic Origins of Cleanliness in the Dutch Golden Age." *Past & Present* 205, no. 1 (2009): 41–69.

Manors and Markets: Economy and Society in the Low Countries, 500–1600. Oxford: Oxford University Press, 2010.

Van Bavel, Bas, Daniel R. Curtis, Jessica Dijkman et al. *Disasters and History: The Vulnerability and Resilience of Past Societies*. Cambridge: Cambridge University Press, 2020.

Van Bavel, Bas, Daniel R. Curtis, and Tim Soens. "Economic Inequality and Institutional Adaptation in Response to Flood Hazards: A Historical Analysis." *Ecology and Society* 23, no. 4 (2018): 30.

Van Bavel, Bas and Jan Luiten van Zanden. "The Jump-Start of the Holland Economy during the Late-Medieval Crisis, c.1350–c.1500." *The Economic History Review* 57, no. 3 (2004): 503–32.

Van Bemmel, A. A. B. *De Lekdijk van Amerongen naar Vreeswijk: Negen eeuwen bescherming van Utrecht en Holland*. Hilversum: Verloren, 2009.

Van Berkel, Klaas. "Science in the Service of Enlightenment, 1700–1795." In *A History of Science in the Netherlands: Survey, Themes and Reference*, edited by K. van Berkel, A. Van Helden, and L. C. Palm, 68–94. Leiden: Brill, 1999.

Van Bochove, Christiaan. *The Economic Consequences of the Dutch: Economic Integration around the North Sea, 1500–1800*. Amsterdam: Aksant, 2008.

"The 'Golden Mountain': An Economic Analysis of Holland's Early Modern Herring Fisheries." In *Beyond the Catch: Fisheries of the North Atlantic, the*

North Sea and the Baltic, 900–1850, edited by Louis Sicking and Darlene Abreu-Ferreira, 209–44. Leiden: Brill, 2009.

Van Brakel, Albert. "De paalworm in Hollandse zeedijken." *Tijdschrift voor Waterstaatsgeschiedenis* 24, no. 2 (2015): 70–81.

Van Bunge, Wiep, ed. *The Early Enlightenment in the Dutch Republic, 1650–1750*. Leiden: Brill, 2003.

"Introduction: The Early Enlightenment in the Dutch Republic, 1650–1750." In *The Early Enlightenment in the Dutch Republic, 1650–1750*, edited by Wiep van Bunge, 1–18. Leiden: Brill, 2003.

Van Creveld, Martin. *Supplying War: Logistics from Wallenstein to Patton*. Cambridge: Cambridge University Press, 2004.

Van Cruyningen, Piet. "Behoudend maar buigzaam: boeren in West-Zeeuws-Vlaanderen, 1650–1850." PhD diss., Wageningen University, 2000.

"Dealing with Drainage: State Regulation of Drainage Projects in the Dutch Republic, France, and England during the Sixteenth and Seventeenth Centuries." *The Economic History Review* 68, no. 2 (2015): 420–40.

"Sharing the Cost of Dike maintenance in the South-Western Netherlands: Comparing 'Calamitous Polders' in Three 'States', 1715–1795." *Environment and History* 23, no. 3 (2017): 363–83.

"State, Property Rights and Sustainability of Drained Lands along the North Sea Coast, Sixteenth-Eighteenth Centuries." In *Rural Societies and Environments at Risk Ecology, Property Rights and Social Organisation in Fragile Areas*, edited by Bas van Bavel and Erik Thoen, 181–207. Turnhout: Brepols, 2013.

Van Dam, Petra J. E. M. "An Amphibious Culture: Coping with Floods in the Netherlands." In *Local Places, Global Processes: Histories of Environmental Change in Britain and Beyond*, edited by David Moon, Paul Warde, and Peter Coates, 78–93. Oxford: Windgather Press, 2016.

"Ecological Challenges, Technological Innovations: The Modernization of Sluice Building in Holland, 1300–1600." *Technology and Culture* 43, no. 3 (2002): 500–20.

"Sinking Peat Bogs: Environmental Change in Holland, 1350–1550." *Environmental History* 6, no. 1 (2001): 32–45.

Van Amsterdams Peil naar Europees referentievlak: De geschiedenis van het NAP tot 2018. Hilversum: Verloren, 2018.

Van Dam, Petra J. E. M. and Harm Pieters. "Enlightened ideas in Commemoration Books of the 1825 Zuiderzee Flood in the Netherlands." In *Navigating History: Economy, Society, Science and Nature. Essays in Honor of Prof. Dr. C.A. Davids*, edited by Pepijn Brandon, Sabine Go, and Wybren Verstegen, 275–97. Leiden: Brill, 2018.

Van de Ven, G. P. *Aan de wieg van Rijkswaterstaat: Wordingsgeschiedenis van het Pannerdens Kanaal*. Zutphen: Walburg Pers, 1976.

Man-made Lowlands: History of Water Management and Land Reclamation in the Netherlands. Utrecht: Uitgeverij Matrijs, 2004.

Van de Ven, G. P., A. M. A. J. Driessen, W. Wolters, H. J. Wasser, and T. Stol. *Niets is bestendig-: De geschiedenis van de rivieroverstromingen in Nederland*. Utrecht: Matrijs, 1995.

Van Deinse, Jacobus Joannes, ed. "Uit het dagboek van Aleida Leurink te Losser, 1698–1754." In *Uit het land van katoen en heide: Oudheidkundige en folkloristische schetsen uit Twente*, 530–72. Enschede: van der Loeff, 1975.

Van den Brink, Paul. *"In een opslag van het oog": De Hollandse rivierkartografie en waterstaatszorg in opkomst, 1725–1754*. Alphen aan den Rijn: Canaletto/Repro-Holland, 1998.

——— "Rijnland en de rivieren: Inrichting en vormgeving van de Hollandse rivierzorg in de achttiende eeuw." *Tijdschrift voor Waterstaatsgeschiedenis* 12 (2003): 69–78.

——— "River Landscapes: The Origin and Development of the Printed River Map in the Netherlands, 1725–1795." *Imago Mundi* 52 (2000): 66–78.

Van den Dool, H. M., H. J. Krijnen, and C. J. E. Schuurmans. "Average Winter Temperatures at De Bilt (The Netherlands): 1634–1977." *Climatic Change* 1, no. 4 (1978): 319–30.

Van den Heuvel, Francien. "'s-Hertogenbosch, een onneembare stad midden in een meer: Hulpverlening en preventie tijdens watersnoden, 1740–1795." *Tijdschrift voor Waterstaatsgeschiedenis* 29, no. 1 (2020): 3–9.

Van der Bijl, M. "De Franse politieke agent Helvetius over de situatie in de Nederlandse Republiek in het jaar 1706." *Bijdragen en Mededelingen van het Historisch Genootschap* 80 (1966): 152–94.

Van der Hoek, J. J. Spahr. *Geschiedenis van de Friese landbouw*. Drachten: Friesche Maatschappij van Landbouw, 1952.

Van der Meer, Theo. "The Persecutions of Sodomites in Eighteenth-Century Amsterdam." *Journal of Homosexuality* 16, no. 1–2 (1988): 263–307.

——— "Sodom's Seed in the Netherlands." *Journal of Homosexuality* 16, no. 1–2 (1988): 1–16.

——— "Sodomy and Its Discontents: Discourse, Desire, and the Rise of a Same-Sex Proto-Something in the Early Modern Dutch Republic." *Historical Reflections/Réflexions Historiques* 33, no. 1 (2007): 41–67.

Van der Poel, J. M. G. "Het Noord Hollandse weidebedrijf om de 19e eeuw." *Holland: Regionaal-historische Tijdschrift* 3 (1986): 148–57.

Van der Schans, Ton. "Als een straf van God, weekdiertje bedreigde onze dijken in de achttiende eeuw." *Kleio* 55, no. 4 (2014): 4–8.

Van der Schrier, G. and R. Groenland. "A Reconstruction of 1 August 1674 Thunderstorms over the Low Countries." *Natural Hazards Earth Systems Science* 17, no. 2 (2017): 157–70.

Van der Steur, A. G. "Een Warmondse boerenboekhouding uit de tijd van de veepest 1742–1749." *Leids Jaarboekje* 62 (1970): 161–87.

Van der Wall, Ernestine. "The Religious Context of the Early Dutch Enlightenment: Moral Religion and Society." In *The Early Enlightenment in the Dutch Republic, 1650–1750*, edited by Wiep van Bunge, 39–57. Leiden: Brill, 2003.

Van der Woude, A. M. *Het Noorderkwartier: een regionaal historisch onderzoek in de demografische en economische geschiedenis van westelijk Nederland*

van de late middeleeuwen tot het begin van de negentiende eeuw. Wageningen: H. Veenman en Zonen, 1972.
Van Deursen, A. T. "The Dutch Republic, 1588–1780." In *History of the Low Countries*, edited by J. C. H. Blom and Emiel Lamberts, 143–220. New York: Berghahn Books, 2006.
Van Dillen, J. G. "Omstandigheden en psychische factoren in de economische geschiedenis van Nederland." In *Mensen en achtergronden*, edited by J. G. Van Dillen, 53–79. Groningen: J. B. Wolters, 1964.
Van Duin, W. E., H. Jongerius, A. Nicolai, J. J. Jongsma, A. Hendriks, and C. Sonneveld. "Friese en Groninger kwelderwerken: Monitoring en beheer 1960–2014." *Wettelijke onderzoekstaken Natuur & Milieu* 68 (2016): 1–98.
Van Duivenvoorde, Wendy. *Dutch East India Company Shipbuilding: The Archaeological Study of Batavia and Other Seventeenth-Century VOC Ships*. College Station: Texas A&M University Press, 2015.
Van Eijnatten, Joris. "Getting the Message: Towards a Cultural History of the Sermon." In *Preaching, Sermon and Cultural Change in the Long Eighteenth Century*, edited by Jorisvan Eijnatten, 343–88. Leiden: Brill, 2009.
 God, Nederland en Oranje: Dutch Calvinism and the Search for the Social Centre. Kampen: Kok, 1993.
Van Engelen, A. F. V., J. Buisman, and F. IJnsen. "A Millennium of Weather, Winds, and Water in the Low Countries." In *History and Climate: Memories of the Future?*, edited by P. D. Jones, A. E. J. Ogilvie, T. D. Davies, and K. R. Briffa, 101–24. Boston: Springer, 2001.
Van Essen, Gea. *Bouwheer en bouwmeester: Bouwkunst in Groningen, Stad en Lande (1594–1795)*. Assen: Koninklijke Van Gorcum, 2010.
Van Iterson, P. D. J. "Havens." In *Maritieme geschiedenis der Nederlanden*, Vol. 3, edited by G. Asaert, J. van Beylen, and H. P. H. Jansen, 59–91. Brussum: De Boer Maritiem, 1977.
Van Lieburg, Fred A. *Living for God: Eighteenth-Century Dutch Pietist Autobiography*. Trans. Annemie Godbehere. Lanham, MD: Scarecrow Press, 2006.
Van Lottum, Jelle. *Across the North Sea: The Impact of the Dutch Republic on International Labour Migration, c.1550–1850*. Amsterdam: Aksant Academic Publishers, 2007.
Van Minnen, J., W. Ligtvoet, L. van Bree, et al. *The Effects of Climate Change in the Netherlands: 2012*. The Hague: Netherlands Environmental Assessment Agency, 2013.
Van Nierop, Henk. "Romeyn de Hooghe and the Imagination of Dutch Foreign Policy." In *Ideology and Foreign Policy in Early Modern Europe (1650–1750)*, edited by David Onnekink and Gijs Rommelse, 197–214. Surrey: Ashgate, 2011.
Van Rooden, Peter. "Dissenters en bededagen. Civil Religion ten tijde van De Republiek." *BMGN: Low Countries Historical Review* 107, no. 4 (1992): 703–12.
Van Roosbroeck, Filip. "To Cure Is to Kill?: State Intervention, Cattle Plague and Veterinary Knowledge in the Austrian Netherlands, 1769–1785." PhD diss., University of Antwerp, 2016.

Van Roosbroeck, Filip and Adam Sundberg. "Culling the Herds? Regional Divergences in Rinderpest Mortality in Flanders and South Holland, 1769–1785." *TSEG/ Low Countries Journal of Social and Economic History* 14, no. 3 (2018): 31–55.
Van Rossum, Matthias. "'Vervloekte goudzugt': De VOC, slavenhandel en slavernij in Azië." *TSEG/Low Countries Journal of Social and Economic History* 12, no. 4 (2015): 29–57.
Van Sas, N. C. F. *De metamorfose van Nederland.* Amsterdam: Amsterdam University Press, 2005.
"De vaderlandse imperatief. Begripsverandering en politieke conjunctuur, 1763–1813." In *Vaderland: Een gescheidenis van de vijftiende eeuw tot 1940,* edited by N. C. F. van Sas, 275–308. Amsterdam: Amsterdam University Press, 1999.
"The Netherlands: A Historical Phenomenon." In *Dutch Culture in a European Perspective: Accounting for the Past, 1650–2000,* edited by Douwe Fokkema and Frans Grijzenhout, 41–66. New York: Palgrave Macmillan, 2004.
Van Schaik, K. *Overlangbroek op de kaart gezet: Dorp, landschap en bewoners, waaronder een familie De Cruijff.* Hilversum: Uitgeverij Verloren BV, 2008.
Van Strien, C. D. *British Travellers in Holland During the Stuart Period: Edward Browne and John Locke as Tourists in the United Provinces.* Leiden: E. J. Brill, 1993.
Van Tielhof, Milja. "After the Flood. Mobilising Money in Order to Limit Economic Loss (the Netherlands, Sixteenth–Eighteenth Centuries)." In *Atti delle "Settimane di Studi" e altri Convegni,* Vol. 49, 393–411. Florence: Firenze University Press, 2018.
Consensus en conflict: Waterbeheer in de Nederlanden, 1200–1800. Hilversum: Verloren, 2020.
"Forced Solidarity: Maintenance of Coastal Defenses along the North Sea Coast in the Early Modern Period." *Environment and History* 21 (2015): 319–50.
The "Mother of All Trades": The Baltic Grain Trade in Amsterdam from the Late 16th to the Early 19th Century. Boston: Brill, 2002.
"Regional Planning in a Decentralized State: How Administrative Practices Contributed to Consensus-Building in Sixteenth-Century Holland." *Environment and History* 23, no. 3 (2017): 431–53.
Van Tielhof, Milja and Petra J. E. M. van Dam. *Waterstaat in stedenland: Het hoogheemraadschap van Rijnland voor 1857.* Utrecht: Matrijs, 2006.
Van Zanden, Jan Luiten. "De economie van Holland in de periode 1650–1805: Groei of achteruitgang? Een overzicht van bronnen, problemen en resultaten." *BMGN: Low Countries Historical Review* 102, no. 4 (1987): 562–609.
The Rise and Decline of Holland's Economy: Merchant Capitalism and the Labour Market. Manchester: Manchester University Press, 1993.
Van Zellem, J. "'Nooyt gehoorde hooge waeteren': bestuurlijke, technische en sociale aspecten, in het bijzonder de hulpverlening, van de overstromingsramp in de Over-Betuwe in 1740–1741." *Tijdschrift voor Waterstaatsgeschiedenis* 12 (2003): 11–20.

Velema, Wyger R. E. *Republicans: Essays on Eighteenth-Century Dutch Political Thought.* Leiden: Brill, 2007.
Verhoeff, J. M. *De oude Nederlandse maten en gewichten,* 1982. www.meertens.knaw.nl/mgw/
Vleer, W. T. *"Sterf sodomieten!": Rudolf de Mepsche, de homofielenvervolging, het Faanse zedenproces en de massamoord te Zuidhorn.* Norg: Veja, 1972.
Vrolik, W., P. Harting, D. J. Storm Buysing, J. W. L. van Oordt, and E. H. von Baumhauer. *Verslag over den paalworm: Uitgegeven door de Natuurkundige afdeeling der Koninklijke Akademie van Wetenschappen.* Amsterdam: C. G. van der Post, 1860.
Walker, Timothy D. "Enlightened Absolutism and the Lisbon Earthquake: Asserting State Dominance over Religious Sites and the Church in Eighteenth-Century Portugal." *Eighteenth-Century Studies* 48, no. 3 (2015): 307–28.
Wallace-Wells, David. *The Uninhabitable Earth: Life after Warming.* New York: Crown/Archetype, 2019.
Walsham, Alexandra. *Providence in Early Modern England.* Oxford: Oxford University Press, 1999.
  The Reformation of the Landscape: Religion, Identity, and Memory in Early Modern Britain and Ireland. Oxford: Oxford University Press, 2011.
Walsmit, Erik, Hans Kloosterboer, Nils Persson, and Rinus Ostermann, eds. *Spiegel van de Zuiderzee: Geschiedenis en cartobibliografie van de Zuiderzee en het Hollands Waddengebied.* Utrecht: HES & De Graaf, 2009.
Wanner, Heinz, Christoph Beck, Rudolf Brázdil et al. "Dynamic and Socioeconomic Aspects of Historical Floods in Central Europe." *Erdkunde* 58, no. 1 (2004): 1–16.
Wanner, Heinz, Jürg Beer, Jonathan Bütikofer et al. "Mid-to Late Holocene Climate Change: An Overview." *Quaternary Science Reviews* 27, no. 19–20 (2008): 1791–828.
Warde, Paul. "Global Crisis or Global Coincidence?" *Past & Present* 228, no. 1 (2015): 287–301.
Weigelt, Ronny, Heike Lippert, Ulf Karsten, and Ralf Bastrop. "Genetic Population Structure and Demographic History of the Widespread Common Shipworm *Teredo navalis* Linnaeus 1758 (Mollusca: Bivalvia: Teredinidae) in European Waters Inferred from Mitochondrial COI Sequence Data." *Frontiers in Marine Science* 4 (2017): 1–12.
Wesselink, Anna, Jeroen Warner, M. Abu Syed et al. "Trends in Flood Risk Management in Deltas around the World: Are We Going 'Soft'?" *International Journal of Water Governance* 3, no. 4 (2015): 25–46.
Wheeler, Dennis, Ricardo Garcia-Herrera, Clive W. Wilkinson, and Catharine Ward. "Atmospheric Circulation and Storminess Derived from Royal Navy Logbooks: 1685 to 1750." *Climatic Change* 101, no. 1–2 (2010): 257–80.
White, Sam. "Animals, Climate Change, and History." *Environmental History* 19, no. 2 (2014): 319–28.
  The Climate of Rebellion in the Early Modern Ottoman Empire. Cambridge: Cambridge University Press, 2011.

A Cold Welcome: The Little Ice Age and Europe's Encounter with North America. Cambridge: Harvard University Press, 2017.

"A Model Disaster: From the Great Ottoman Panzootic to the Cattle Plagues of Early Modern Europe." In *Plague and Contagion in the Islamic Mediterranean*, edited by N. Varlik, 91–116. Kalamazoo: Arc Humanities Press, 2017.

"The Real Little Ice Age." *Journal of Interdisciplinary History* 44, no. 3 (2014): 327–52.

Wilkinson, Lise. *Animals and Disease: An Introduction to the History of Comparative Medicine*. Cambridge: Cambridge University Press, 1992.

Wilson, Charles. *The Dutch Republic and the Civilisation of the Seventeenth Century*. New York: McGraw-Hill, 1968.

"Taxation and the Decline of Empires, an Unfashionable Theme." In *Economic History and the Historian*, edited by Charles Wilson, 114–27. London: Weidenfeld, 1969.

Wilson, Robert J. S., Brian H. Luckman, and Jan Esper. "A 500 Year Dendroclimatic Reconstruction of Spring–Summer Precipitation from the Lower Bavarian Forest Region, Germany." *International Journal of Climatology* 25, no. 5 (2005): 611–30.

Wisner, Ben, Jean-Christophe Gaillard, and Ilan Kelman. "Framing Disaster: Theories and Stories Seeking to Understand Hazards, Vulnerability and Risk." In *Handbook of Hazards and Disaster Risk Reduction*, edited by Ben Wisner, Jean-Christophe Gaillard, and Ilan Kelman, 18–33. New York: Routledge, 2012.

Wolff, Wim J. "Non-Indigenous Marine and Estuarine Species in the Netherlands." *Zoologische Mededelingen Leiden* 79, no. 1 (2005): 1–116.

Woodshole Oceanographic Institution. "Marine Fouling and Its Prevention." Prepared for the Bureau of Ships, Navy Department. Annapolis, MD: U.S. Naval Institute, 1952.

The Works of Virgil Translated into English Prose. London: Geo B. Whittaker et al., 1826.

Worster, Donald. *Dust Bowl: The Southern Plains in the 1930s*. Oxford: Oxford University Press, 1979.

"Transformations of the Earth: Toward an Agroecological Perspective in History." *The Journal of American History* 76, no. 4 (1990): 1087–106.

Wouda, Bertus. *Een stijgende stand met een zinkend land: Waterbeheersingssystemen in polder Nieuw-Reijerwaard, 1441–1880*. Hilversum: Verloren, 2009.

Xoplaki, Eleni, Panagiotis Maheras, and Jürg Luterbacher. "Variability of Climate in Meridional Balkans during the Periods 1675–1715 and 1780–1830 and Its Impact on Human Life." *Climatic Change* 48, no. 4 (2001): 581–615.

Zhang, David D., Harry F. Lee, Cong Wang et al. "The Causality Analysis of Climate Change and Large-Scale Human Crisis." *Proceedings of the National Academy of Sciences* 108, no. 42 (2011): 17296–301.

Zhu, Zhenchang, Vincent Vuik, Paul J. Visser et al. "Historic Storms and the Hidden Value of Coastal Wetlands for Nature-Based Flood Defence." *Nature Sustainability* 3 (2020): 1–10.

Zomer, Jeroen. "Middeleeuwse veenontginningen in het getijdenbekken van de Hunze: Een interdisciplinair landschapshistorisch onderzoek naar de paleogeografie, ontginning en waterhuishouding (ca 800–ca 1500)." PhD diss., Rijksuniversiteit Groningen, 2016.

Index

's Gravenhage (The Hague), 51, 53–4, 115, 218, 233, 240
's Gravesande, Willem Jacob, 49, 144, 184–5
's Hertogenbosch (Den Bosch), 165, 173, 177, 179, 190

Aalst, 139
Adaptation. *See also* Natural disasters
   cattle plague, 69, 78–86
   in disaster scholarship, 2, 21
   flooding (coastal), 97–100, 144
   flooding (river), 232
   room for the rivers, 274
   shipworms, 15, 148, 157–62
Africa, 34
Agriculture. *See also* Little Ice Age; Maunder Minimum; cattle; decline; reclamation
   Arcadian ideal, 55, 58–60, 64, 124, 258
   in art, 59–62, 64–5, 124, 218–20
   cattle holding, 87
   climate/weather, 55–6, 199–202, 224, 265
   importance in Dutch Republic, 15, 32, 62, 262
   recession, 9, 33, 64, 87, 176, 202, 215, 258, 261, 267
   reclamation, 15, 63, 95, 111, 148, 181, 262
   regional variation, 62, 170, 221–3, 256
   specialization, 12, 63–4, 265

Alblasserwaard, 179, 184, 191, 197, 199, 207
Alkmaar, 239
All Saints' Day Flood (1570), 46, 99, 270
Allerheiligenvloed. *See* All Saints' Day Flood (1570)
Alta, Eelko, 245–7
Ameland, 105
Amphibious culture, 98
Amstelland, 184. *See also* Water boards
Amsterdam, 6, 27, 30, 33–4, 65–6, 104, 144–6, 161–2, 184–5, 204–5, 233, 241, 258–9
Amsterdamse Peil, 185
Anglo-Dutch Wars, 33. *See* also *Rampjaar*
Arnhem, 171, 190–1, 197
Asia. *See* Dutch East Indies
Asperen, 197
Assendelft, 84
Austrian Netherlands, 215, 225, 228

Baltic Grain Trade, 33, 63
Bankoff, Greg, 19, 97
Beels, Leonard, 155
Belkmeer, Cornelis, 148
Betuwe (region), 179
Beverwijk, 236, 238
Beverwijk Trials, 236–41, 243, 245, 247
Bicker-Raye, Jacob, 200
Biesbosch, 172, 181, 193
Bishop of Münster, 26
Blom, Koenraad, 216

325

Bodegraven, 43
Boerhaave, Herman, 49, 182. *See also* Expertise; Leiden University
Bogaert, Adriaan, 102
Bolstra, Melchior, 179, 183–7, 193, 208
Bommelerwaard, 188, 193
Bommenee, Adriaan, 139
Bosch, Toon, 188
Brazil, 34, 131
Brusse, Paul, 198
Bunschoter Dike, 162
Buytendijk, Leendert Jans, 233

Calvinism. *See* Reformed Church
Cattle. *See also* Agriculture; cattle plague
  cultural significance, 12, 59–61, 63–5, 217
  Danish oxen trade, 57, 64–6, 80, 82–3, 225–7, 232
  economic and social significance, 60–4, 68–9
Cattle plague. *See* continuous crisis management, cattle, agriculture, physico-theology, inoculation
  in 1651, 219
  in 1682, 219
  certification, 83–4, 86, 227–9
  climate/weather, 38, 55–6, 199–202, 223–4
  cultural meaning, 58, 67, 69–71, 73–8, 88
  economic consequences, 69
  Enlightenment, 25, 218, 236–41, 243–4, 247–50, 256
  folk remedies, 217, 234
  import/export restrictions, 20, 78–83, 227
  mortality, 57–9, 68, 217–23, 233, 238, 253, 262, 265
  origins, 23, 57, 253
  physicians, 50, 83, 217, 234–5, 238, 243, 245–6, 248–9
  stamping out, 83–6
  symptoms, 57, 75, 235
  theories of disease, 80–2, 85–6, 227, 244–5
Causal stories, 22, 254, 260
  cattle plague, 23, 59, 71–3, 77, 79, 86, 248, 268
  cultural memory, 265
  definition, 21
  flooding (coastal), 94, 106

  flooding (river), 180, 185, 205
  hybridity, 263–4
  shipworms, 148, 153, 155, 163, 183
  uses, 21
Christmas Flood (1717), 23–4, 124, 152–3, 161, 180, 206, 220, 253–4, 264–5, 268. *See also* Flooding (coastal); cultural memory; *Hulpkarspelen*
  dike adaptation, 94, 115–19
  financing dike repair, 115, 121
  flood cultures, 95–101
  Groningen, 89–93
  international impacts, 91–2
  mortality, 103–4, 113
  origins, 89
  providential interpretation, 101–6, 117–18
  relief, 91, 93, 101
Cleves, 177
Climate change. *See also* Little Ice Age; Maunder Minimum
  epizootics, 23, 223–6
  evidence, 109–10, 174–6
  experience/perception, 1–2, 39–41
  expressions, 37–9
  flooding (coastal), 109–11
  flooding (river), 174–6
  hybridity, 261–3
  shipworms, 136–8
Coastal flooding. *See* Flooding (coastal)
Collegiants, 205
Cologne, 26, 177
Columbus, Christopher, 131
Comets
  changing interpretation, 248–9
  Great Comet (1744), 212–14, 248
  Halley's Comet (1758), 248
  physico-theological interpretations, 212–14
  providential interpretations, 71, 212
Continuous crisis management, 83, 228, 232–3. *See also* Cattle plague
Costerus, Bernard, 43
Crous, Albert Ebbo, 94–5, 101, 117
Cruquius, Nicolaas, 49, 150, 182–4, 208
Culemborg, 174, 179, 190–2, 198
Cultural memory, 2, 19, 22, 215, 266
  flooding (coastal), 17, 24, 93–4, 120, 254, 265
  flooding (river), 179

Index

memoria, 95, 100–1, 104–5, 117, 176, 210, 254, 265, 269–73
shipworms, 24, 126, 162

Dalton Minimum, 134, 175. *See also* Climate change; Little Ice Age
Dams. *See* Water infrastructure
Days of Thanksgiving, Fasting, Prayer, 76, 152, 203. *See also* Providence
 cattle plague, 77
 history, 76
 uses, 77–8
De Beer, Zacheus, 199
De Hooghe, Romeyn, 28–32, 41, 47, 253, 257
De Leth, Hendrik, 122–5
De Mepsche, Rudolphe, 153
De Missy, Jean Rousset, 147
De Robles, Caspar, 272
De With, Witte, 131
De Witt, Johan, 28, 35, 42
Decline
 cattle plague, 58–9, 78, 87–8, 218, 247–8
 Christmas Flood, 106
 economic and political indicators, 46–7, 215–16, 238, 257
 economic turn, 257
 and Enlightenment, 48–50, 205–7, 236–43, 256–7
 environmental history, 12–14
 environmental indicators, 47, 49
 flooding (river), 169–70, 203, 205
 historiography, 9–11
 moral indicators, 47–8, 253–5, 268
 proto-national narratives, 207, 252, 256
 Rampjaar, 45–6
 rural paradigm, 257–8
 shipworms, 126, 151–7, 163–4
 universal characterization, 9, 251–2, 255–6
Degroot, Dagomar, 40
Deknatel, Johannes, 97
Delfland, 184, 187, 207. *See also* Water boards
Delfshaven, 134
Delft, 134, 200, 215, 218, 233
Denmark. *See also* Cattle plague; oxen trade, Danish; Christmas Flood (1717)
Dief dike, 179
Diemer dike, 162, 262
Dijkrechten. See Water boards

Dikes. *See* Water infrastructure
Disaster cascades, 22, 24–5, 169, 202, 207, 217, 223–7, 242, 249, 255–6, 259, 265–7. *See also* Continuous crisis management
 definition, 19
Disaster *memoria*. *See* Cultural memory
Disaster novelty. *See also* Shipworms, shipworm epidemic; natural disasters
 and cultural memory, 126, 162–3
 exceptionality, 17, 21, 93, 101, 103–5, 165, 175, 220, 223, 249, 254, 256, 268–9
 perception, 24, 163
Disaster Year (1672). *See Rampjaar*
Disasters. *See* Natural disasters
Divine providence. *See* Providence
Dollard Sea Arm, 97, 108
Doniawerstal, 229
Dordrecht, 190, 193
Drechterland, 139, 144, 148
Drechterlandse noorderdijk, 139
Drenthe, 62, 68, 222
Droogmakerijen. See Reclamation
Drought, 2, 16, 28, 42, 136–8, 216, 223, 265. *See also* Climate change; Little Ice Age; Maunder Minimum; shipworm epidemic; mice plagues
Dutch decline. *See* Decline
Dutch East India Company (VOC), 6, 34, 44, 131, 158
Dutch East Indies, 33–4, 131–2
Dutch Enlightenment. *See* Enlightenment
Dutch Golden Age
 characterization of, 6–9
 myth of, 11
 perception, 12, 14, 22, 32, 34, 45, 47–8, 59–62, 67, 70, 78, 96–7, 99, 102, 154–7, 205, 249, 258
Dutch Republic
 economic development, 12, 32–4, 215–16
 environmental conditions, 36–41
 founding, 7–9
 geopolitical, 33, 214–15
 political organization, 6–7, 35–6
 religious pluriformity, 35
Dutch Revolt. *See* Eighty Years' War
Dutch West India Company (WIC), 6, 34, 44

East Friesland, 75, 91, 103, 132, 206
Eighty Years' War, 33

Index

Elden, 178
Engelhardt, Henricus, 149
Engelman, Jan, 244
England. *See also* Anglo-Dutch Wars; Rampjaar; cattle plague
   cattle plague, 83–4
   climate, 137
   conflict with, 10, 26, 33, 42
   flooding (coastal), 91, 103
   inoculation, 240–1, 244
Enkhuizen, 159
Enlightenment. *See also* Van Effen, Justus; physico-theology; Holland Society of Sciences
   cattle plague, 25, 218
   decline discourses, 14, 48, 236
   disaster response, 48–9
   flooding (river), 205–7
   shipworms, 124, 147, 156–7, 163
   sociability, 237–41
Estates General, 51, 60, 76. *See also* Days of Thanksgiving, Fasting, Prayer; Dutch Republic; continuous crisis management
   cattle plague, 82, 228
   conflict mediation, 112, 115
   incentivizing plague remedies, 234
   shipworms, 146, 152
Estates of Friesland, 69, 82, 114
Estates of Groningen
   cattle plague, 80
   Christmas Flood (1717), 114–15, 121
Estates of Holland. *See also* Days of Thanksgiving, Fasting, Prayer; continuous crisis management; expertise
   cattle plague, 80, 238
   flooding (river), 184, 187, 191–2, 197, 203, 206–7
   flooding (river) relief, 191
   Rampjaar, 26
   shipworm epidemic, 144, 149–50
Estates of Utrecht, 78, 235. *See also* continuous crisis management
Estates of Zeeland, 138
Europische Mercurius, 145–8, 174, 179, 198
Exclusive providentialism, 244–7
Expertise. *See* Leiden University; Utrecht University; Hydraulic Department (Holland)

citizen-science, 238–41, 245–7
   hydraulic, 20, 48, 116, 149–50, 162, 169, 180–3, 193, 207–10, 255
   medical, 217, 234–5
   natural historical, 147–8
   role of civic engagement, 148–9

Fabricius, Isbrandus, 69, 105–6
Famine
   increasing likelihood, 223
   providential interpretation, 71, 155, 247
Flanders, 103
Flooding (coastal). *See also* Storm surges; All Saints' Day Flood (1570); St. Martin's Flood (1686); Christmas Flood (1717); cultural memory; adaptation; vulnerability; shipworm epidemic
   in 1492 (St. Elizabeth's Flood), 104, 172
   in 1675 (Second All Saints' Day Flood), 23, 30, 44
   in 1702, 44, 53
   in 1714, 87
   in 1715, 87, 105
   in 1716, 115
   in 1719, 119, 121
   in 1720, 121
   in 1953, 272
   climate/weather, 38–41, 89, 110
   documentation, 39
   environmental change, 110–12
   fear of total inundation, 124, 126, 144, 161, 163, 258
   flood cultures, 94–101, 108, 124
   interpretation of dike breaches, 71
   providential interpretation, 101–4
   relief in Groningen, 112
   risk of, 15–17, 108–12, 143
Flooding (river). *See also* Sedimentation; subsidence; reclamation; expertise; vulnerability
   in 1598, 183
   in 1653, 166
   in 1658, 179
   in 1709, 165
   in 1726, 165–8, 176, 179, 182–3, 187, 190, 198–9, 207
   in 1740–41, 173–5, 177–80, 187, 216, 225, 259
   in 1744, 165, 207, 216, 232
   in 1747, 165, 207, 216, 232

Index

in 1751, 165, 207, 210, 232, 236
in 1753, 165, 207, 232, 236
in 1754, 243
in 1757, 165–6, 207, 232
in 1926, 272
changing governance, 182
climate/weather, 173–5, 177, 199–202
documentation, 39, 175
environmental change, 36–7
fear of total inundation, 169, 199, 207–8, 258
flood frequency and severity, 17, 21–2, 24, 166, 177, 255, 261–3, 269, 273
ice dams, 173–4, 180–1, 198, 232
perception of risk, 24, 169, 177, 181, 185, 202
providential interpretation, 169, 203, 210
relief, 20, 25, 169, 187–98, 204–7, 210, 255, 264, 268
Rhine-Meuse morphology, 172
river cartography, 184–7, 191–7, 208, 255
social vulnerability, 37, 176
solidarity, 188, 205–7, 210, 266, 269
structural conditions, 172–3, 180
Fokkert, Jan Barte, 233
Forty Years' War, 46
France. *See also* Louis XIV (of France); *Rampjaar*; Forty Years' War; War of the Spanish Succession; Treaty of Utrecht (1713); War of the Austrian Succession
conflict with, 10, 42, 46, 51, 54, 181
Francken, Aegidius, 74
Frederick II (Prussia), 224
Friesland, 270, 272. *See also* Estates of Friesland; Alta, Eelko; cattle plague; continuous crisis management
cattle breeding, 63
cattle plague, 58, 66–9, 77, 80, 82, 84, 220–4, 228–30, 233, 245
cattle holding, 68, 85
celebrating peace, 53
Christmas Flood (1717), 102, 105, 116
flooding (coastal), 87, 267
mob violence, 215
shipworms, 122, 127, 152, 158, 162, 267
Further Reformation, 34, 154

Gabbema, Simon, 100, 104
Gay, Peter, 48
Gecommittteerden van Stad Groningen en Ommelanden Tot de Dijken cum plena, 112
Gelderland. *See also* Rhine River; water boards
cattle plague, 69, 82, 198, 228, 233
drainage issues, 171
flooding (river), 166, 176–9, 193, 207, 225
flooding (river) relief, 190, 197
Rampjaar, 43
river improvement, 181, 185
river landscape, 170
Generality Lands
cattle plague, 222, 228
flooding (river), 166, 179
Rampjaar, 43
river landscape, 170–1
War of the Austrian Succession, 215
Golden Age. *See* Dutch Golden Age
Gorinchem, 43, 184, 190, 199
Gouda, 43, 105, 198
Groningen (city). *See also Ommelanden*; Groningen (province)
conflict with *Ommelanden*, 24, 94, 114–15, 254, 268
flooding (coastal) relief, 91, 113–14
Rampjaar, 26, 42
Groningen (province). *See also* Water infrastructure; Estates of Groningen; *Ommelanden*
cattle holding, 68
cattle plague, 69, 74, 77, 81–2, 84, 224
Christmas Flood (1717) mortality, 91, 102–3
coastal vulnerability, 93
dike adaptation, 115–19, 180
early coastal settlement, 107–8
early history, 108–9
flood risk, 108–11
flooding (coastal) exposure, 89
increasing social vulnerability, 111–13, 254
mice plagues, 87
Pachtersoproer, 215
public debt, 54
reclamation, 97
sodomy persecution, 153, 155
St. Martin's Flood (1686), 52
water management, 96, 108
Grootebroek, 267
Groynes. *See* Water infrastructure

Haarlem, 161, 200, 215, 236, 238–9
Halma, François, 53, 67, 73
Harkenroht, Jacob, 75, 153
Harlingen, 270
Harvest failures, 40, 117, 169, 232, 259, 265
  climate/weather, 55–6, 200–2, 223
  likelihood, 39
  mice plagues, 87, 210, 216, 242
  military inundations, 43
  price spikes, 53
Hattem, 233
Hedikhuizen, 197
Heiloo, 75
Hekelius, Johann Christian, 91
Helvetius, Adrianus Engelhard, 46
Hemelumer Oldeferd, 222
Herring Fishery, 9, 33, 215
Heusden, 177, 197
Holland. *See also* Water Infrastructure; water boards; West Friesland
  agricultural recession, 33, 163
  Catholicism, 75–6
  cattle breeding, 63
  cattle holding, 62, 67
  cattle plague, 56, 66, 74, 79–81, 84–6, 220–3, 233
  Christmas Flood (1717), 102, 105
  coastal dike designs, 143
  conflict with House of Orange, 35–6
  Danish oxen trade, 57, 64, 225
  deurbanization, 10
  dike adaptation, 157–62, 180
  economic development, 32
  flood risk, 183–7, 207–8
  flooding (coastal), 53, 87, 267
  flooding (river), 166–70, 173, 178–80, 198–9, 225
  flooding (river) relief, 190–8, 204–5
  incentivizing plague remedies, 234
  influence in Dutch Republic, 7, 11
  land abandonment, 230, 266
  mice plagues, 216
  Pachtersoproer, 215
  public debt, 54
  Rampjaar, 32–3, 41–4
  reclamation, 63, 96
  river management, 181–2
  shipworms, 126–7, 132, 134, 139–40, 144, 158, 267
  water boards, 96

Holland Society of Sciences
  as Enlightenment institution, 239
  prizes for cattle plague, 239–40
  prizes for river hydraulics, 239
  Transactions (journal), 244
Hollandsche Maatschappij der Wetenschappen. *See* Holland Society of Sciences
Hollandsche Spectator, 48, 156–7. *See also* Van Effen, Justus
Hollandse waterlinie. *See* Water lines
Hoofden. *See* Water infrastructure
Hoogheemraadschappen. *See* Water boards
Hoorn, 225, 236, 241
Huekelom, 197
Huissen, 179
Hulpkarspelen, 113–17
Hünniger, Dominik, 83
Hydraulic Department (Holland). *See also* Flooding (river); Water management
  professionalization, 182
  water management centralization, 181–2

Iconoclasm (1566), 99
IJssel River, 136, 171, 180, 207
Inlaagdijk. *See* Water infrastructure
Inoculation. *See also* Enlightenment; cattle plague; Holland Society of Sciences; exclusive providentialism; Alta, Eelko
  definition, 236
  early Dutch trials, 236–7, 256
  enlightened sociability, 238–41
  opposition, 244–7, 249
  smallpox analogy, 236
  third eighteenth-century cattle plague epizootic, 248

Kadijk. *See* Water infrastructure
Kedichem, 199
Keeken (Germany), 177
Kemner, Johannes, 106
Kerstvloed. *See* Christmas Flood (1717)
Kjærgaard, Thorkild, 226
Kool, Agge Roskam, 237–9, 244
Kraamer, Simon Jacobszoon, 214, 216, 220
Krammaten. *See* Water infrastructure
Krimpenerwaard, 198
Kuypers, François, 203, 205
Kwelders. *See* Voorland

Index

L'Epie, Zacharias, 148, 159
Lakenman, Seger, 144, 159
Lancisi, Giovanni, 83
Land abandonment, 229–30, 266
Land of Heusden and Altena, 193
Landaanwinning. *See* Reclamation
Langedijk, 230
Layard, Daniel Peter, 241
Leens, 89, 118
Leiden, 10, 76, 161, 185, 207, 215, 233, 238. *See also* Leiden University; Siege of Leiden (1573–4)
Leiden University, 144, 182–3, 208, 235, 255
Lek River, 179, 183–5, 262. *See also* Flooding (river); Bolstra, Melchior; sedimentation; *Rijnland*
  in 1740–41, 177, 179, 188, 198
  in 1744, 207
  in 1747, 207
  flood risk, 180, 187, 262
  Rhine–Meuse River System, 183–5
Leurink, Aleida, 55–6, 199
Limburg, 26
Linge River, 179, 191, 198–9
Little Ice Age. *See also* Climate change; North Atlantic Oscillation (NAO); Maunder Minimum; Grindelwald Fluctuation; Dalton Minimum
  characterization of, 37–8
  and Dutch Republic, 7, 40–1, 262–3
  flooding (river), 174
  Little Ice Age–type impacts, 223
  shipworms, 134–6
  social impacts, 38–40
  storminess, 110
Lopikerwaard, 198
Louis XIV (France), 26, 51
Lulofs, Johan, 208
Lustigh, Lambert Rijckszoon, 73, 81, 85

Malaria, 216
Marchant, Jan, 234
Maunder Minimum, 38, 40, 42, 53, 55, 134. *See also* Climate change; Little Ice Age; North Atlantic Oscillation (NAO)
  flooding (river) frequency, 175
  periodization, 37
  storminess, 109–10
  westerlies, 110
Mel, Koenraad, 155

Merwede River, 27, 172, 181–2, 188
Meuse River
  in 1740–1, 166, 177–80, 216
  in 1757, 166
  discharge, 136, 173
  flood risk, 262
  interactions with Rhine, 170–3, 181, 255
  salinity, 134
Mice plagues, 87, 210, 216, 225–6, 232, 236, 242, 259, 265
Military inundations. *See* Water lines
Mobachius, Joachim, 242
Mobachius, Johannes, 70, 73, 101–2, 104, 151
Montfoort, 198
Moubach, Abraham, 67, 71
Muizenjaren. See Mice plagues

Natural disasters. *See also* Adaptation; vulnerability; novelty; cultural memory; disaster cascades
  climate/weather, 38–41
  Dutch decline, 12–15
  environmental history, 4–6
  events and processes, 18–20, 264–7
  historiography, 2–4
  history and memory, 21–2, 269–75
  hybridity, 18, 261–4
  scale, 20–1, 267–9
  trends in, 1–2
  as trials of faith, 15, 31, 50, 252–3, 260
Natural theology, 71, 243. *See also* Providence; physico-theology
Nederasselt, 177, 183
Nederlandsche jaerboeken, 235
Nederrijn River, 171, 180, 183, 185, 207
Nen, Gerrit Jacobszoon, 200
New Netherland, 34
Nieuwentijdt, Bernard, 243
Nijmegen, 193
Nine Years' War, 46
North Atlantic Oscillation (NAO). *See also* Climate change; Little Ice Age; Maunder Minimum
  flooding (river), 175
  storminess, 110
North Brabant. *See* Generality Lands
North Sea. *See also* Wadden Sea; flooding (coastal); water management; water infrastructure
  Christmas Flood (1717), 23, 90, 251, 253, 265

North Sea. (cont.)
  Danish oxen trade, 64–5
  discharge, 171
  Dutch prosperity, 12
  flood cultures, 93–101
  flood disasters, 91
  physical geography of the Netherlands, 36, 107, 170
  salinity, 136
  shipworms, 126, 133
  storminess, 108–10
Novelty. See Disaster novelty
Nozeman, Cornelis, 237–9, 244
Ommelanden
  Christmas Flood (1717), 91, 103
  conflict with Groningen (city), 94, 114–16, 254, 268
  dike responsibility, 107, 264
  political marginalization, 112

Oud Loosdrecht, 258–9, 262
Outhof, Gerardus, 88, 90, 100, 104
Over-Betuwe (region), 197–8
Overijssel
  cattle plague, 69–70, 74, 80, 233
  flooding (river) in 1753, 207
  public debt, 54
  Rampjaar, 43
  river modification, 181
  winter in 1739-40, 199
Oxen trade, Danish. *See also* Cattle; cattle plague
  cultural significance, 64
  early history, 63–4
  economic and ecological consequences, 64–6

Paaldijken. See Water infrastructure
Paalschermen. See Water infrastructure
Pachtersoproer, 215
Pamphlets. *See also* Cultural memory
  cattle plague, 25, 65–7, 71, 81, 88, 218, 234, 237, 246
  Christmas Flood (1717), 101, 104–6
  flooding (river), 165, 176, 188, 202, 205, 207, 210
  Rampjaar, 43–5
  shipworms, 123, 138, 144, 148, 150, 152, 158–62
  sodomy persecution, 155
  Treaty of Utrecht (1713), 53–4

Pannerdens Canal, 177, 179, 181, 185
Peace of Münster, 33, 35
Peat
  climate proxies, 109
  drainage and subsidence, 95–6, 183–4
  extraction, 36, 63, 95, 108, 183
  flooding (coastal), 108
  lake formation, 63, 96
  lake reclamation, 32, 62, 96, 262
  landscape feature, 36, 63, 108, 170
Periwig Period, 9–11
physico-theology, 243–4, 248
Pierlinck, Jacob, 169, 175–7, 207
Pietism. *See also* Reformed Church
  in Dutch Republic, 34
  Further Reformation, 154
  interpretations of disaster, 242
  moral critiques, 24, 47, 74–5, 163, 255
Polders. See Reclamation
Poot, Hubert Korneliszoon, 56, 105
Pos, Jacob, 236, 258–62, 265
Propositie of 1751, 12, 258
Providence. *See also* Physico-theology; exclusive providentialism; Days of Thanksgiving, Fasting, Prayer
  Book of Nature, 71, 212, 247
  cattle plague, 87, 217, 242–4
  causal story, 58–9, 71, 86, 150, 236
  Christmas Flood (1717), 102–6
  civic rituals, 76–8
  decline, 49, 88, 151–2, 163, 169
  dike adaptation, 121
  documentation, 71–2
  explanation for environmental change, 71, 202–3
  extreme weather, 202
  flooding (coastal), 100–1, 117–18, 150–1
  general providence, 71
  historiography, 72–3
  hybridity, 263–4
  interpretations of comets, 212–14
  physico-theology, 248
  providential favor, 58, 60, 94, 99–100
  rhetoric of inversion, 58–9, 67, 70, 78, 88, 99, 106, 188, 219, 258
  secular responses, 78
  shipworms, 123, 151
  social utility, 73–6
  sodomy persecution, 154–6, 255
  solidarity, 210, 230–1
  special providence, 103

Pruikentijd. See Periwig Period
Psiloteredo megotara. See Shipworms

Quarter of Nijmegen (region), 197
Quint, Thadeus François, 200, 224

Ram, Godefridus, 198, 212, 220, 226, 229
Rampjaar. *See also* Romeyn de Hooghe; France; England; Decline; Forty Years' War; water lines
  comparison to 1744, 214
  cultural consequences, 44–6
  decline, 22, 30–2, 46–7, 58, 203
  documentation, 41
  economic consequences, 44, 51, 215
  environmental consequences, 26–8, 42–4, 53
  hybridity, 30
  military consequences, 26–7
  political consequences, 28, 35–6, 41–2
  social consequences, 28–9
Reclamation. *See also* Subsidence; peat; agriculture; water infrastructure; *Voorland*; water management; water boards
  agricultural productivity, 15
  cattle holding, 62–3
  cultural interpretation, 99
  ecological crisis, 63
  flood risk, 262
  flooding (coastal), 95, 108
  flooding (river), 170–1, 181
  foreign perceptions, 97
  Groningen, 97, 107–8, 110–12, 118–19
  Holland, 96, 222
  as shipworm adaptation, 148
  urban investment, 64
  Zeeland, 62, 97
Reformed Church. *See also* Pietism; Further Reformation
  condemnation of Catholic rituals, 75
  decline, 47
  dissenters, 244
  pamphlet literature, 45
  pluriformity, 35
  as Public Church, 34–5
Reynvaan, Edualdus, 127–8
Rhine River. *See also* Meuse River; IJssel River; Lek River; Merwede River; Waal River; Nederrijn River; Pannerdens Canal; flooding (river); water infrastructure; Little Ice Age; North Atlantic Oscillation (NAO)
  in 1740–1, 166, 177–80, 216
  anthropogenic changes, 170–1
  discharge, 136
  distributaries, 136, 171–2
  drought, 136–8
  extreme weather, 173–4
  fear of total inundation, 183–4
  geophysical influence, 170
  increasing flooding (river) risk, 185
  interactions with Meuse, 172–3
  water management challenges, 180–1
Rijnland, 44, 183–7, 207, 233. *See also* Water boards
Rinderpest. *See* Cattle plague
River flooding. *See* Flooding (rivers)
Rotterdam, 134, 184–5, 204, 258

Schama, Simon, 31, 99
Schieland, 184, 187, 207. *See also* Water boards
Schoock, Martin, 100
Schoonhoven, 43
Schwenke, Thomas, 240
Second Orangist Revolution, 215
Sedimentation, 19, 36, 118, 265
  ice dams, 181
  IJssel River, 136, 171, 180
  Lek River, 171
  Rhine River, 36
  water management challenge, 166–9, 181, 183, 198
Sellius, Gottfried, 147–8
Sermons. *See also* Providence; Pietism; Days of Thanksgiving, Fasting, Prayer; cultural memory
  cattle plague, 69–70, 77–8, 101
  Christmas Flood (1717), 101–5, 254
  civic virtue, 249
  disaster *memoria*, 210
  disaster relief, 205
  flooding (coastal), 100–1
  flooding (river), 203–4
  Great Comet of 1744, 214
  printing culture, 44
  punishment/disaster sermons, 71–6
  shifts in disaster sermons, 242–4
  shipworms, 124, 154–5
  sodomy persecution, 154–5
  solidarity, 206–7

Sermons. (cont.)
  Treaty of Utrecht (1713), 54
  War of the Spanish Succession, 101
Seventeenth-century crisis, 7
Shipworm epidemic. *See also* Shipworms; water infrastructure; providence; decline; disaster novelty
  dike adaptation, 138–9, 143–4, 148–50, 157–62, 254, 266
  discovery in Holland, 139–40
  early records in Europe, 132
  fear of total inundation, 123, 144, 184, 187, 199, 208, 272
  land abandonment, 230
  misinformation, 144–7
  origins, 134–8, 140–3, 262–3
  perception, 17, 21, 24, 126, 162–4, 254, 256
  providential interpretation, 150–2, 203
  sodomy persecution, 155–7, 255, 266
  water management centralization, 182
Shipworms. *See also* Shipworm epidemic
  in 1751, 236
  climate change, 273
  discovery on Walcheren, 122
  early science, 147–8
  ecological tolerances, 127–30, 261
  geographic origins, 132–5
  maritime experience, 129–32
Siege of Leiden (1573–4), 99
Sleeper dikes. *See* Water infrastructure
Sluices. *See* Water infrastructure
Sluiter, Johannes, 70, 75
Smids, Ludolph, 104
Smit, Jan, 191–7, 218–19
Social unrest, 251, 255, 259, 265. *See Pachtersoproer*
  in 1747-8, 236
  during Little Ice Age, 40
  during *Rampjaar*, 18, 28
  in Groningen, 117
  in Haarlem, 238
Sodomy persecution, 24. *See also* Shipworm epidemic; Pietism
  in 1730s, 153
  Catholicism, 156
  decline, 154–7, 163, 255, 266
  novelty, 124–6
  origins, 154
Soens, Tim, 93, 113
Southern Sea, 27, 98, 122, 132, 171, 262

reclamation, 148
salinity gradient, 136
Spanish Empire, 7, 27, 34. *See also* Eighty Years' War
St. Martin's Flood (1686), 52, 93, 112, 114, 118, 151
Stad Groningen. *See* Groningen (city)
Staketwerken. *See* Water infrastructure
Staten van Friesland. *See* Estates of Friesland
Staten van Holland en West-Friesland. *See* Estates of Holland
Staten van Stad en Lande. *See* Estates of Groningen
Staten van Utrecht. *See* Estates of Utrecht
Staten van Zeeland. *See* Estates of Zeeland
Staten-Generaal. *See* Estates General
Storm surges, 12, 15, 17, 267, 273. *See also* Little Ice Age; Maunder Minimum
  in 1719, 121
  in 1720, 121
  Christmas Flood (1717), 89–90, 265
  climatic trends, 109–10
  coastal erosion, 140
  documentation, 39
  flood cultures, 99
  river morphology, 36
  settlement in Groningen (province), 107–9
Straat, Pieter, 158–62
Subsidence. *See also* Flooding (river); flooding (coastal); peat; reclamation; water management; water infrastructure
  definition, 95
  drainage, 36
  ecological crisis, 63
  flooding (coastal), 108, 140
  flooding (river), 166, 185
  hybridity, 263
  land loss, 96
  long-term driver of disaster, 19, 264
  technological lock-in, 273
  threat to water infrastructure, 12

Tak, Jan, 236–8, 244
Taxation. *See also* Continuous crisis management; water management; decline
  cattle plague, 58, 68–9, 82, 84
  debt burden, 46

extraordinary taxation, 215, 259, 262
foreign perceptions, 47
Holland's responsibility, 35
horned animals, 68
inefficiencies, 9
Pachtersoproer, 215
protectionism, 64
Rampjaar, 52
remission, 111, 114, 143, 182, 197–8, 208, 229–30, 268
War of the Spanish Succession, 87
water management, 96, 267
Temple, William, 45, 253
Tenant Revolt. *See* Pachtersoproer
Teredo navalis. *See* Shipworms
Teredo norvagica. *See* Shipworms
Texel, 134
Tiele, Alardus, 74
Tielerwaard, 179
Toulon, 132
Transverse dike. *See* Water infrastructure
Treaty of Utrecht (1713), 214, 216, 253. *See also* War of the Spanish Succession
peace celebrations, 51–4
perception, 54–5, 105

Uithuizermeeden, 113
Ulrum, 91
United Provinces of the Netherlands. *See* Dutch Republic
Utrecht (city), 30, 53, 153, 207
Utrecht (province). *See also* Continuous crisis management
cattle plague, 66, 78–80, 84, 220, 233
extreme weather, 226
flooding (river), 185, 193, 198–9, 225
mice plagues, 216
Utrecht University, 235

Van Barn-in 't-Loo, Hendrik, 214
Van Byler, Henricus Carolinus, 70, 76, 155
Van Claerbergen, Johan Vegelin, 229
Van Dam, Petra, 98
Van den Berge, Maria, 243
Van den Honert, Johan, 242
Van der Deure, Pieter, 158–62
Van der Spek, Paulus, 200, 218, 220, 233
Van der Vinne, Isaac Vincentsz., 178
Van Effen, Justus. *See also* Enlightenment; sodomy persecution; *Hollandsche Spectator*

decline, 156–7
shipworms, 157
significance, 48
Van Eijnatten, Joris, 248
Van Gyzen, Jan, 71, 88
Van Hardeveldt, Anthonie, 206
Van Linschoten, Jan Huygen, 131
Van Seeratt, Thomas. *See also* Christmas Flood (1717); expertise; water infrastructure
as commies provinciaal, 90, 106–7
flood interpretation, 119–20
technocratic optimism, 119, 121, 161
Variolation. *See* Inoculation
Velema, Wyger, 157
Velsen, Cornelis, 49, 150, 160–1, 181–3, 199, 207–10, 255
Vereenigde Oostindische Compagnie (VOC). *See* Dutch East India Company (VOC)
Verhoefslaging. *See* Water infrastructure
Vianen, 193
Vier Noorder Koggen, 122, 139. *See also* Water boards
Vierhuizen, 91
Vierlingh, Andries, 99
Voorland. *See also* Vulnerability; water infrastructure
dike adaptation, 159
erosion, 5, 12, 108, 140
flooding (coastal), 265
in Holland, 140
reclamation, 111–12
storm surge buffer, 111
Vulnerability, 2
dike adaptation, 140
disaster outcomes, 20, 265
early modern interpretation, 100, 255
flooding (coastal), 254
flooding (river), 37, 176
marginalization, 23, 113, 262, 265

Waal River. *See also* Rhine River
in 1740–41, 177, 188, 191
in 1757, 166–7
Rhine-Meuse River System, 171–3, 181
Wadden Sea, 89, 93, 107, 118, 270, *See* North Sea; Groningen (province); Christmas Flood (1717)
coastal morphology, 108
coastal vulnerability, 112–13, 254

336 Index

Wadden Sea (cont.)
 erosion, 136, 140
 flood identities, 107
 reclamation, 111–12
Wagenaar, Jan, 202, 204–7
Wageningen, 178, 190
Walcheren. *See also* Water infrastructure; Westkapelle Sea Dike
 shipworm epidemic, 24, 126–8, 138–9
War of the Austrian Succession
 cattle plague, 224–5, 255
 military and political consequences, 214–15
War of the Spanish Succession, 23, 46, 51, 102, 253. *See also* Treaty of Utrecht (1713)
 cattle plague, 57
 economic consequences, 51, 87
 geopolitical consequences, 251
Water boards, 149–50, 162, 180, 271. *See also* Reclamation; water infrastructure; water management; expertise; adaptation
 in 1726, 199
 ad hoc communalization, 267
 cartography, 255
 cooperation, 207
 decentralized management, 179–80
 emergence, 96
 flooding (river) relief, 188–91, 193–7
 flooding (river) risk in Holland, 183–7
 Friesland, 158
 Groningen, 96, 108
 Holland, 96, 122, 158–62
 shipworm adaptation in Holland, 139–44
 Zeeland, 138–9
Water infrastructure, 96, 161, 262. *See also* Water management; water boards; adaptation; flooding (coastal); flooding (river)
 causes of failure, 173–4
 coastal, 111, 119, 127
 cultural meaning, 93–4, 97, 99, 106, 117–18
 depictions, 30
 financial responsibility, 24, 94, 107–8, 112, 114, 117, 254
 Friesland, 116
 Groningen, 115
 improvement, 118–19, 161, 249
 innovation, 24–5, 144, 157–62, 249
 inspection, 90, 122, 127, 132, 139, 145
 military inundation, 27, 43–4
 reclamation, 96–7
 rivers, 171, 179
 shipworm impacts, 122–3, 127–8, 132, 139–40
 types (Holland), 140–3
 types (Zeeland), 126–7
 Zeeland, 158
Water lines
 climate/weather, 42
 consequences, 43–4
 Holland water line, 27
 Pannerdens Canal, 181
 Rampjaar, 26–8, 42
Water management. *See also* Hydraulic department (Holland); water infrastructure; water boards
 adaptation, 50, 252, 256
 centralization, 180–2
 conflict, 112–17, 271
 diversity, 35, 97–9, 165, 178
 in Dutch Republic, 12, 15–16
 early history, 95–7, 170–1
 economic development, 32, 63
 environmental consequences, 36–7, 262
 expertise, 149–50, 169, 181–3, 255
 flood cultures, 93, 95, 124, 151, 207–8, 254
 Groningen, 108
 hard and soft approaches, 274
 increasing costs, 33, 37, 262, 267
 interprovincial, 169, 208
 legacy, 16
 neglect of maintenance, 176, 179
 riverine challenges, 25, 171–3, 180–1, 232, 261
 social inequality, 175–6
West Friesland, 122
 cattle plague, 222
 environmental change, 140–3
 land abandonment, 230, 266
 shipworm epidemic, 140, 143–5, 157–62
West Indies, 131
Westerkwartier
 farmer uprising, 117
Westerzeedijk (Friesland), 270–3
Westgeest, Marijtje Cornelisdr, 233
West-Indische Compagnie (WIC). *See* Dutch West India Company (WIC)
Westkapelle Sea Dike, 126–7
White, Sam, 225

Index

Wierdijken. See Water infrastructure
Wierriemen. See Water infrastructure
Wijk bij Duurstede, 171
William II (Prince of Orange), 35
William III (Prince of Orange), 41–2, 44, 181, 215, 253. *See also Rampjaar*
William IV (Prince of Orange), 215
William of Orange ("The Silent"), 35, 41
Windmills. *See* Water infrastructure
Windstorm of 1674 (Utrecht), 30
Winter. *See also* Little Ice Age; Maunder Minimum
   in 1739–40, 24, 169, 202, 210, 216, 224–5, 230, 259
   in 1708–9, 53, 199, 230
Woerden (town), 212
Woerden (water board), 184
Wopkes, Jan, 66–7, 82
Wormspijkers. See Water infrastructure
Woudrichem, 172, 191, 203

Yarranton, Andrew, 47

Zeeland
   cattle plague, 222, 228
   Christmas Flood (1717), 102
   early evidence of shipworms, 132
   flooding (river) relief, 204
   Rampjaar, 42
   reclamation, 62, 97
   Second Orangist Revolution, 215
   shipworm epidemic, 122–3, 126–8, 138–9, 144, 152–3, 158, 267
   War of the Austrian Succession, 215
Zeelandic Flanders. *See* Generality Lands
Zijlvesten. See Water boards
Zuiderzee. See Southern Sea
Zwammerdam, 43
Zwolle, 233

Lightning Source UK Ltd.
Milton Keynes UK
UKHW010638190122
397376UK00002B/4